T0091494

The Nanotechnology Revolution

The Nanotechnology Revolution

A Global Bibliographic Perspective

Dale A. Stirling

PAN STANFORD PUBLISHING

Published by

Pan Stanford Publishing Pte. Ltd.
Penthouse Level, Suntec Tower 3
8 Temasek Boulevard
Singapore 038988

Email: editorial@panstanford.com
Web: www.panstanford.com

British Library Cataloguing-in-Publication Data
A catalogue record for this book is available from the British Library.

The Nanotechnology Revolution: A Global Bibliographic Perspective

ISBN 978-981-4774-19-2 (Hardback)
ISBN 978-1-315-11083-7 (eBook)

Contents

Foreword

While the concept of manipulating materials at the atomic level is not new per se, doing so only became practical around ten years ago. Since then, scientists and engineers have engaged in an international flurry of research and innovation into nanotechnology and nanomaterials. They have already changed the world, and they seem to just be getting started. Dale Stirling's book *The Nanotechnology Revolution: A Global Bibliographic Perspective* is the best account that I have seen of these early years of exploration.

I've been involved in developing standards for the safe use of nanotechnology since the late 1990s; I've traveled to just about every continent and met with too many brilliant minds to count as we've all tried to grapple with the field's explosive development. I thought that I had tracked it decently well, at least pertaining to safety. Nope. Even though I've been in the thick of the revolution, Dale Stirling's book took me on a ride that I could not have imagined.

I have known Dale to be a careful, thoughtful, and meticulous research historian and librarian. I cannot think of a better researcher to explore the key changes in the field of nanotechnology while still successfully showcasing its sheer scope and diversity. Dale outlines the 115 countries where nanotechnology is researched or used, assesses its disparate economic impacts, probes the sociological impacts it is already causing, provides key sources of human and environmental health work, and lists current legal and regulatory impacts, and considers its ethical implications.

Are you new to the subject, either as a student or as a curious layperson? Start here. Are you writing a manuscript and need a reference on regulations for a nano-object? Use this book. Do you want to understand one of the fundamental drivers of change in the new century? Here's where to look.

Hold on! Nanotechnology is just getting started, and this book is how you'll get up to speed and join the revolution.

Richard C. Pleus, Ph.D., M.S.
Founder, Intertox Inc. and Intertox Decision Science, LLC

Preface

This reference work is intended to accomplish several things. First and foremost, it is a comprehensive and thematically organized guide to nanotechnology related documents. It is designed to be used, frequently, by those who are involved with nanotechnology—whether as academicians, researchers, manufacturers, or any of dozens of disciplines interested in or impacted by the universe of nanotechnology.

However, no reference work should start off with a flurry of information without some context. Therefore, the introduction includes the author's interpretation of the history of nanotechnology, focusing on key events and selected individuals. Also, due to the massive amount of information presented within the pages of this book, a detailed analysis of its organization is presented, as well as information on how the actual bibliography was compiled. Lastly, the author intends this reference work to be a "snapshot in time" and by its content illustrate the state of nanotechnology in the early 21st century.

Acknowledgments

During a nearly forty-year career as a historian working in atypical fields such as natural resource management, environmental site investigations, and toxicology and risk assessment, several constants have sustained me. The first and foremost is my wife Stephanie—she has humored me, supported me, and spoiled me in all my creative endeavors. Also important have been the mentors, supervisors, and company leaders who recognized the value and purpose of history and historical research. Lastly, I want to thank Robert Freitas for his input on the book's organization.

Introduction

What is nanotechnology? Per the *Oxford Dictionary*, nanotechnology (nanō tek′näləjē) is "the branch of technology that deals with dimensions and tolerances of less than 100 nanometers, especially the manipulation of individual atoms and molecules."[a] Although relatively young as a technology, its roots can be traced to the 1850s and the following milestones are key[b]:

- 1857: Michael Faraday discovers colloidal ruby gold
- 1936: Erwin Müller's invents the field emission microscope
- 1947: Bell Lab scientists discover the semiconductor transistor
- 1950: Victor La Mer and Robert Dinegar develop a process for growing monodisperse colloidal materials
- 1951: Erwin Müller' invents the field ion microscope
- 1956: Arthur von Hippel coins the term *molecular engineering*
- 1981: Gerd Binning and Heinrich Rohrer' invent the scanning tunneling microscope, which allowed scientists to see direct spatial images of individual atoms
- 1985: Rice University researchers discover the buckminster-fullerene, a molecule representing a soccer ball in shape and composed entirely of carbon
- 1985: Bell Labs scientist Louis Brus discovers colloidal semi-conductor nanocrystals known as quantum dots
- 1990s: Nanotechnology companies begin to incorporate (Nano-phase Technologies, Helix Energy Solutions Group, Zyvex, etc.)

---

[a]http://www.oxforddictionaries.com/us/definition/american_english/nanotechnology. Accessed online on July 17, 2016.
[b]Information abstracted from http://www.nano.gov/timeline. Accessed online on August 19, 2015.

- 1991: Sumio Iijima of NEC discovers the carbon nanotube, which exhibits valued properties of strength and conductivity

Although these milestones are important, the following events and individuals are essential cornerstones of nanotechnology and are described in detail.

In December 1959, the American physicist Richard Feynman gave a talk at the annual meeting of the American Physical Society at Caltech. A transcript of the talk titled "There's Plenty of Room at the Bottom" was published in the February 1960 issue of *Engineering & Science*. Feynman was at Caltech to "talk about the problem of manipulating and controlling things on a small scale," and he noted that "the principles of physics, as far as I can see, do not speak against the possibility of maneuvering things atom by atom. It is not an attempt to violate any laws; it is something, in principle, that can be done; but in practice, it has not been done because we are too big." What Feynman couldn't know was that his talk was the genesis of nanotechnology as a viable science and his "invitation to enter a new field in physics" was accepted by many.[c]

In July 1974, Tokyo Science University Professor Norio Taniguchi presented a paper titled "On the Basic Concept of Nano-technology" at the International Congress on Production Engineering in Tokyo.[d] It is generally acknowledged that he was the first person to coin the term *nanotechnology*. Unlike Feynman, who proposed manipulation at the atomic level for a variety of purposes, Taniguchi spoke specifically about manufacturing and processing and noted the "intense needs for the processing of materials obtaining the preciseness and fineness of the order of 1 nm in length." His paper's abstract perfectly reflects the paper's content and intent:

'Nano-technology' is the production technology to get the extra high accuracy and ultra fine dimensions, i.e. the preciseness and fineness of the order of 1 nm (nanometer), 10^{-9} m in length. The name of 'Nano-technology' originates from this nanometer. In the processing

---

[c]Feynman R.P. (1960). There's plenty of room at the bottom. *Engineering and Science.* **23**(5):22–36.

[d]Norio T. (1974). On the basic concept of nano-technology. In *Proceedings of the International Conference on Production Engineering, Part I.* Tokyo. pp. 18–23.

*of materials, the smallest bit size of stock removal, accretion or flow
of materials probably of one atom or one molecule, namely 0. 1 ~ 0.
2 nm in length. Therefore, the expected limit size of fineness would
be of the order of 1 nm.*

Taniguchi concluded his paper with a hopeful but realistic
statement:

*In the 'Nano-technology' in materials processing, the processing by
one atom or one molecule should be fully utilized. However, on the
processings due too thermal, chemical, electro-chemical reactions,
there remain also several problems to solve, i.e. how to supply
the process energy impulsively and with high power density and
also through very fine path. In contrast, the processings due to
dynamic reaction of ion beam or molecular beam seem to be more
promising for the technology, in spite of the difficulties of control and
measurement for nanometer.*

Feynman and Taniguchi are important for introducing nanotechnology concepts, applications, both theoretical and realized, and
potential future uses. But it wasn't until the 1980s that the ideas they
fostered truly took shape. The persistence, research, and writings
of a handful of key individuals were responsible for promoting and
propelling nanotechnology from the late 20th century into the new
millennium.

Perhaps the best known of these is K. Eric Drexler. He earned
multiple undergraduate degrees from the Massachusetts Institute
of Technology (MIT) in the 1970s, and his 1986 book *Engines
of Creation: The Coming Era of Nanotechnology* (Anchor Books)
envisioned a future that uses molecular-scale machines for a
variety of purposes. In the same year, he founded the Foresight
Institute, a nonprofit organization dedicated to nanotechnology.
He earned a Ph.D. in molecular nanotechnology from MIT in
1991. His dissertation ("Molecular Machinery and Manufacturing
with Applications to Computation") was published as *Nanosystems:
Molecular Machinery Manufacturing and Computation* by John Wiley
& Sons, Inc., in 1992. In the 2000s, he served as a technical advisor to
Nanorex, a company that developed software for molecular systems

design. Most recently, he was an academic visitor in residence at Oxford University and his book *Radical Abundance: How a Revolution in Nanotechnology Will Change Civilization* (Public Affairs [Perseus Books Group]) was published in 2013.[e]

Organization

I had no specific organization in mind for this book when I began conducting literature searches in 2012. It seemed obvious to have a few general subject headings, but as search results flowed, a thematic structure began to appear. The structure became more formalized as I sorted through thousands of journal articles, books, conference papers, government reports, and gray literature abstracts.

Each of the nine sections begins with a short introductory paragraph that examines its key features. In many cases, there is a single subject heading (Section 2: Economics)—this reflects the fact that there is not a plethora of related literature. However, some sections have multiple headings to reflect the complexity and amount of related literature. In many sections the first subheading is titled "General Overviews," which includes literature of a broad nature relating to the main heading. Then, as dictated by the variety of related literature, additional subheadings are used as required. An example of this is the largest section of the book, Section 9: Nanotechnology Applications, and its subheading "Nanomedicine," which includes the following subheadings (a sample partial listing only):

Nanomedicine: General Topics

(a) Environmental and Health Risks
(b) Ethics
(c) History, Trends, and Future Directions
(d) Informatics

---

[e]Information abstracted from Drexler's website at http://e-drexler.com/p/idx04/00/0404drexlerBioCV.html. Accessed online on July 16, 2015.

Nanomedicine: Applications and Concepts

- (e) Alternative Medicine
- (f) Aptamers
- (g) Biosensors
- (h) Cardiology
- (i) Clinical nanomedicine
- (j) Dendrimers
- (k) Dentistry
- (l) Dermatology
- (m) Diagnostics

Perhaps the most important theme of the book is its global perspective. Although the United States is generally recognized as a leader in nanotechnology, many other countries and regions can claim equal interest and commitment to nanotechnology and all it promises. Therefore, Section 1 is devoted to literature relating to 115 countries and regions. Not surprisingly, nanotechnology literature is common in Europe, much of Asia and other developed countries, but its reach is even documented in such countries as Bhutan, Burkina Faso, and Senegal. However, some 93 countries are not the subject of such research and publication (Refer to Appendix 1 for a list). Section 2 is devoted to the literature of economics and its interface with nanotechnology. The majority of documents are from journals, but there are some pivotal works published by government agencies as well. Section 3 cites literature that examines nanotechnology as it serves or is the subject of education and educators. It cites literature from a nano-cultural perspective and the sociological impact of nanotechnology. Section 4 may be the most pertinent as it cites literature that examines documented and potential impacts that nanotechnology has on our environment and human health.

In the headlong rush towards real-world applications of nano-technology and related manufactured products there is real concern that its impact is not being studied to the extent that it should. Similarly, ethical, legal, and regulatory concerns in nanotechnology are the subject of literature cited in Section 5.

The retrospective and futurespective examination of nanotech-nology is reflected in the literature cited in Section 6. However,

Section 7 may well be the most interesting as it cites literature that studies the literature of nanotechnology. Such introspection allows us to gauge the progress of research and writing on all aspects of nanotechnology. Through such examination, we understand how context and content changes over time. Sections 1 through 7 cite literature that provides researchers with a solid foundation on the history, promise, future, and impact of nanotechnology on today's world. If no other literature was consulted, the reader would be informed. However, Sections 8 and 9 focus on the essence of what nanotechnology is as revealed in its key concepts and theories as well as real-world applications. Accordingly, Section 8 cites literature on nanostructured devices, materials, and systems. These are the technologies that are changing how the world uses nanotechnology in adaptive ways. Section 9 cites literature relating to the real-world application of nanotechnology to common every-day activities—ranging from its use in aerospace, to construction, to transportation, and perhaps most significantly, nanomedicine and the use of nanotechnology in nearly every subdiscipline of medicine.

To facilitate the best use of this reference work, each citation is numbered and is keyed into two important indexes. The first index (Appendix 2) is subject oriented and is based on main and subsection headings. However, the index is broader than that with all 4621 citations scanned for key words. The second index (Appendix 3) is geographic and allows the reader to quickly locate which countries and regions are cited in this book.

Research Methods

Between 2012 and 2016, I conducted exhaustive searches of nanotechnology utilizing publicly available and fee-based databases, web portals, and online search engines. During the process of conducting literature searches, specific themes and subject areas were discerned, and those provide the structure of this refer-ence guide. The bibliography is global in scope; it reflects the broad interest in nanotechnology from developed and developing nations.

The following parameters were used to determine which documents would be cited in this bibliography. Review articles were the preferred document source overall because they are comprehensive in scope, timely, and peer-reviewed. However, all documents, including journal articles, were chosen with the following criteria in mind:

- They focus on the history and development of nanomedicine
- They focus on trends and emerging technologies
- They focus on ethical, regulatory, and legal issues related to nanomedicine
- They focus on the relationship between nanotechnology and its various disciplines and sub-disciplines

Also of critical interest were documents that indexed, abstracted, or provided bibliographic citations and references on the literature of nanotechnology. Finally, I sought a balance that reflects the multidisciplinary nature of nanotechnology. Therefore, documents from many branches of science and other disciplines are cited in this bibliography. Most importantly, documents were included only if abstracted by a database provider, available as a digital download, or acquired from a research portal or document delivery service so they could be reviewed for context and content.

I also sought a balance between spectrums of information. I chose to include scholarly works because they represent the penultimate of research and presentation, but I also chose to include popular literature such as trade journals and magazines because they represent the interests of the nanotechnology business world. I also sought a global balance—that is, to focus not just on what the United States is doing, but also on what the entire world is doing.

Where possible, I have used full journal titles. This despite the internationally recognized journal abbreviation conventions posited by the U.S. National Library of Medicine (NLM). This is due to the explosion of open-source publishing and titles that are not indexed by the NLM and that have not been assigned a recognized journal abbreviation.

The new open-source environment is both a blessing and a curse in scientific research. It offers a field of plenty for authors, but the actual intent and aim of some publishers is suspect and the quality of the science may be in question. Indeed, the NLM's Literature Selection Technical Review Committee is charged with reviewing and recommending journals for inclusion in MEDLINE, and whether a journal is added is based on the following critical elements:

- Scope and coverage
- Quality of content
- Quality of editorial work
- Production quality
- Audience

However, I leave it to the reader to decide if a journal article, or indeed any document, appearing in this bibliography provides information of value and substance as it applies to his or her involvement in nanotechnology.

Section 1

Countries and Regions

The purpose of this section is to highlight the global reach of nanotechnology. However, for many developing countries little to no attention is paid to nanotechnology. Indeed, for some countries the only source of nanotech information are articles written about nanotechnology by scientists or others from one of these countries or regions or, an article or report is written about the application of nanotechnology in that country. (Please refer to Appendix 1 for a list of these countries.)

Africa and Related Regions

1. Al-Rawashdeh M.M. et al. (2013). Development highlights of micro-nano technologies in the MENA region and pathways for initiatives to support and network. *Green Processing and Synthesis*. **2**(2):91–100.
2. Boshoff N. (2009). Neo-colonialism and research collaboration in Central Africa. *Scientometrics*. **81**(2):413–434.
3. Ezema I.C. et al. (2014). Initiatives and strategies for development of nanotechnology in nations: a lesson for Africa and other least developed countries. *Nanoscale Research Letters*. **9**(1):1–8.

The Nanotechnology Revolution: A Global Bibliographic Perspective
Dale A. Stirling
Copyright © 2018 Pan Stanford Publishing Pte. Ltd.
ISBN 978-981-4774-19-2 (Hardcover), 978-1-315-11083-7 (eBook)
www.panstanford.com

4. Gastrow M. (2011). Open innovation in South Africa: case studies in nanotechnology, biotechnology, and open source software development. *Journal for New Generation Sciences.* **9**(1).

5. Limaye M.V. et al. (2015). Bogolanfini dyeing: a traditional nanotechnology from West Africa. Stockholm University, Faculty of Science, Department of Materials & Environmental Chemistry.

6. Ma'moun Al-Rawashdeh B. et al. (2012). Development highlights of micro-nano technologies in the MENA region and pathways for initiatives to support and network. *Green Processing and Synthesis.* **2**:91–100.

7. Masoka X. et al. (2012). Nanotechnology research & occupational health and safety in Africa. *African Newsletter.* **22**(3):56.

8. Musee N. et al. (2013). Relevance of nanotechnology to Africa: synthesis, applications, and safety. In *Chemistry for Sustainable Development in Africa.* Berlin Heidelberg: Springer, pp. 123–158.

9. Oukil M.S. (2011). A development perspective of technology-based entrepreneurship in the Middle East and North Africa. *Annals of Innovation & Entrepreneurship.* **2**(1).

10. Rosei F. et al. (2008). Materials science in the developing world: challenges and perspectives for Africa. *Advanced Materials.* **20**(24):4627–4640.

11. Saidi T. (2009). The expectations and challenges in the development of nanotechnology in sub-Saharan Africa. Thesis: Maastricht Graduate School of Governance.

12. Simate G.S. et al. (2013). Biotechnology and nanotechnology: a means for sustainable development in Africa. In *Chemistry for Sustainable Development in Africa.* Berlin Heidelberg: Springer, pp. 159–191.

13. Tobin L. (2009). First annual report on nanoscience and nanotechnology in Africa. Stirling: Institute of Nanotechnology.

14. Wetter K.J. (2010). Big continent and tiny technology: nanotechnology and Africa. Pambazuka News.

15. Yawson R.M. (2011). Africa's nanofuture: the importance of regionalism. *International Journal of Nanotechnology.* **8**(6–7):420–436.

Andorra

16. Arguimbau L. and Alegret S. (2011). Chemical research in the Catalan Countries: a brief quantitative assessment of the agents, resources, and results. *Contributions to Science.* **6**(2):215–232.

Antarctica

17. Hemmings A.D. (2009). From the new geopolitics of resources to nanotechnology: emerging challenges of globalism in Antarctica. *The Yearbook of Polar Law Online.* **1**(1):55–72.
18. Hemmings A.D. and Jabour J. (2008). Regulating nanotechnology in Antarctica. In *International Conference on Nanoscience and Nanotechnology* (ICONN 2008), vol. 25, p. 29.

Arab and Arabic States

19. Alfeeli B. et al. (2011). Current status and prospects of nanotechnology in Arab States. *Advances in Synthesis, Processing, and Applications of Nanostructures: Ceramic Transactions.* **238**:77–92.
20. Alfeeli B. et al. (2013). A review of nanotechnology development in the Arab world. *Nanotechnology Review.* doi:https://doi.org/10.1515/ntrev-2012-0070.
21. Haik Y. (2008). Nanotechnology in the United Arab Emirates: hype and hope. *Nanotechnology Law & Business.* **5**:219.
22. Terdman M. (2010). Nanotechnology and the environment in the Arab world. *Muslim Environment Watch.* **1**(1), http://muslimenvironment.wordpress.com.

Argentina

23. Andrini L. and Figueroa S.J. (2007). Governmental encouragement of nanosciences and nanotechnologies in Argentina. *Nanotechnologies in Latin America,* 27.
24. Aydogan-Duda N. (2012). Nanotechnology in Argentina. In *Making It to the Forefront.* New York: Springer, pp. 69–75.

25. Balseiro C.A. (2013). Physics in Argentina: the case of nanoscience and nanotechnology. In *APS March Meeting Abstracts*, vol. 1, p. 9004.

26. Baranao L. (2011). [Science in Argentina. Where do we go from here?]. *Medicina*. **72**(4):339–349 (Spanish).

27. Kaiser J. and Marshall E. (2008). Lino Barañao interview: new minister raises expectations for science in Argentina. *Science*. **321**(5889):622–622.

28. Malsch I. (2000). Nanotechnology in Argentina. NanoforumEULA, University of Twente, Netherlands. Available at http://www. mesaplus.utwente.nl/nanoforumeula

Armenia

29. Arzumanyan T. (2006). Current issues of research, development and innovation in Armenia. *International Journal of Foresight and Innovation Policy*. **2**(2):133–145.

30. Gabrielyan B. and Arzumanyan T. (2004). Problems and approaches in national innovation policy in Armenia. In Filho W. L. (ed.), *Supporting the Development of R&D and the Innovation Potential of Post-Socialist Countries*. NATO Science Series, vol. 42, Amsterdam: IOS Press, pp. 49–57.

31. Khnkoyan A. (2011). Development of scientific and innovation policy in Armenia since 2000s. Armenian State Committee of Science of the Republic of Armenia.

32. Khnkoyan A. (2012). National innovation system and the development of scientific and innovation policy in the Republic of Armenia. *Наука та інновації*. (3):75–83.

33. Markkula M. (2008). The science, technology and innovation policy of the Republic of Armenia. UNESCO Armenian STI Mission.

Asia and Related Regions

34. Choi K.H. (2004). Ethical issues on nanotechnology in Asia. *Key Engineering Materials*. **277**:945–949.

35. Karim M.E. (2014). Nanotechnology in Asia: a preliminary assessment of the existing legal framework. *KLRI Journal of Law and Legislation*. **4**(2):169–223.

36. Liu L. (2009). *Emerging Nanotechnology Power: Nanotechnology R&D and Business Trends in The Asia Pacific Rim.* World Scientific, Singapore.

37. Liu L. (2015). Overview on nanotechnology R&D and commercialization in the Asia Pacific region. In *The Nano-Micro Interface: Bridging the Micro and Nano Worlds.* Wiley-VCH Verlag GmbH & Co. KGaA, Weinheim, Germany, pp. 37–54.

38. Nguyen T.V. and Pham L.T. (2011). Scientific output and its relationship to knowledge economy: an analysis of ASEAN countries. *Scientometrics.* **89**(1):107–117.

39. Plusnin N.I. and Lazarev G.I. (2008). Project of international science-education center and integration problems of nano science education in far eastern region of Asia. *Journal of Physics.* **100**(5):052030.

Australia

40. Alford K.J. et al. (2009). Creating a spark for Australian science through integrated nanotechnology studies at St. Helena secondary college. *Journal of Nano Education.* **1**(1):68–74.

41. Australian Academy of Science (2009). Nanotechnology in Australia: trends, applications and collaborative opportunities. Canberra.

42. Australian Academy of Science (2012). National nanotechnology research strategy. Canberra.

43. Bartholomaeus N.F.A.A. (2011). Regulation of nanotechnologies in Australia and New Zealand. *International Food Risk Analysis Journal.* **1**(2):33–40.

44. Batley G. and McLaughlin M.J. (2007). Fate of manufactured nanomaterials in the Australian environment. Centre for Environmental Contaminants Research, CSIRO Land and Water, Bangor, NSW, Australia.

45. Bowman D.M. and Bennett M.G. (2012). Current state of Australia's evolving approach to regulating nanotechnologies. *Nanotechnology Law and Business.* **9**:330.

46. Bowman D.M. and Fitzharris M. (2007). Too small for concern? Public health and nanotechnology. *Australian and New Zealand Journal of Public Health.* **31**(4):382–384.

47. Bowman D.M. and Hodge G. (2007). Engaging in small talk: nanotechnology down under: getting on top of regulatory matters. *Nanotechnology Law and Business.* **4**(2):225.

48. Braach-Maksvytis V. (2002). Nanotechnology in Australia–towards a national initiative. *Journal of Nanoparticle Research.* **4**(1):1–7.

49. Cortie M.B. et al. (2013). Nanomedical research in Australia and New Zealand. *Nanomedicine.* **8**:12.

50. Edwards J. et al. (2007). Current OHS best practices for the Australian nanotechnology industry. A position paper by the nanosafe Australia network. *Journal of Occupational Safety & Health, Australia & New Zealand.* **23**(4):315–331.

51. Faunce T.A. (2007). Nanotherapeutics: new challenges for safety and cost-effectiveness regulation in Australia. *Medical Journal of Australia.* **186**(4):189.

52. Faunce T. et al. (2008). Sunscreen safety: the precautionary principle, the Australian therapeutic goods administration and nanoparticles in sunscreens. *NanoEthics.* **2**(3):231–240.

53. Fletcher N. (2008). Nanotechnology in Australia–a network of interactions. *IEEE Nanotechnology Magazine.* **2**(3):19–22.

54. Gorjiara T. and Baldock C. (2014). Nanoscience and nanotech-nology research publications: a comparison between Australia and the rest of the world. *Scientometrics.* **100**(1):121–148.

55. Harwood J. and Schibeci R. (2008). Community participation in Australian science and technology policy: the case of nanotechnology. *Prometheus.* **26**(2):153–163.

56. Kane D.M. et al. (2011). *Nanotechnology in Australia: Showcase of Early Career Research.* Pan Stanford Publishing, Singapore.

57. Katz E. et al. (2005). *Citizens Panel on Nanotechnology Report to Participants (DMR-2673).* CSIRO, Clayton South, Victoria.

58. Katz E. et al. (2009). Evolving scientific research governance in Australia: a case study of engaging interested publics in nanotechnology research. *Public Understanding of Science.* **18**(5):531–545.

59. Ludlow K. (2007). Readiness of Australian Food Regulation for the Use of Nanotechnology in Food and Food Packaging. *University of Tasmania Law Review.* **26**:177.

60. Ludlow K. et al. (2007). A review of possible impacts of nanotechnology on Australia's regulatory framework. Monash Centre for Regulatory Studies, Faculty of Law, Monash University and Institute for Environmental and Energy Law, Faculty of Law, KU Leuven.

61. Lyons K. and Scrinis G. (2009). Under the regulatory radar? Nanotechnologies and their impacts for rural Australia. Tracking rural change: community, policy and technology in Australia, New Zealand and Europe. Australian National University E Press, Canberra.

62. Lyons K. and Whelan J. (2010). Community engagement to facilitate, legitimize and accelerate the advancement of nanotechnologies in Australia. *NanoEthics.* **4**(1):53–66.

63. Martinez-Fernandez C. and Leevers K. (2004). Knowledge transfer and industry innovation: the discovery of nanotechnology by South-West Sydney organisations. *International Journal of Technology Management.* **28**(3–6):560–581.

64. NanoSafe Australia Network (2007). Current OHS best practices for the Australian nanotechnology industry. Victoria.

65. Nicolau D. (2004). Challenges and opportunities for nanotechnology policies: an Australian perspective. *Nanotechnology Law & Business.* **1**:446.

66. Nicolau D. (2005). Innovation and knowledge transfer in emerging fields: the case of nanotechnology in Australia. *Nanotechnology Law and Business.* **2**:386.

67. Peterson A. et al. (2010). Communicating with citizens about nanotechnologies: views of key stakeholders in Australia: final report. Monash University, Clayton, Victoria.

68. Priestly B. and Stebbing M. (2008). Risk perception and risk communication: is nanotechnology at the crossroads in Australia? In *2008 International Conference on Nanoscience and Nanotechnology*, Melbourne, pp. 238–240.

69. Prime Minister's Science, Engineering & Innovation Council (2005). Nanotechnology: enabling technologies for Australian innovative industries. Canberra.

70. Pritchard J. et al. (2007). Nanotechnology: the next wave of commercial development for health and medical devices: an Australian story. *Nanomedicine.* **2**(2):255–260.

71. Stebbing M. et al. (2006). Affective evaluation, trust, perceived risk and acceptability of new technology: the case of nanotechnology in Australia. *Australasian Epidemiologist.* **13**(3): 95.

72. Stevens M.G. et al. (2002). Nanotechnology in society. *Australian Science Teachers Journal.* **48**(3):22.

73. Tegart G. (2008). An Australian viewpoint of nanotechnology issues. *International Journal of Technology Transfer and Commercialisation.* **7**(2–3):123–128.

74. Tegart G. (2009). Energy and nanotechnologies: priority areas for Australia's future. *Technological Forecasting and Social Change.* **76**(9):1240–1246.

75. Tolstoshev A. (2006). Nanotechnology: assessing the environmental risks for Australia. Earth Policy Centre, University of Melbourne, Australia.

76. Warris C. (2004). Nanotechnology benchmarking project. Australian Academy of Science, Canberra.

Austria

77. Eisenberger I. et al. (2011). Nano regulation in Austria (1): chemical and product safety. Nano Trust Dossiers No. 018en.

78. Fuchs D. and Gazsó A. (2015). Nano risk governance: the Austrian case. *International Journal of Performability Engineering.* **11**(6):569–576.

79. Metag J. and Marcinkowski F. (2014). Technophobia towards emerging technologies? A comparative analysis of the media coverage of nanotechnology in Austria, Switzerland and Germany. *Journalism.* **15**(4):463–481.

Bangladesh

80. Palash M.L. and Mozumder M. (2013). Nanotechnology and Governance in Bangladesh. *Nanotechnology Law and Business.* **10**:146.

81. Parvin M. (2004). Need for governing nanotechnology research and development in Bangladesh. *International Journal of Scientific & Engineering Research.* **4**(11):179–185.

Belarus

82. Berezkina N.Y. et al. (2013). The use of citation databases for the assessment of the research activities of organizations in Belarus. *Scientific and Technical Information Processing.* **40**(4):190–194.

83. Vitiaz P.A. et al. (2006). State-of-the-art and prospects of nano-materials, nanotechnologies and nanodevices development in Belarus. *International Journal of Nanotechnology.* **3**(1):8–28.

Belgium

84. Goorden L. et al. (2008). Nanotechnologies for tomorrow's society: a case for reflective action research in Flanders, Belgium. In *Presenting Futures.* Amsterdam, Netherlands: Springer, pp. 163–182.

Bolivia

85. Del Barco R. and Foladori G. (2014). Nanotechnology and lithium: a window of opportunity for Bolivia? Advances and challenges. A paper presented at the *6th International Scientific Conference on Economic and Social Development and 3rd Eastern European ESD Conference.* Business Continuity, Vienna, April 24–25.

86. Gamarra R.D.B. and Foladori G. (2010). Nanotecnología y litio, ¿ una ventana de oportunidad para Bolivia? *Biociencias y Nanociencias.* MT5.

Brazil

87. Arcuri A. et al. (2009). Developing strategies in Brazil to manage the emerging nanotechnology and its associated risks. In *Nanomaterials: Risks and Benefits*. Netherlands: Springer, pp. 299–307.

88. Aydogan-Duda N. (2012). Nanotechnology in Brazil. In *Making It to the Forefront*. New York: Springer, pp. 63–67.

89. da Silva Sant'Anna L. et al. (2014). Nanomaterials patenting in Brazil: some considerations for the national regulatory framework. *Scientometrics*. **100**(3):675–686.

90. Dos Santos D.M. et al. (2005). Future studies in Brazil: CGEE approach for bio-and nanotechnology. *Journal of Business Chemistry*. **2**(3):126–137.

91. Eguchi E.S. et al. (2013). The way of nanotechnology in Brazil animal production. *PUBVET*. **7**(8): Article #1528.

92. Engelmann W. and Von Hohendorff R. (2014). Current scenario of nanotechnology in Brazil. *Journal of Hazardous, Toxic, and Radioactive Waste*. 10.1061/(ASCE)HZ.2153-5515.0000253.

93. Fonseca P. F. and Pereira T. S. (2014). The governance of nanotechnology in the Brazilian context: entangling approaches. *Technology in Society*, **37**:16–27.

94. Gosain R. (2000). *Nanotechnology Patent Protection: Brazil and the World Market*. Sao Paulo: Daniel Advogados.

95. Kay L. and Shapira P. (2011). The potential of nanotechnology for equitable economic development: the case of Brazil. In *Nanotechnology and the Challenges of Equity, Equality and Development*. Netherlands: Springer, pp. 309–329.

96. Kay L. et al. (2009). The role of Brazilian firms in nanotechnology development: science and innovation policy. A paper presented at the *Science and Innovation Policy Conference*. Atlanta, Georgia, October 2–3.

97. Lima M. and de Almeida M. (2012). Articulation of texts on nanoscience and nanotechnology for the initial training of physics teachers. *Revista Brasileira de Ensino de Física*. **34**(4):1–9.

98. Macnaghten P. and Guivant J.S. (2011). Converging citizens? Nanotechnology and the political imaginary of public engagement in Brazil and the United Kingdom. *Public Understanding of Science.* **20**(2):207–220.

99. Mateus C.F.R. and DaSilva A.C. (2006). The Brazilian initiative in MNT for aerospace applications. In *CANEUS 2006: MNT for Aerospace Applications.* American Society of Mechanical Engineers, pp. 347–354.

100. Peixoto F.J. (2011). Nanotechnology innovation policy in Brazil: an analysis of the economic subvention program. A paper presented at the *7th International Ph.D. School on National Systems of Innovation and Development.* Tampere, Finland, May 16–26.

101. Peixoto F.J. M. (2011). Understanding innovation in nanotechnology: a proposal for improving innovation policy for nanotechnology in Brazil. Paper apresentado na *DIME-DRUID Academy Winter Conference*, Aalborg, Denmark.

102. Rediguieri C.F. (2009). Study on the development of nanotechnology in advanced countries and in Brazil. *Brazilian Journal of Pharmaceutical Sciences.* **45**(2):189–200.

103. Thome A. et al. (2015). Review of nanotechnology for soil and groundwater remediation: Brazilian perspective. *Water, Air and Soil Pollution.* **226**(4):2243.

104. Toma H.E. (2005). Interfaces e organização da pesquisa no Brasil: da Química à Nanotecnologia. *Química Nova.* **28**:S48–S51.

105. Zanchet D. et al. (2002). Nanoscience and nanotechnology research at the Brazilian National Synchrotron Laboratory (LNLS). *Physica Status Solidi B Basic Research.* **232**(1):24–31.

Bulgaria

106. Simeonova K. (2006). Research and innovation in Bulgaria. *Science and Public Policy.* **33**(5):351–363.

Burkina Faso

107. Paul S. et al. (2013). Plants used in traditional beekeeping in Burkina Faso. *Open Journal of Ecology*. **3**(05):354.

Cameroon

108. Fokunang C.N. et al. (2011). Traditional medicine: past, present and future research and development prospects and integration in the National Health System of Cameroon. *African Journal of Traditional, Complementary and Alternative Medicines*. **8**(3):284–295.

Canada

109. Beaudry C. and Schiffauerova A. (2011). Impacts of collaboration and network indicators on patent quality: the case of Canadian nanotechnology innovation. *European Management Journal*. **29**(5):362–376.

110. Beaudry C. and Schiffauerova A. (2011). Is Canadian intellectual property leaving Canada? A study of nanotechnology patenting. *The Journal of Technology Transfer*. **36**(6):665–679.

111. Bergeron S. and Archambault É. (2005). Canadian stewardship practices for environmental nanotechnology. *Science*. **514**:6505.

112. Council of Canadian Academies (2008). Small is different: a science perspective on the regulatory challenges of the nanoscale. Report of the Expert Panel on Nanotechnology, Ottawa.

113. Einsiedel E. (2005). In the public eye: the early landscape of nanotechnology among Canadian and U.S. publics. *Journal of Nanotechnology*. Available online at http://www.azonano.com/article.aspx?ArticleID=1468.

114. Hu G. et al. (2012). Visualizing nanotechnology research in Canada: evidence from publication activities, 1990–2009. *The Journal of Technology Transfer*. **37**(4):550–562.

115. Mehta M.D. (2005). Regulating biotechnology and nanotechnology in Canada: a post-normal science approach for inclusion of the fourth helix. *International Journal of Contemporary Sociology.* **42**(1):107–120.

116. Moazami A. (2012). A network perspective of nanotechnology innovation: a comparison of Quebec, Canada and the United States. Doctoral Dissertation: Quebec, Canada, Concordia University Montreal.

117. Riggsby W.S. (1985). Some recent developments in the molecular biology of medically important Canada. *Microbiological Sciences.* **2**(9):257–263.

118. Roseman M. (2005). An overview of nanotechnology in Canada: a review and analysis of foreign nanotechnology strategies developed for the Prime Minister's advisory council on science and technology (PMACST). Toronto, Canada.

119. Tyshenko M.G. (2014). Nanotechnology framing in the Canadian national news media. *Technology in Society.* **37**:38–48.

120. Yegul M.F. et al. (2009). Delineation of landscapes of emerging sciences through publication data: a case of nanoscience in Canada. *2009 International Conference on Application of Information and Communication Technologies*, Baku, pp. 1–5.

Chile

121. Floody M.C. et al. (2009). Natural nanoclays: applications and future trends–a Chilean perspective. *Clay Minerals.* **44**(2):161–176.

122. Foladori G. and Fuentes V. (2007). Nanotechnology in Chile: towards a knowledge economy? In *Nanotechnologies in Latin America*. Berlin: Karl Dietz Verlag, p. 68.

123. Krauskopf M. et al. (2007). Low awareness of the link between science and innovation affects public policies in developing countries: the Chilean case. *Scientometrics.* **72**(1):93–103.

124. Pozo J. et al. (2011). Nanotechnology survey Chile 2011. Available at http://nanotecnologiasurvey.blogspot.com

125. Pozo J. et al. (2012). Perception of risks in nanotechnology: determining key aspects in Chile. *Journal of Risk Analysis and Crisis Response.* **2**(1):34–43.

126. Schnettler B. et al. (2013). Preferences for sunflower oil produced conventionally, produced with nanotechnology or genetically modified in the Araucanía Region of Chile. *Ciencia e Investigación Agraria.* **40**(1):17–29.

127. Schnettler B. et al. (2014). Acceptance of nanotechnology applications and satisfaction with food-related life in southern Chile. *Food Science and Technology (Campinas).* **34**(1):157–163.

China

128. Appelbaum R.P. and Parker R.A. (2008). China's bid to become a global nanotech leader: advancing nanotechnology through state-led programs and international collaborations. *Science and Public Policy.* **35**(5):319–334.

129. Appelbaum R.P. et al. (2011). Developmental state and innovation: nanotechnology in China. *Global Networks.* **11**(3):298–314.

130. Asian Technology Information Program (2003). Nanotechnology programs and organizations in China (ATIP03.054r). Albuquerque, New Mexico.

131. Bai C. (2005). Ascent of nanoscience in China. *Science.* **309**(5731): 61–63.

132. Bhattacharya S. et al. (2012). China and India: the two new players in the nanotechnology race. *Scientometrics.* **93**(1):59–87.

133. Cai N. et al. (2002). Research on the lifecycle and the emphasis of supporting policy of the current development of nanotechnology in China. *Studies in Science of Science.* **2**: 013.

134. Guan J. and Ma N. (2007). China's emerging presence in nanoscience and nanotechnology: a comparative bibliometric study of several nanoscience 'giants'. *Research Policy.* **36**(6):880–886.

135. Hu X. and Rousseau R. (2013). Are Chinese nanoscience citation curves converging towards their American counterparts? *Malaysian Journal of Library & Information Science.* **18**(3):49–56.

136. Huang S. et al. (2015). Nanotechnology in agriculture, livestock, and aquaculture in china: a review. *Agronomy for Sustainable Development.* **35**(2):369–400.

137. Jarvis D.S. and Richmond N. (2011). Regulation and governance of nanotechnology in China: regulatory challenges and effectiveness. *European Journal of Law and Technology.* **2**(3). http://ejlt. org/article/view/94/155

138. Jia L. et al. (2011). Fast evolving nanotechnology and relevant programs and entities in China. *Nano Today.* **6**(1):6–11.

139. Li Z.Y. (2012). Nanophotonics in China: overviews and highlights. *Frontiers of Physics.* **7**(6):601–631.

140. Liang L.M. and Xie C.X. (2003). Investigation of China's nanotechnology study based on frequency analysis of key words. *Studies in Science of Science.* **2**:005.

141. Lin M.W. and Zhang J. (2007). Language trends in nanoscience and technology: the case of Chinese-language publications. *Scientometrics.* **70**(3):555–564.

142. Liu L. and Zhang L.D. (2005). Nanotechnology in China: now and in the future. *Nanotechnology Law & Business.* **2**:399.

143. Song Y. and Ning B. (2014). Focus on study of nanotoxicology in China. *Zhonghua yu fang yi xue za zhi* [Chinese journal of preventive medicine]. **48**(7):552 (Chinese).

144. Weiss P.S. (2008). A conversation with Dr. Chunli Bai: champion of Chinese nanoscience. *ACS Nano.* **2**(7):1336–1340.

145. Wu S. et al. (2014). Nanoeducation in China: current status. *Journal of Nanoparticle Research.* **16**(5):1–5.

146. Ye X. et al. (2012). International collaborative patterns in China's nanotechnology publications. *International Journal of Technology Management.* **59**(3/4):255–272.

147. Zhao Y. and Ma N. (2012). Portrait of China's R&D activities in nano-science and nanotechnology in bibliometric study. *Advanced Materials Research.* **535**:505–510.

Colombia

148. Carpenter C. (2014). Application of a nanofluid for asphaltene inhibition in Colombia. *Journal of Petroleum Technology.* **66**(02):117–119.

149. Correa H.C.M. (2011). Aplicación de nanotecnología en la industria textil Colombiana. Bogota: National University of Colombia.

150. Fernández M.C. (2012). Regulación de la nanotecnología y aplicación en el sistema general de seguridad social en salud en Colombia. http://hdl.handle.net/10946/747

151. Gaviria M.I. et al. (2013). Red tecnoparque sena: una puerta tangible a la nanotecnologia (Tecnoparque sena network: an open door to nanotechnology). *Revista de Fiscia.* **46E**:115–125.

152. Martelo C. and Vinck D. (2009). Redes sociotécnicas de cogestión de conocimiento en nanotecnologías en Colombia: ¿ entre la visibilidad internacional y la apropiación local? *REDES, Revista de Estudios Sociales de la Ciencia*, **15**(29):113–137.

153. Martinez H. et al. (2014). Biotechnology profile analysis in Colombia. *Scientometrics.* **101**(3):1789–1804.

154. Martinez V. et al. (2010). Nanotecnología para Colombia. Una mirada histórica, pasando por el contexto global, latinoamericano y las regiones. *Revista Nano Ciencia y Tecnologia.* **2**(1):49–64 (Spanish).

155. Méndez Naranjo K.C. et al. (2008). Tendencias investigativas de la nanotecnología en empaques envases para alimentos. *Revista Lasallista de Investigación.* **11**(2):18–28.

156. Perez-Martelo C.B. and Vinck D. (2009). Redes sociotécnicas de cogestión de conocimiento en nanotecnologías en Colombia: entre la visibilidad internacional y la apropiación local? *REDES, Revista de Estudios Sociales de la Ciencia.* **15**(29):113–137.

157. Petersen G.M.Z. et al. (2014). Introduction of concepts of nanoscience and nanotechnology in the high school public system of Bogotá. A paper presented at the *9th Latin American and Caribbean Conference for Engineering and Technology.* Medellin, August 3–5, 2011.

158. Rojas-Sola J.I. and De San-Antonio-Gomez C. (2010). Bibliometric analysis of Colombian scientific publications in engineering, multidisciplinary subject category in web of science database (1997–2009). *Dyna*. **77**(164):9–17.

159. Sanabria L.E. and Quintero L.S. (2012). Analysis of Colombian bitumen modified with a nanocomposite. *Journal of Testing and Evaluation*. **40**(7):1–6.

160. Silva F. (2010). Nanotecnología En Colombia, Retos Y Oportunidades. Dissertation: Bogota, University of Colombia.

161. Vinck D. and Pérez Marteló C. (2008). Redes sociotécnicas de cogestión de conocimiento en nanotecnología en Colombia: ¿ entre la visibilidad internacional y la apropiación local? *Redes*. **15**(29):113–137.

Costa Rica

162. Rivera-Álvarez A. and Vega-Baudrit J.R. (2015). Divulgación de la nanociencia y nanotecnología en Costa Rica 2013–2014. *Revista de Física*. **49E**:59–66.

163. Vargas I.A. and Álvarez E.M. (2014). Mapeo de la industria de biotecnología y nanotecnología en Costa Rica (IC-IM-02-2014).

164. Vega-Baudrid J.R. (2007). La Nanotecnología en Costa Rica: la experiencia en el LANOTEC. IX Congreso Nacional de Ciencias Exploraciones fuera y dentro del aula. Costa Rica Institute of Technology, Cartago, Costa Rica, August 24–25.

165. Vega-Baudrid J. et al. (2012). Sustainable nanotechnology policies for innovation in Costa Rica. *JONPI*. **3**(2):21–38.

Croatia

166. Čatić I. (2010). *Hrvatska znanost na prekretnici. Strojarstvo: časopis za teoriju i praksu u strojarstvu*. **52**(3):260–260.

Cuba

167. Aguiar J.D. et al. (2012). Producción científica cubana sobre nanociencias y nanotecnología. *Ciencias de la Información*. **43**(1):5–14 (Spanish).

168. Castellanos C.R. (2014). Nanotechnologies in Cuba: popularization and training. In *The History of Physics in Cuba*. Netherlands: Springer, pp. 323–328.

169. Castillo A. et al. (2013). Modelación económico-financiera de la gestión para empresas biotecnológicas en Cuba. *Biotecnología Aplicada*. **30**(4):290–298 (Spanish).

170. Castillo F.E. (2013). Retos de este siglo: nanotecnología y salud. *Revista Cubana de Hematología, Inmunología y Hemoterapia*. **29**(1).

171. Cedeño J.D.A. et al. (2011). Propuesta del sitio Web para la Gestión de Contenidos sobre Nanociencias y Nanotecnologías del Centro de Estudios Avanzados de Cuba. *Ciencias de la Información*. **42**(2):61–70 (Spanish).

172. Quiñones A.A. (2012). La percepción social de las nanotecnologías en Cuba en un grupo de actores sociales relevantes: ¿ un tema para después? (The social perception of nanotechnologies in Cuba in a group of relevant social actors: a topic for next?). *Revista Congreso Universidad*. **1**(1).

173. Stone R. (2015). Fidel Castro's first-born son foments a nanotech revolution. *Science*. **348**(6236):748–749.

174. Thorsteinsdóttir H. et al. (2004). Cuba: innovation through synergy. *Nature Biotechnology*. **22**(Suppl):DC19–DC24.

Czech Republic

175. Abrhám J. and Herget J. (2013). Identification, current state and development of clusters in the Czech Republic. *Mediterranean Journal of Social Sciences*. **4**(11):144.

176. Bärtl S. (2013). There's plenty of room at the bottom but is there room at the top? Nanotechnology in the Czech Republic: policy and barriers to innovation. Student Paper: Lund University, School of Economics and Management, Lund, Sweden.

177. Czech Invest (2008). Nanotechnology in the Czech Republic. Available online at http://www.czechinvest.org/data/files/nanotechnologies-preview-1232-en.pdf

178. Luňáček J. et al. (2004). Physics status in the nanotechnology education project at the VŠB-Technical University of

Ostrava. A paper presented at the *International Conference on Engineering Education and Research "Progress Through Partnership."* Technical University of Ostrava, Ostrava-Poruba, Czech Republic.

179. Voves J. (2007). Trends and skills needs in the field of nanotechnology–the state of affairs in the Czech Republic in the European context. Skill needs in emerging technologies. Prague: Czech Technical University, Faculty of Electrical Engineering, Department of Microelectronics.

Denmark

180. Andersen M.M. and Molin M.J. (2007). Nanobyg: a survey of nanoinnovation in Danish construction (Risø-R-1602). Roskilde, Denmark: Risø National Laboratory.

181. Andersen M.M. and Rasmussen B. (2010). Nanotechnology development in Denmark—environmental opportunities and risk (Risø-R-1550). Roskilde, Denmark: Risø National Laboratory.

182. Dannemand Andersen P. et al. (2005). Technology foresight on Danish nano-science and nano-technology. *Foresight.* **7**(6):64–78.

183. Hansen S.F. and Kristensen H.V. (2014). Horizon-scanning and identification of emerging risks among nanotech companies in Denmark. *Nanotechnology Law and Business.* **11**:82–92.

184. Jamison A. and Mejlgaard N. (2010). Contextualizing nanotechnology education: fostering a hybrid imagination in Aalborg, Denmark. *Science as Culture.* **19**(3):351–368.

185. Jensen J.S. and Koch C. (2007). Accelerating nano-technological innovation in the Danish construction industry. In Boyd D. (ed.) *Procs 23rd Annual ARCOM Conference.* Belfast, UK, Association of Researchers in Construction Management, September 3–5, pp. 609–618.

186. Kjærgaard R.S. (2008). Making a small country count: nanotechnology in Danish newspapers from 1996 to 2006. *Public Understanding of Science.* **19**(1):80–97.

187. Kristensen H.V. et al. (2009). Adopting eco-innovation in Danish polymer industry working with nanotechnology:

drivers, barriers and future strategies. *Nanotechnology Law and Business.* **6**:416.

188. Munch-Andersen M. et al. (2007). NanoByg: a survey of nanoinnovation in Danish construction. Danmarks Tekniske Universitet, Risø Nationallaboratoriet for Bæredygtig Energi.

189. Søgaard Jørgensen M. et al. (2006). Green technology foresight about environmentally friendly products and materials: the challenges from nanotechnology, biotechnology and ICT (Working Report No. 34). Copenhagen: Danish Ministry of the Environment.

190. Stipp S.L.S. (2006). New Danish research program applying nanotechnology to improve oil recovery. *Elements.* **2**(3).

191. Tønning K. and Poulsen M.D. (2007). Nanotechnology in the Danish industry: survey on production and application. Copenhagen: Environmental Protection Agency.

192. Vestergård G.L. and Nielsen K.H. (2015). From the preserves of the educated elite to virtually everywhere: a content analysis of Danish science news in 1999 and 2012. *Public Understanding of Science.* doi:10.1177/0963662515603272.

Developing and Emerging Countries

193. Bradley E.L. et al. (2011). Applications of nanomaterials in food packaging with a consideration of opportunities for developing countries. *Trends in Food Science & Technology.* **22**(11):604–610.

194. Brame J. et al. (2011). Nanotechnology-enabled water treatment and reuse: emerging opportunities and challenges for developing countries. *Trends in Food Science & Technology.* **22**(11):618–624.

195. Buürgi B.R. and Pradeep T. (2006). Societal implications of nanoscience and nanotechnology in developing countries. *Current Science.* **90**(5):645–658.

196. Çetindamar D. et al. (2009). Does technology management research diverge or converge in developing and developed countries? *Technovation.* **29**(1):45–58.

197. Chaudhry Q. and Castle L. (2011). Food applications of nanotechnologies: an overview of opportunities and challenges for developing countries. *Trends in Food Science & Technology.* **22**(11):595–603.

198. Cozzens S. (2012). The distinctive dynamics of nanotechnology in developing nations. In *Making It to the Forefront.* New York: Springer, pp. 125–138.

199. Daar A. et al. (2007). How can developing countries harness biotechnology to improve health? *BMC Public Health.* **7**(1):346.

200. El Naschie M.S. (2006). Nanotechnology for the developing world. *Chaos, Solitons & Fractals.* **30**(4):769–773.

201. Grimshaw D.J., Gudza L.D. and Stilgoe J. (2011). How can nanotechnologies fulfill the needs of developing countries? In *Nanotechnology and the Challenges of Equity, Equality and Development.* Amsterdam, Netherlands: Springer, pp. 379–391.

202. Hassan M. H. (2005). Small things and big changes in the developing world. *Science.* **309**(5731):65–66.

203. Invernizzi N. and Foladori G. (2005). Nanotechnology and the developing world: will nanotechnology overcome poverty or widen disparities. *Nanotechnology Law & Business Journal.* **2**(3): Article 11.

204. Jamal H. (2006). Roadmap to nanoelectronics for developing countries: a realistic approach. *WSEAS Transactions on Electronics.* **3**(4):214.

205. Maclurcan D. (2005). Nanotechnology and developing countries–Part 1: What possibilities? *Online Journal of Nanotechnology.* **1**. Available at http://www.azonano.com/article.aspx? ArticleID =1428

206. Maclurcan D. (2005). Nanotechnology and developing countries–Part 2: What realities? *Online Journal of Nanotechnology.* **1**. Available at http://www.azonano.com/article. aspx? ArticleID =1429

207. Meyer M. (2000). Patent citations in a novel field of technology: what can they tell about interactions between emerging communities of science and technology? *Scientometrics.* **48**(2):151–178.

208. Naseri R. and Davoodi R. (2011). Commercialization of nanotechnology in developing countries. *IPEDR*. **12**:85–389.

209. Niosi J. et al. (2012). The international diffusion of biotechnology: the arrival of developing countries. *Journal of Evolutionary Economics*. **22**(4):767–783.

210. Núñez-Mujica G.D. (2006). Employing geoethics to avoid negative nanotechnology scenarios in developing countries. *The Journal of Geoethical Nanotechnology*. **1**(4):2–8.

211. Ramani S.V. (ed.) (2014). *Nanotechnology and Development: What's in It for Emerging Countries*. New York: Cambridge University Press.

212. Rodrigues R. et al. (2007). Nanotechnology and the global poor: United States policy and international collaborations. *Technical Proceedings of the 2007 Nanotechnology Conference and Trades Show*. **1**:593–596.

213. Romig A.D. et al. (2007). An introduction to nanotechnology policy: opportunities and constraints for emerging and established economies. *Technological Forecasting and Social Change*. **74**(9):1634–1642.

214. Salamanca-Buentello F. et al. (2005). Nanotechnology and the developing world. *PLoS Medicine*. **2**(5):383.

215. Schummer J. (2007). Impact of nanotechnologies on developing countries. In Allhoff F., Lin P., Moor J. and Weckert J. (eds.), *Nanoethics: The Ethical and Social Implications of Nanotechnology*. Hoboken, NJ: Wiley, pp. 291–307.

216. Thakur D. (2011). Open access nanotechnology for developing countries: lessons from open source software. In *Nanotechnology and the Challenges of Equity, Equality and Development*. Amsterdam, Netherlands: Springer, pp. 331–347.

217. Uskokovic V. et al. (2010). Strategies for the scientific progress of the developing countries in the new millennium. *Science, Technology and Innovation Studies*. **6**(1):33–62.

218. Woodson T.S. (2014). Nanotechnology and development: what's in it for emerging countries? *The Journal of Development Studies*. **50**(11):1590–1591.

Ecuador

219. Loayza Villa M.F. (2010). Nueva estrategia terapéutica basada en nanotecnología contra leishmaniasis tegumentaria en Ecuador (New pharmacological strategy based in nanotechnology against cutaneous leishmaniasis in Ecuador). Ph.D. Thesis: Universidad San Francisco de Quito.

220. Páez L. and Moreno M. (2014). Financiamiento para la innovación en los nuevos sectores estratégicos del Ecuador. In *Economía y las Oportunidades de Desarrollo: Desafíos en América-Latina*. ECORFAN, pp. 223–244.

221. Perugachi R. et al. (2006). Las Nanoarcillas y sus potenciales aplicaciones en el Ecuador. *Revista Tecnológica-ESPOL*. **19**(1):121–124.

222. Terrones M. (2010). Nanomaterials research: a big step for Ecuador. *Nature Materials*. **9**:704–705.

Egypt

223. Denney D. (2014). Nanotechnology applications for challenges in Egypt. *Journal of Petroleum Technology*. **66**(02):123–126.

224. El-Diasty A.I. and Ragab A.M.S. (2013). Applications of nanotechnology in the oil & gas industry: latest trends worldwide & future challenges in Egypt. A paper presented at the *North Africa Technical Conference and Exhibition*. Society of Petroleum Engineers, Cairo.

225. Ibrahim El-Diasty A. (2015). The potential of nanoparticles to improve oil recovery in bahariya formation, Egypt: an experimental study. A paper presented at the *SPE Asia Pacific Enhanced Oil Recovery Conference*. Kuala Lumpur, Malaysia.

226. Ramadan A.B.A. (2009). Air pollution monitoring and use of nanotechnology based solid state gas sensors in greater Cairo area, Egypt. In *Nanomaterials: Risks and Benefits*. Netherlands: Springer, pp. 265–273.

227. Selim S.A.S. et al. (2015). Integrating nanotechnology concepts and its applications into the secondary stage physics curriculum in Egypt. *European Scientific Journal*. **11**(12):193–212.

El Salvador

228. Roberto Alegría Coto J. (2004). La nanotecnologia y su impacto en la educación superior universitaria. *Realidad y Reflexion.* **4**(12):19–30 (Spanish).

Estonia

229. Kattel R. (2004). Governance of innovation policy: the case of Estonia. *Trames.* (4):419–427.
230. Tiits M. (2003). Towards modern STI policy making in Estonia. *Trames.* (1):53–62.
231. Tiits M. (2007). Technology-intensive FDI and economic development in a small country: the case of Estonia. *Trames.* (3):324–342.
232. Vissak T. (2007). The emergence and success factors of fast internationalizers: four cases from Estonia. *Journal of East-West Business.* **13**(1):11–33.

Europe and European Union

233. Abicht L. et al. (2006). Identification of skill needs in nanotechnology. In *Cedefop Panorama Series 120.* Luxembourg: European Communities.
234. Ahmadi M. and Ahmadi L. (2013). European patent law framework regarding nanotechnology applications in stem cells. *Nanotechnology Law and Business.* **10**:65.
235. Amenta V. et al. (2015). Regulatory aspects of nanotechnology in the agri/feed/food sector in EU and non-EU countries. *Regulatory Toxicology and Pharmacology.* **73**(1):463–476.
236. Baraton M.I. et al. (2006). European activities in nanoscience education and training. *MRS Proceedings.* **931**:093.
237. Bochon A. (2011). Evolution of the European commission recommendation for a code of conduct for responsible nanosciences and nanotechnology research. *Nanotechnology Law and Business.* **8**:117.

238. Brosset E. (2013). The law of the European Union on nanotechnologies: comments on a paradox. *Review of European, Comparative & International Environmental Law.* **22**(2):155–162.

239. Calignano G. and Quarta C.A. (2015). The persistence of regional disparities in Italy through the lens of the European Union nanotechnology network. *Regional Studies, Regional Science.* **2**(1):469–478.

240. Chesneau A. et al. (2008). Teaching nanoscience across scientific and geographical borders: a European master programme in nanoscience and nanotechnology. *Journal of Physics: Conference Series.* **100**(3):032002.

241. Coles D. and Frewer L.J. (2013). Nanotechnology applied to European food production: a review of ethical and regulatory issues. *Trends in Food Science & Technology.* **34**(1):32–43.

242. Collin J. (2006). European commission action plan on nanotechnologies: a brief presentation with a view on intellectual property. *Nanotechnology Law and Business.* **3**:80.

243. Colombelli A. et al. (2014). The emergence of new technology-based sectors in European regions: a proximity-based analysis of nanotechnology. *Research Policy.* **43**(10):1681–1696.

244. Cunningham S.W. and Werker C. (2012). Proximity and collaboration in European nanotechnology. *Papers in Regional Science.* **91**(4):723–742.

245. Dalton-Brown S. (2012). Global ethics and nanotechnology: a comparison of the nanoethics environments of the EU and China. *NanoEthics.* **6**(2):137–150.

246. Del Castillo A.M.P. (2013). The European and member states' approaches to regulating nanomaterials: two levels of governance. *NanoEthics.* **7**(3):189–199.

247. Dorbeck-Jung B. and Shelley-Egan C. (2013). Meta-regulation and nanotechnologies: the challenge of responsibilisation within the European Commission's code of conduct for responsible nanosciences and nanotechnologies research. *NanoEthics.* **7**(1):55–68.

248. Dorocki S. and Kula A. (2009). Przestrzenne zróżnicowanie rozwoju nanotechnologii w Europie [Spatial diversity of nanotechnology development in Europe]. *Studies of the Industrial Geography Commission of the Polish Geographical Society.* **29**(1):27–41 (Polish).

249. Ehmann F. et al. (2013). Next-generation nanomedicines and nanosimilars: EU regulator's initiatives relating to the development and evaluation of nanomedicines. *Nanomedicine (Lond).* **8**(5):849–856.

250. Eisenberger I. et al. (2010). Nano regulation in the European Union. NanoTrust-Dossiers. Institut für Technikfolgen-Abschätzung (ITA), Wien.

251. Escoffier L. (2010). The European Union consultation on nanotechnology. *Nanotechnology Law and Business.* **7**:97.

252. Esslinger A. (2007). Patenting nanotechnology inventions in Europe. *Nanotechnology Law and Business.* **4**:495.

253. Euronano (2009). Nanotechnology in Europe: assessment of the current state, opportunities, challenges and socio-economic impact (Phase 1, Framework Service Contract 150083-2005-02-BE). Grenoble.

254. European Commission (2010). Communicating nanotechnology: why, to whom, saying what and low? Brussels: Communication Unit, Directorate General for Research, European Commission.

255. European Union (2012). Monitoring policy and research activities on science in society in Europe (MASIS): a final synthesis reports. Brussels: Directorate General for Research and Innovation.

256. Gabellieri C. and Frima H. (2011). Nanomedicine in the European Commission policy for nanotechnology. *Nanomedicine: Nanotechnology, Biology and Medicine.* **7**(5):519–520.

257. Galiay P. (2011). Situation in Europe and the World: a code of conduct for responsible European research in nanoscience and nanotechnology. In *Nanoethics and Nanotoxicology.* Berlin Heidelberg: Springer, pp. 497–509.

258. Gaskell G. et al. (2005). Imagining nanotechnology: cultural support for technological innovation in Europe and the United States. *Public Understanding of Science.* **14**(1):81–90.

259. Godman M. and Hansson S.O. (2009). European public advice on nanobiotechnology: four convergence seminars. *NanoEthics.* **3**(1):43–59.

260. Gyalog T. (2007). Nanoscience education in Europe. *Euro Physics News.* **38**(1):13–15.

261. Heinze T. (2004). Nanoscience and nanotechnology in Europe: analysis of publications and patent applications including comparisons with the United States. *Nanotechnology Law and Business.* **1**:427.

262. Hellsten E. (2007). The European nanotechnology strategy: environmental and health aspects. *Journal of Nanoscience and Nanotechnology.* **6**(2):502–513.

263. Hofmann-Amtenbrink M. et al. (2014). Nanotechnology in medicine: European research and its implications. *Swiss Medical Weekly.* **144**:14044.

264. Hullmann A. (2008). European activities in the field of ethical, legal and social aspects (ELSA) and governance of nanotechnology. DG Research, Brussels: European Commission.

265. Hydzik P. (2011). [Risks associated with nanotechnology based on European Union legislation]. *Przeglad Lekarski.* **69**(8):490–491 (Polish).

266. Jemala M. (2015). Systemic insights into nanotechnology patenting in EU countries. *International Journal of Agile Systems and Management.* **8**(1):1–22.

267. Jurewicz M. (2014). Cosmetic products with the use of nanotechnology in terms of the European Union law. *Przemysl Chemiczny.* **93**(12):2138–2140.

268. Justo-Hanani R. and Dayan T. (2015). European risk governance of nanotechnology: explaining the emerging regulatory policy. *Research Policy.* **44**(8):1527–1536.

269. Kallinger C. et al. (2008). Patenting nanotechnology: a European patent office perspective. *Nanotechnology Law and Business.* **5**:95.

270. Kanama D. (2007). EU nanoroadmap: issues and outlook for technology roadmaps in the nanotechnology field. *Science and Technology Trends: Quarterly Review.* **23**:55–64.

271. Kelly B. and Bogaert P. (2008). Medical nanotechnology in Europe-Brian Kelly and Peter Bogaert discuss the regulatory and legal implications of medical nanotechnology in the EU. *Regulatory Affairs Journal-Devices.* **16**(4):243.

272. Kiškis M. (2010). Regulation of nanotechnology in the European Union. *Societal Studies.* **4**(8):357–372.

273. Kozhukharov V. and Machkova M. (2013). Nanomaterials and nanotechnology: European initiatives, status and strategy. *Journal of Chemical Technology and Metallurgy.* **48**(1):3–11.

274. Laurent B. (2012). Responsible agreements: constructing markets for nanotechnology through the European regulation. *Nanotechnology Law and Business.* **9**:267.

275. Laurent B. (2012). Science museums as political places. Representing nanotechnology in European science museums. *JCOM: Journal of Science Communication.* **11**(04):6.

276. Lee M. (2010). Risk and beyond: EU regulation of nanotechnology. *European Law Review.* **35**(6):799–821.

277. Malsch I. (1999). Nanotechnology in Europe: scientific trends and organizational dynamics. *Nanotechnology.* **10**(1):1.

278. Marrani D. (2013). Nanotechnologies and novel foods in European law. *NanoEthics.* **7**(3):177–188.

279. Michelson E. (2004). Analyzing the European approach to nanotechnology. Washington, D.C.: Woodrow Wilson International Center for Scholars.

280. Mielke S.K. (2013). Regulating in thin air: nanotechnology regulation in the European union. *Review of European, Comparative & International Environmental Law.* **22**(2):146–154.

281. Nano Forum (2003). Third nanoforum report: nanotechnology and its implications for the health of the EU citizen.

282. Nano Forum (2005). Sixth nanoforum report: European nanotechnology infrastructure and networks.

283. O'loughlin R.C. (2006). Antitrust or antitrade-self-assessment of market share chills the incentive to license nanotechnology

patents in the European Union. *Washburn Law Journal.* **46**: 345.

284. Ovalle-Perandones M.A. et al. (2013). The influence of European framework programmes on scientific collaboration in nanotechnology. *Scientometrics.* **97**(1):59–74.

285. Pandza K. et al. (2011). Collaborative diversity in a nanotechnology innovation system: evidence from the EU framework programme. *Technovation.* **31**(9):476–489.

286. Papadaki M. (2003). Research on medical applications of nanotechnology in the European Union [Cellular/tissue engineering]. *IEEE Engineering in Medicine and Biology Magazine.* **22**(1): 88.

287. Pfitzner L. et al. (2007). Metrology, analysis and characterization in micro-and nanotechnologies: a European challenge. *ECS Transactions.* **10**(1):35–49.

288. Rickerby D.G. and Morrison M. (2007). Nanotechnology and the environment: a European perspective. *Science and Technology of Advanced Materials.* **8**(1):19–24.

289. Schellekens M. (2010). Patenting nanotechnology in Europe: making a good start? An analysis of issues in law and regulation. *The Journal of World Intellectual Property.* **13**(1):47–76.

290. Scheufele D.A. et al. (2009). Religious beliefs and public attitudes toward nanotechnology in Europe and the United States. *Nature Nanotechnology.* **4**(2):91–94.

291. Sinner F. et al. (2010). European center for nanotoxicology-nanoscale materials: a new challenge for toxicology. *Scientia Pharmaceutica.* **78**:580.

292. Soldatenko A. (2011). An overview of activities related to nanotechnologies in central and eastern Europe, Caucasus and central Asia (White Paper). University of Strasbourg.

293. Talesnik M. et al. (2012). Report on best practices in nanotechnology education at the secondary school level (NMP4-SA-2012-319054). Brussels: NanoEIS Consortium.

294. ten Have H. (2014). Nanotechnology and ethics–European public policies. In *Pursuit of Nanoethics.* Netherlands: Springer, pp. 193–208.

295. Tolles W.M. (1996). Nanoscience and nanotechnology in Europe. *Nanotechnology.* **7**(2):59.

296. Tolles W.M. (1994). Nanoscience and nanotechnology in Europe (NRL/FR/1003-94-9755). Washington, D.C.: Naval Research Laboratory.

297. Tomellini R. and Faure U. (2003). Research on nanosciences and nanotechnologies within the European unions 6th framework programme. *Reviews on Advanced Materials Science.* **5**:1–5.

298. Tomellini R. and Monk R. (2005). A European approach to nanotechnologies. *GAIA-Ecological Perspectives for Science and Society.* **14**(1):28–29.

299. van Broekhuizen P. and Reijnders L. (2011). Building blocks for a precautionary approach to the use of nanomaterials: positions taken by trade unions and environmental NGOs in the European nanotechnologies debate. *Risk Analysis.* **31**(10):1646–1657.

300. van Calster G. (2006). Regulating nanotechnology in the European Union. *Nanotechnology Law and Business.* **3**:359.

301. van Calster G. (2009). Simply swallow: the application of nanotechnologies in European food law. *European Food and Feed Law Review.* **4**(3):167–171.

Finland

302. Koponen P. et al. (2007). Nanotechnology in Finnish Industry 2006 survey results. Espoo, Finland: Spinverse Consulting.

303. Most F.V. (2009). Research councils facing new science and technology: the case of nanotechnology in Finland, the Netherlands, Norway and Switzerland. Thesis: University of Twente. doi:10.3990/1.9789036528979

304. Nikulainen T. (2010). Identifying nanotechnological linkages in the Finnish economy: an explorative study. *Technology Analysis & Strategic Management.* **22**(5):513–531.

305. Nikulainen T. and Kulvik M. (2009). How general are general purpose technologies? Evidence from Nano-, Bio- and ICT-technologies in Finland (ETLA Discussion Paper No. 1208).

306. Nikulainen T. and Palmberg C. (2008). Nanotechnology and industrial renewal in Finland: a synthesis of key findings. Helsinki: Finnish Funding Agency for Technology & Innovation.

307. Palmberg C. and Nikulainen T. (2006). Industrial renewal and growth through nanotechnology? An overview with focus on Finland (No. 1020). Helsinki: The Research Institute of the Finnish Economy.

308. Palmberg C. and Nikulainen T. (2008). Nanotechnology and industrial renewal in Finland: a synthesis of key findings. Helsinki: Etalatieto Ltd.

309. Palmberg C. et al. (2012). Transferring science-based technologies to industry—does nanotechnology make a difference? (No. 1064). Helsinki: The Research Institute of the Finnish Economy.

310. Raivio T. et al. (2008). Nanosafety in Finland: a summary report (Tekes Review No. 224). Helsinki: Tekes, The Finnish Funding Agency for Technology & Innovation.

311. van der Most F. (2000). Research councils facing new science and technology the case of nanotechnology in Finland, The Netherlands, Norway and Switzerland. PhD Dissertation: University of Twente.

France

312. André J.C. (2005). [Réflexions autour de la nano-éthique et de la nanonormalisation]. *Environment, Risques & Santé.* **4**(6):411–415 (French).

313. Bieberstein A. et al. (2011). Consumers between indifference and mistrust: French and German behaviour regarding the potential introduction of nanotechnologies in food. *INRA Sciences Sociales.* http://purl.umn.edu/149768

314. Bieberstein A. et al. (2013). Consumer choices for nano-food and nano-packaging in France and Germany. *European Review of Agricultural Economics.* **40**(1):73–94.

315. Bordé J. (2011). Situation in France: ethical reflection on research in nanoscience and nanotechnology. In *Nanoethics*

and Nanotoxicology. Berlin Heidelberg: Springer, pp. 437–453.

316. Desmoulin S. (2008). French and European Community Law on the nanometric forms of chemical substances: questions about how the law handles uncertain risks. *Nanotechnology Law and Business.* **5**:341.

317. Fontaine A. et al. (2008). The C'nano competence centres. *International Journal of Nanotechnology.* **5**(6–8):571–573.

318. Hesto P. and Lourtioz J.M. (2016). Research in nanoscience and nanotechnology: the French research system. In *Nanosciences and Nanotechnology.* Switzerland: Springer International Publishing, pp. 315–324.

319. Jouvenet M. (2013). Boundary work between research communities: culture and power in a French nanosciences and nanotechnology hub. *Social Science Information.* **52**(1):134–158.

320. Laurent B. (2013). Nanomaterials in political life: in the democracies of nanotechnology. In *Nanomaterials: A Danger or a Promise?* London: Springer, pp. 379–399.

321. Nel A. et al. (2013). Implications of the French registry for engineered nanomaterials. *ACS Nano.* **7**(6):4694–4696.

322. Ogasawara A. (2002). Trends in French science, technology, and innovation policy: the MINATEC industry-academia-government nanotechnology innovation center project. *Science and Technology Trends: Quarterly Review.* **4**:64–71.

G-7 Countries (Canada, France, Germany, Italy, Japan, United Kingdom, United States of America)

323. Yang L.Y. et al. (2012). A comparison of disciplinary structure in science between the G7 and the BRIC countries by bibliometric methods. *Scientometrics.* **93**(2):497–516.

324. Zhao Q. and Guan J. (2010). International collaboration of three 'giants' with the G7 countries in emerging nanobiopharmaceuticals. *Scientometrics.* **87**(1):159–170.

G-15 Countries (Algeria, Argentina, Brazil, Chile, Egypt, India, Indonesia, Iran, Jamaica, Kenya, Malaysia, Mexico, Nigeria, Peru, Senegal, Sri Lanka, Venezuela, and Zimbabwe)

325. Karpagam R. et al. (2011). Publication trend on nanotechnology among G15 countries: a bibliometric study. *COLLNET Journal of Scientometrics and Information Management.* **5**(1):61–80.

G-20 Countries (Argentina, Australia, Brazil, Canada, China, France, Germany, India, Indonesia, Italy, Japan, Republic of Korea, Mexico, Russia, Saudi Arabia, South Africa, Turkey, United Kingdom, United States, and the European Union)

326. Karpagam R. (2014). Literature in nanotechnology among G20 Countries: a scientometrics study based on scopus database. PhD Dissertation: Chennai, Anna University.

Germany

327. Blind K. and Gauch S. (2009). Research and standardisation in nanotechnology: evidence from Germany. *The Journal of Technology Transfer.* **34**(3):320–342.
328. Bundesministerium für Bildung und Forschung (2011). Status quo der Nanotechnologie in Deutschland. Bonn, Berlin: BMBF.
329. Donk A. et al. (2012). Framing emerging technologies: risk perception of nanotechnology in the German press. *Science Communication.* **34**(1):5–29.
330. Federal Ministry of Education and Research (2004). Nanotechnology conquers markets: German innovation initiative for nanotechnology. Berlin: The Ministry, Publications and Website Division.
331. Guenther L. and Ruhrman G. (2013). Science journalists' selection criteria and depiction of nanotechnology in German media. *Journal of Science Communication.* **12**(3):1–17.

332. Heinze T. and Kuhlmann S. (2008). Across institutional boundaries? Research collaboration in German public sector nanoscience. *Research Policy*. **37**(5):888–899.

333. Henke S. (2004). Financing of MEMS/MOEMS and nanotechnology in Germany. In *MicroNano Integration*. Berlin Heidelberg: Springer, pp. 37–39.

334. Henn S. (2007). Evolution of regional clusters in nanotechnology: empirical findings from Germany. *Nanotechnology Law and Business*. **4**:501.

335. Lambauer J. et al. (2008). Nanotechnology and its impact on the German energy sector. *Advanced Engineering Materials*. **10**(5):423–427.

336. Rieke V. and Bachmann G. (2004). German innovation initiative for nanotechnology. *Journal of Nanoparticle Research*. **6**(5):435–446.

337. Wald A. (2007). Effects of 'mode 2'-related policies on the research process: the case of publicly funded German nanotechnology. *Science Studies*. **20**(1).

338. Zweck A. et al. (2008). Nanotechnology in Germany: from forecasting to technological assessment to sustainability studies. *Journal of Cleaner Production*. **16**(8):977–987.

Guatemala

339. Salazar F.G. (2001). Propuesta para la creación de un parque tecnológico en nanoctecnológia en guatemala. *Revista Electrónica Ingeniería Primero*. **18**:112–135.

Hong Kong

340. Soliman A. and Yip H. (2013). Nanotechnology regulation in Hong Kong: a comparative legal study. *City University of Hong Kong Law Review*. **4**(1).

Hungary

341. Kelemen L. et al. (2012). Scientific publications of the institute of biophysics of the Hungarian academy of sciences. *European Polymer Journal*. **48**:1745–1754.

India

342. Ali A. and Sinha K. (2014). Exploring the opportunities and challenges in nanotechnology innovation in India. *Journal of Social Science for Policy Implications.* **2**(2):227–251.

343. Ali A. and Sinha K. (2015). Policy on risk governance for nanotechnology development in India. *Nanotechnology Law and Business.* **12**:60.

344. Barpujari I. (2011). Attenuating risks through regulation: issues for nanotechnology in India. *Journal of Biomedical Nanotechnology.* **7**(1):85–86.

345. Barpujari I. (2011). Incentivizing innovation and serving the public good: extending the patent regime to nanotechnology in India and Sri Lanka. http://www.thesnet.net/wp-content/uploads/9_Barpujari.pdf.

346. Barpujari I. (2011). Public engagement in emerging technologies: issues for India. In Zulsdorf T.C., Coenan A., Ferrari A., Fiedeler U., Milburn C. and Wienroth M. (eds.), *Quantum Engagements: Social Reflections of Nanoscience and Emerging Technologies.* Heidelberg, Germany: Akademische Verlagsgesellschaft, pp. 123–137.

347. Beumer K. and Bhattacharya S. (2013). Emerging technologies in India: developments, debates and silences about nanotechnology. *Science and Public Policy.* **40**(5):628–643.

348. Chowdhury N. (2006). Regulatory supervision of emerging technologies: a case for nanotechnology in India. *Economic and Political Weekly.* **41**(46):4730–4733.

349. Gupta V.K. (2009). Indian publications output in nanotechnology during 1990–2008. *Advanced Science Letters.* **2**(3):402–404.

350. Jain A., Hallihosur S. and Rangan L. (2011). Dynamics of nanotechnology patenting: an Indian scenario. *Technology in Society.* **33**(1):137–144.

351. Jayanthi A.P., Beumer K. and Bhattacharya S. (2012). Nanotechnology: risk governance in India. *Economic and Political Weekly.* **47**(4):34–40.

352. Karpagam R. et al. (2011). Mapping of nanoscience and nanotechnology research in India: a scientometrics analysis, 1990–2009. *Scientometrics.* **89**(2):501–522.

353. Kumar A. (2009). Nanotechnology development in India: an overview (RIS Discussion Paper #193). New Delhi: Research and Information System for Developing Countries.

354. Kumar A. and Desai P.N. (2011). Mapping the Indian nanotechnology innovation system. *World Journal of Science, Technology & Sustainable Development.* **11**(1):53–65.

355. Kumar M.J. (2014). Quantum computing in India: an opportunity that should not be missed. *IETE Technical Review.* **31**(3):187–189.

356. Madhulika B. et al. (2009). Nanotechnology development in India: quantitative analysis. Turkey: Istanbul Bilgi University, September 20–23, 338.

357. Mazumder S. et al. (2014). Nanotechnology commercialization: prospects in India. *Journal of Materials Science & Nanotechnology.* **2**(2):1.

358. Mohan L. et al. (2010). Research trends in nanoscience and nanotechnology in India. *DESIDOC Journal of Library & Information Technology.* **30**(2):40–58.

359. Patra D. et al. (2009). Nanoscience and nanotechnology: ethical, legal, social and environmental issues. *Current Science.* **96**(5):651–657.

360. Patra D. (2011). Nanoscience, nanotechnology, or nanotechnoscience: perceptions of Indian nanoresearchers. *Public Understanding of Science.* **22**(5):590–605.

361. Purushotham H. (2012). Transfer of nanotechnologies from R&D institutions to SMEs in India. *Carbon.* **16**:47.

362. Sarma S.D. (2011). How resilient is India to nanotechnology risks? *European Journal of Law and Technology.* **2**(3):1–15.

363. Sarma S.D. and Anand M. (2012). Status of nano science and technology in India. *Proceedings of the National Academy of Sciences, India Section B: Biological Sciences.* **82**(1):99–126.

364. Sastry R.K. et al. (2011). Nanotechnology for enhancing food security in India. *Food Policy.* **36**(3):391–400.

365. Sen P. (2008). Nanotechnology: the Indian scenario. *Nanotechnology Law & Business.* **5**:225.

366. Singh D.N. (2007). Nanotechnology: an Indian perspective. *IETE Technical Review.* **24**(1):43–49.

367. Singh Y. (2010). Two decades of nanoscience and nanotechnology research in India: a scientometric analysis. *Nano Trends.* **8**(1):15–28.

368. Subramanian K.S. and Tarafdar J.C. (2011). Prospects of nanotechnology in Indian farming. *Indian Journal of Agricultural Sciences.* **81**(10):887–893.

369. Subramanian V. et al. (2012). Nanotechnology in India: inferring links between emerging technologies and development. In *Making It to the Forefront.* New York: Springer, pp. 109–124.

370. The Energy and Resources Institute (2010). Nanotechnology development in India: building capability and governing the technology (Briefing Paper). New Delhi.

371. Thirumagal A. (2012). Bibliometric study of nanotechnology in India: an analysis. *SRELS Journal of Information Management.* **49**(5):577–587.

Indonesia

372. Ratnawati R. et al. (2006). Nanotechnology: an emerging new technology for Indonesia. Part I. Nanotechnology in general. *Reaktor.* **10**(1):46–53.

Iran

373. Baradar R. et al. (2009). Mapping the Iranian ISI papers on nanoscience and nanotechnology: a citation analysis study. *Malaysian Journal of Library & Information Science.* **14**(3):95–107.

374. Baradar R. et al. (2012). The growth and interdisciplinary patterns of nanoscale research in Iran. *International Journal of Information Science and Management.* **8**(2):29–37.

375. Farshchi P. et al. (2011). Nanotechnology in the public eye: the case of Iran, as a developing country. *Journal of Nanoparticle Research.* **13**(8):3511–3519.

376. Ghazinoory S. et al. (2009). A model for national planning under new roles for government: case study of the National Iranian Nanotechnology Initiative. *Science and Public Policy.* **36**(3):241–249.

377. Ghazinouri R. and Ghazinoory S. (2009). Nanotechnology and sociopolitical modernity in developing countries; case study of Iran. *Technological and Economic Development of Economy.* (3):395–417.

378. Hassanzadeh M. and Khodadust R. (2012). Dimensions of Iranian international co-authorship network in the field of nanotechnology. *Journal of Science & Technology Policy.* **5**(1):31–44.

379. Hosseini S.J.F. and Eghtedari N. (2013). A confirmatory factorial analysis affecting the development of nanotechnology in agricultural sector of Iran. *African Journal of Agricultural.* **8**(16):1401–1404.

380. Hosseini S.M. and Rezaei R. (2009). Factors affecting the attitudes of Iranian agricultural faculty members towards nanotechnology. *World Applied Sciences Journal.* **2**:197–202.

381. Lemańczyk S. (2014). Science and national pride: the Iranian press coverage of nanotechnology, 2004–2009. *Science Communication.* **36**(2):194–218.

382. Mirdamadi S.M. et al. (2005). Analysis of socio-cultural obstacles for dissemination of nanotechnology from Iran's agricultural expert's perspective. *International Scholarly and Scientific Research & Innovation.* **5**(1):341–345.

383. Nasiri A. and Nasiri M. (2009). Customizing conventional patent-based financial instruments for financing nanotechnology firms: an Iranian perspective. *IPCBEE.* **2**(2011):95–99.

384. Rahimpour M. et al. (2012). Public perceptions of nanotechnology: a survey in the mega cities of Iran. *NanoEthics.* **6**(2):119–126.

385. Rezaeifar A. et al. (2005). Activities in Iran for standardization of nanotechnology. In *ASME 4th Integrated Nanosystems Conference*. American Society of Mechanical Engineers, pp. 87–88.

386. Sarkar S. and Beitollahi A. (2009). An overview on nanotechnology activities in Iran. *Iranian Journal of Public Health*. **38**(Suppl 1):65–68.

Iraq

387. Althabhawi N.M. and Zainol Z.A. (2013). Patentable novelty in nanotechnology inventions: a legal study in Iraq and Malaysia. *NanoEthics*. **7**(2):121–133.

Ireland

388. Hanford C.E. et al. (2014). Nanotechnology in the agri-food industry on the island of Ireland: applications, opportunities and challenges. Belfast: Queen's University, Institute for Global Food Security and Dublin: Teagasc Ashtown Food Research Centre.

Israel

389. Rosenbaum B. et al. (2007). Israel's nanotechnology research landscape: a survey of Israeli nanotechnology capabilities and technology transfer policies. *Nanotechnology Law and Business*. **4**:109.

Italy

390. Arnaldi S. (2014). Exploring imaginative geographies of nanotechnologies in news media images of Italian nanoscientists. *Technology in Society*. **37**:49–58.

391. Arnaldi S. (2014). Who is responsible? Nanotechnology and responsibility in the Italian daily press. In *Responsibility in Nanotechnology Development: The International Library of*

Ethics Law and Technology. Amsterdam, Netherlands: Springer, pp. 175–188.

392. Calignano G. (2014). Italian organisations within the European nanotechnology network: presence, dynamics and effects. *Journal of the Geographical Society of Berlin.* **145**(4):241–259.

393. Calignano G. and Quarta C.A. (2015). The persistence of regional disparities in Italy through the lens of the European Union nanotechnology network. *Regional Studies, Regional Science.* **2**(1):469–478.

394. Cassia L. and De Massis A. (2010). The market for nanotechnology applications and its managerial implications: an empirical investigation in the Italian landscape. In Fuerstner I. (ed.), *Products and Services: From R&D to Final Solutions*. Rijeka, Croatia: Sciyo, pp. 199–210.

395. Chiesa V. and De Massis A. (2006). *La Nanoindustria: Analisi Dei Principali Player Italiani Nelle Nanotecnologie*. Roma: Aracne Editrice.

396. Chiesa V. et al. (2007). How to sell technology services to innovators: evidence from nanotech Italian companies. *European Journal of Innovation Management.* **10**(4):510–531.

397. Chiodo E. et al. (2015). Consumer perceptions of nanotechnology applications in Italian wine. *Italian Journal of Food Science.* **27**(2):93–107.

398. Escoffier L. (2007). Brief review of nanotechnology related activities in Italy. *Nanotechnology Law and Business.* **4**:385.

399. Lisotti A. et al. (2014). The NANOLAB project: educational nanoscience at high school. *MRS Proceedings.* **1657**. doi:10.1557/opl.2014.374

400. Mantovani E. and Porcari A. (2006). Nanotechnology in Italy: national programs, players and activities. *Nano Science and Technology Institute.* **1**:576–581.

401. Mirabile M. et al. (2014). Workplace exposure to engineered nanomaterials: the Italian path for the definition of occupational health and safety policies. *Health Policy.* **117**(1):128–134.

Jamaica

402. Clayton A. (2001). Developing a bioindustry cluster in Jamaica: a step towards building a skill-based economy. *Social and Economic Studies.* 1–37.

403. Jones P.W. (2006). Jamaica and the nanotechnology approach to economic development. *Economics Bulletin.* **28**(7). Available at SSRN: http://dx.doi.org/10.2139/ssrn.895168

Japan

404. Bobowski S. (2013). Knowledge cluster initiatives by MEXT–case of Tokai region nanotechnology manufacturing cluster in Japan. *Prace Naukowe Uniwersytetu Ekonomicznego we Wrocławiu.* (295):27–40.

405. Bu-ran W. and Bang-chao Y. (2005). The publicity state and analysis of patents for nano science and technology in Japan. *Micronanoelectronic Technology.* **1**:004.

406. Fujita Y. et al. (2006). Perception of nanotechnology among the general public in Japan—of the NRI nanotechnology and society survey project. *Asia Pacific Nanotech Weekly.* **4**:1–2.

407. Grey F. (1993). STM-based nanotechnology: the Japanese challenge. *Advanced Materials.* **5**(10):704–710.

408. Ishizu S. et al. (2008). Toward the responsible innovation with nanotechnology in Japan: our scope. *Journal of Nanoparticle Research.* **10**(2):229–254.

409. Kanama D. (2013). Multimodal evaluations of Japan's nanotechnology competitiveness. *International Journal of Innovation and Technology Management.* **10**(02):1340003.

410. Kanama D. and Kondo A. (2007). Analysis of Japan's nanotechnology competitiveness-concern for declining competitiveness and challenges for nano-systematization. *Science and Technology Trends: Quarterly Review.* **25**:36–49.

411. Katao K. (2006). Nanomaterials may call for a reconsideration of the present Japanese chemical regulatory system. *Clean Technologies and Environmental Policy.* **8**(4):251–259.

412. Kishi T. (2004). Nanotechnology R&D policy of Japan and nanotechnology support project. *Journal of Nanoparticle Research.* **6**(6):547–554.

413. Kishi T. and Bando Y. (2004). Status and trends of nanotechnology R&D in Japan. *Nature Materials.* **3**(3):129–131.

414. Obayashi M. (1986). Origins of molecular biology in Japan. *Journal of UOEH.* **8**(2):251–256.

415. Ohtsu M. (2007). Nanophotonics in Japan. *Journal of Nanophotonics.* **1**(1):011590–011590.

416. Okuyama S. et al. (2008). Comparative view of nanotechnology patents in Japan and the US: a case study of two patents. *Nanotechnology Law and Business.* **5**:455.

417. Ross L. (2004). A cursory look at commercializing nanotechnology in Japan. *Nanotechnology Law and Business.* **1**(2): 213.

418. Sugi M. (1987). Molecular engineering in Japan: a prospect of research on Langmuir-Blodgett films. *Thin Solid Films.* **152**(1):305–326.

419. Takemura M. (2008). Japan's engagement in health, environmental and societal aspects of nanotechnology. *Journal of Cleaner Production.* **16**(8):1003–1005.

420. Yamada I. and Toyoda N. (2007). Nano-scale surface modification using gas cluster ion beams: a development history and review of the Japanese nano-technology program. *Surface and Coatings Technology.* **201**(19):8579–8587.

421. Žagar A. (2014). Nanotech cluster and industry landscape in Japan. *EU-Japan Centre for Industrial Cooperation.* http://www.eu-japan.eu/sites/eu-japan.eu/files/NanotechInJapan.pdf

Kazakhstan

422. Blank N. and Smagulov B. (2013). "Kazakh national nanotechnological initiative" is a key instrument of technological modernization for Kazakhstan. *Scientific Israel-Technological Advantages.* **15**(2):119–120.

423. Figovsky O.L. et al. (2013). Kazakh national nanotechnological initiative: key instrument of technological modernization for Kazakhstan. *International Letters of Social and Humanistic Sciences.* **06**:13–23.

424. Kaigorodtsev A. and Bordiyanu I. (2014). SWOT-analysis of an innovation cluster creation in East Kazakhstan. *Актуальні проблеми економіки.* **6**:284–290.

425. Mansurov Z. (2012). The development of nanotechnology in Kazakhstan. *Chemical Bulletin of Kazakh National University.* **1**:456–457.

426. Shamelkhanova N. et al. (2013). Educational integration for post-graduate training (PGT) systems in the field of nanotechnology: the case of Kazakhstan. *Public Administration and Regional Studies.* **1**(11):5–22.

427. Shamelkhanova N.A. et al. (2013). Methodological bases of innovative training of specialists in nanotechnology field (Kazakhstan's Case) No. 1. *Materials Engineering.* **15**(1):128–129.

Kenya

428. Harsh M. (2011). Equity and participation in decisions: what can nanotechnology learn from biotechnology in Kenya? In *Nanotechnology and the Challenges of Equity, Equality and Development.* Netherlands: Springer, pp. 251–269.

Korea, Republic of

429. Bae S.-H. et al. (2013). The innovation policy of nanotechnology development and convergence for the new Korean government. *Journal of Nanoparticles Research.* **15**(11):1–15.

430. Chon M.C. (2004). Nanotechnology and nanomaterials in Korea. *Journal of Structural Chemistry.* **45**(S1):S6.

431. Chou K.T. and Liou H.M. (2011). Risk and ethical governance of nano-convergence technology: an initial comparison of the technological impact assessment between South Korea and Taiwan. *Asian Journal of WTO & International Health Law and Policy.* **6**(1):235–280.

432. Kim H.S. The research life cycle and innovation through grey literature in nanotechnology in Korea. Dongdaemun: Korea Institute of Science & Technology Information.

433. Kim S. et al. (2010). Korean experience in nanotechnology industrialization. *Tech Monitor.* 21–29.

434. Kostoff R. et al. (2008). Relation of seminal nanotechnology document production to total nanotechnology document production: South Korea. *Scientometrics.* **76**(1):43–67.

435. Lee C.J. et al. (2013). Factors influencing nanotechnology commercialization: an empirical analysis of nanotechnology firms in South Korea. *Journal of Nanoparticle Research.* **15**(2): 1–17.

436. Lee J. et al. (1986). Public and expert's perception about nanotechnology hazards in Korea. *Journal Environmental Toxicology.* **23**(4):247–256.

437. Lee J. et al. (2006). Synthesis of new nanostructured carbon materials using silica nanostructured templates by Korean research groups. *International Journal of Nanotechnology.* **3**(2–3):253–279.

438. Lee J.W. (2002). Overview of nanotechnology in Korea–10 years' blueprint. *Journal of Nanoparticle Research.* **4**(6):473–476.

439. Lee Y.-G. and Song Y.-I. (2007). Selecting the key research areas in nanotechnology field using technology cluster analysis: a case study based on national R&D programs in South Korea. *Technovation.* **27**:57–64.

440. Lee Y.S. (2006). Current research status of biomedical micro and nano technologies in Korea. *Nanomedicine: Nanotechnology, Biology and Medicine.* **2**(4):298.

441. Lim D. (2009). Biotechnology industry, statistics and policies in Korea. *Asian Biotechnology and Development Review.* **11**(2):1–27.

442. Lim J.M. (2004). The present and future of nanotechnology in medicine. *Korean Journal of Hepatology.* **10**(3):185–190.

443. Lim J.S. et al. (2015). Study of US/EU national innovation policies based on nanotechnology development, and implications

for Korea. *Journal of Information Science Theory and Practice.* **3**(1):50–65.

444. So D.S. et al. (2012). Nanotechnology policy in Korea for sustainable growth. *Journal of Nanoparticle Research.* **14**(6):1–11.

445. Uskoković V. and Uskoković D.P. (2011). Extrapolating strategies for the scientific and technological development of underdeveloped societies from the examples of South Korea, Slovenia and Serbia. *International Journal of Technology Management & Sustainable Development.* **10**(2):125–145.

446. Wieczorek I. (2007). Nanotechnology in Korea—actors and innovative potential. In *Innovation and Technology in Korea.* Berlin: Physica-Verlag, pp. 205–231.

Kosovo

447. Bumajdad A. (2013). Nanoscience and nanotechnology research highlights at Kuwait University. *Green Processing and Synthesis.* **2**(2):181–183.

Latin America (including Andean countries, Iberoamerica, Pan America)

448. Argentinian Centre of Scientific and Technological Information. N.D. Nanotechnology in Ibero-America current situation and trends. A report prepared for the Ibero-American Observatory of Science, Technology, Society & Innovation.

449. Foladori G. (2006). Nanotechnology in Latin America at the crossroads. *Nanotechnology Law and Business.* **3**:205.

450. Foladori G. (2013). Nanotechnology policies in Latin America: risks to health and environment. *NanoEthics.* **7**(2):135–147.

451. Foldari G. and Invernizzi N. (eds.) (2007). *Nanotechnologies in Latin America.* Berlin: Karl Dietz Verlag.

452. Foladori G. and Invernizzi N. (2013). Inequality gaps in nanotechnology development in Latin America. *Journal of Arts and Humanities.* **2**(3):35–45.

453. Foladori G. et al. (2012). Nanotechnology: distinctive features in Latin America. *Nanotechnology Law and Business.* **9**:88.

454. Foladori G. et al. (2013). Nanotechnology: risk management and regulation for health and environment in Latin America and in the Caribbean. *Trabalho, Educação e Saúde.* **11**(1):145–167.

455. Invernizzi N. et al. (2015). Nanotechnology for social needs: contributions from Latin American research in the areas of health, energy and water. *Journal of Nanoparticle Research.* **17**(5):1–19.

456. Kay L. and Shapira P. (2009). Developing nanotechnology in Latin America. *Journal of Nanoparticle Research.* **11**(2):259–278.

457. Mendoza G. and Rodgriguez J.L. (2007). La nanociencia y la nanotecnología: una revolución en curso. *Perfiles Latinoamericanos.* **29**:162–86 (Spanish).

458. Pardo-Guerra J.P. (2011). Mapping emergence across the Atlantic: some (tentative) lessons on nanotechnology in Latin America. *Technology in Society.* **33**(1):94–108.

459. Pastrana H.F. et al. (2012). Nanotecnología, patentes y la situación en América Latina. *Mundo Nano. Revista Interdisciplinaria en Nanociencia y Nanotecnología.* **5**(9).

460. Pohlmann A.R. and Beck R.C. (2012). A special issue on the developments in biomedical nanotechnology in Latin America. *Journal of Biomedical Nanotechnology.* **8**(2):191.

461. Prieto P. (2009). Physics in the Andean countries: a perspective from condensed matter, novel materials and nanotechnology. *APS April Meeting Abstracts.* **1**:7003.

462. Sánchez J.T. and Serena P. (2011). Situación de la divulgación y la formación en nanociencia y nanotecnología en Iberoamerica. *Mundo Nano. Revista Interdisciplinaria en Nanociencia y Nanotecnología.* **4**(2):12–17.

Latvia

463. Štaube T. et al. (2014). The origins of nanotechnology in Latvia. *Advanced Materials Research.* **1025**:1083–1087.

Libya

464. Elmarzugi N.A. et al. (2014). Awareness of Libyan students and academic staff members of nanotechnology. *Journal of Applied Pharmaceutical Science.* **4**(06):110–114.

Lithuania

465. Grinius L. et al. (2007). Perspectives for safe use and application of modern biotechnology in Lithuania (UNEP-GEF Project No. GFL-2328-2716-4935). Lithuania Ministry of the Environment.
466. Snitka V. (2011). Nanotechnology research in Lithuania. *MNT Bulletin.* **7**(1).
467. Streimikiene D. (2014). The impact of research and development for business innovations in Lithuania. *The Amfiteatru Economic Journal.* **16**(37).
468. Zverev Y.M. et al. (2011). The research and technology development in Lithuania and the prospects of research and technology cooperation between Lithuania and the Russian federation. *Baltijskij Region.* **2**(8):49–55.

Macedonia

469. Nikolov E. et al. (2009). R&D national policy in the Republic of Macedonia according to the security related and generally R&D scene: status and shortfalls. In *Proceedings of the International Conference–"Business and Science for Security and Defence R&D"*, Bulgaria, Sofia.

Malawi

470. Utembe W. and Gulumian M. (2013). Questioning the adequacy of the regulatory regime for nanotechnology in Malawi. *Malawi Law Journal.* **7**:9–34.

Malaysia

471. Hashim U. et al. (2009). Nanotechnology development status in Malaysia: industrialization strategy and practices. *International Journal of Nanoelectronics and Materials.* **2**(1):119–134.

Malta

472. Surface N. et al. (2007). University of Malta to offer European master's degree in heat treatment and surface engineering. *Transactions of the Institute of Metal Finishing.* **1**(2):49.

Mexico

473. Barrañón A. (2010). Women in Mexican nanotechnology. In *Proceedings of the 2010 American Conference on Applied Mathematics.* World Scientific and Engineering Academy and Society (WSEAS), pp. 239–243.

474. Barrañón A. and Juanico A. (2010). Major issues in designing an undergraduate program in nanotechnology: the Mexican case. *WSEAS Transactions on Mathematics.* **9**(4):264–274.

475. Delgado-Ramos G.C. (2007). Sociología política de la nanotecnología en el hemisferio occidental: el caso de Estados Unidos, México, Brasil y Argentina. *Revista de Estudios Sociales.* **27**:164–181 (Spanish).

476. Delgado-Ramos G.C. (2013). Ethical, social, environmental, and legal aspects of nanotechnologies. *International Journal of Innovation & Technology Management.* **10**(02):1340001.

477. Foladori G. and Lau E.Z. (2007). Tracking nanotechnology in Mexico. *Nanotechnology Law & Business.* **4**:213.

478. Foladori G. and Záyago Lau E. (2014). The regulation of nanotechnologies in Mexico. Nanotechnology Law & Business. **11**(2):164–171.

479. Foladori G. et al. (2012). Mexico-US scientific collaboration in nanotechnology. *Frontera Norte*. **24**(48):145–164.

480. Foladori G. et al. (2015). Nanotechnology in Mexico: key findings based on OECD criteria. *Minerva*. **53**(3):279–301.

481. Fuentes L. et al. (2010). Introducing the materials world modules in Mexico: the Chihuahua project. *Journal of Materials Education*. **32**(5–6):245–254.

482. Lau E.Z. (2011). Nanotech cluster in nuevo Leon, Mexico: reflections on its social significance. *Nanotechnology Law and Business*. **8**:49.

483. Lau E.Z. et al. (2012). Toward an inventory of nanotechnology companies in Mexico. *Nanotechnology Law and Business*. **9**:283.

484. Lau E.Z. et al. (2014). Twelve years of nanoscience and nanotechnology publications in Mexico. *Journal of Nanoparticle Research*. **16**(1):1–10.

485. Lazos-Martínez R.J. (2014). Standards for nanotechnology in Mexico. *CENAM, Centro Nacional de Metrología*. http://archive.nrc-cnrc.gc.ca/obj/inms-ienm/doc/mexico-nanotechnology.pdf

486. Lazos-Martínez R.J. and González-Rojano N. (2013). Nanometrology in emerging economies: the case of Mexico. *Mapan*. **28**(4):299–309.

487. López-Vázquez E. et al. (2012). Perceived risks and benefits of nanotechnology applied to the food and packaging sector in México. *British Food Journal*. **114**(2):197–205.

488. Robles-Belmont E. and Vinck D. (2011). A panorama of nanoscience developments in Mexico based on the comparison and crossing of nanoscience monitoring methods. *Journal of Nanoscience and Nanotechnology*. **11**(6):5499–5507.

489. Robles-Belmont E. and de Gortari-Rabiela R. (2013). Dynamics of the emergence of micro and nanotechnologies in the

healthcare sector in Mexico. *Nanotechnology, Law and Business Journal.* **10**:54.

490. Suárez M. (2013). Transnational knowledge networks in nanotechnology in Mexico: policy incentives and dynamics. *Synesis: A Journal of Science, Technology, Ethics, and Policy.* **4**(1):G76–G84.

491. Záyago E. (2011). Nanotech cluster in Nuevo León, México: reflections on its social significance. *Nanotechnology, Law and Business Journal.* **8**(1):49–59.

492. Záyago E. and Foladori G. (2010). La nanotecnología en México: un desarrollo incierto. *Economía Sociedad y Territorio.* **10**(32): 143–178.

493. Záyago E. et al. (2012). Toward an inventory of nanotechnology companies in Mexico. *Nanotechnology, Law and Business Journal.* **9**(3):283–292.

494. Záyago Lau E. et al. (2014). Twelve years of nanoscience and nanotechnology publications in Mexico. *Journal of Nanoparticle Research.* **16**(2193):1–10.

495. Záyago Lau E. et al. (2014). Researching risks of nanomaterials in Mexico. *Journal of Hazardous, Toxic, and Radioactive Waste.* doi:10.1061/ (ASCE)HZ.2153-5515.0000247

Monaco

496. Akyildiz I.F. et al. (2012). Monaco: fundamentals of molecular nano-communication networks. *IEEE Wireless Communications.* **19**(5):12–18.

Mongolia

497. Fu W.Y. and Wang Z.C. (2002). Development on current nanotechnology in our country. *Journal of Inner Mongolia University for Nationalities.* **5**:030.

498. Gantsolmon K. et al. (2015). Nanowaste management policy in the capital city, Mongolia. *Applied Mechanics and Materials.* **768**:29–35.

499. Khaisandai L. (2013). Mongolia in Northeast Asia: issues of economic development and cooperation. *Mongolian Journal of International Affairs*. (17):59–63.

Morocco

500. Al Akhawayn University (2007). Book of abstracts: First Moroccan days on nanoscience & nanotechnology (MDNN1). Ifrane, Morocco.
501. Boukharouaa N.E. et al. (2014). The Moroccan diaspora and its contribution to the development of innovation in Morocco. In *The Global Innovation Index*, Chapter 8, pp. 123–131.
502. Moroccan Ministry of Economy and Finances (2009). Nanotechnology sector in Morocco: state and route of development. Direction of investigation and financial forecast. Rabat.

Nepal

503. Bhattarai R.K. (2013). Structural immunity to technology diffusion in Nepal: planned technology and unplanned adaptation. Kathmandu, Nepal: Tribhuvan University.
504. Lamichhane S.K. (2015). Public communication on science and technology. *Himalayan Physics*. **5**:12–16.

Netherlands

505. Est V.Q. et al. (2012). Governance of nanotechnology in the Netherlands-informing and engaging in different social spheres. *International Journal of Emerging Technologies & Society*. **10**(1):6–26.
506. Krabbenborg L. (2012). The potential of national public engagement exercises: evaluating the case of the recent Dutch societal dialogue on nanotechnology. *International Journal of Emerging Technologies and Society*. **10**:27.
507. Moniz A.B. et al. (2008). Application of FTA and the perceived impact on employment policies: the case of nanotechnology in the Netherlands, Germany and Portugal. The *3rd International*

Seville Conference on Future-Oriented Technology Analysis, Seville, Spain.

508. von Raesfeld A. et al. (2012). When is a network a nexus for innovation? A study of public nanotechnology R&D projects in the Netherlands. *Industrial Marketing Management*. **41**(5):752–758.

509. Walhout B. et al. (2010). Nanomedicine in The Netherlands: social and economic challenges (Background Note for First and Second Chamber MPs Visiting the High Tech Campus Eindhoven (April 16, 2010)). The Hague: Rathenau Instituut.

510. Werker C. and Cunningham W. (1999). Policy and concentration of activities: the case of Dutch nanotechnology. Delft: Delft University of Technology, Department of Economics of Innovation.

New Zealand

511. Cook A.J. and Fairweather J.R. (2005). Nanotechnology—ethical and social issues: results from New Zealand focus groups (Research Report No. 281). Canterbury, New Zealand: Lincoln University.

512. Cook A.J. and Fairweather J.R. (2007). Intentions of New Zealanders to purchase lamb or beef made using nanotechnology. *British Food Journal*. **109**(9):675–688.

513. Cronin K. and Hutchings J. (2012). Supergrans and nanoflowers: reconstituting images of gender and race in the promotion of biotechnology and nanotechnology in Aotearoa New Zealand. *New Genetics and Society*. **31**(1):55–85.

514. Hall-McMaster S. Nanotechnology: solving NZ's biggest problems at the smallest level. Dunedin: University of Otago. Available online at http://eureka.blueriver.co.nz/uploaded/SamHall-McMasterNanotechnology.pdf

515. Hickson R. (2009). Setting directions for nanotechnology in New Zealand. *International Journal of Nanotechnology*. **6**(3–4):288–297.

Nigeria

516. Bankole M.T. et al. (2014). A review on nanotechnology as a tool of change in Nigeria. *Scientific Research and Essays.* **9**(8):213–223.

517. Batta H. et al. (2014). Science, nano-science and nanotechnology content in Nigeria's elite and popular press: focus on framing and socio-political involvement. *New Media and Mass Communication.* **31**:9–19.

518. Cosmas G.T. et al. (2012). A predictor of nanotechnology benefits to computing and microelectronics production in Nigeria. *International Journal of Academic Research.* **4**(6):230–236.

519. Siyanbola W.O. et al. (2013). Designing and implementing a science, technology and innovation policy in a developing country: recent experience from Nigeria. *Industry and Higher Education.* **27**(4):323–331.

520. Temitope A.E. et al. (2013). Synthesis of silver nanoparticles using some alcoholic beverages from Nigeria market. *International Journal of Nano & Material Science.* **2**(1):25–35.

North America

521. Priest S. (2006). The North American opinion climate for nanotechnology and its products: opportunities and challenges. *Journal of Nanoparticle Research.* **8**(5):563–568.

Norway

522. Kjølberg K.L. (2009). Representations of nanotechnology in Norwegian newspapers—implications for public participation. *NanoEthics.* **3**(1):61–72.

523. Research Council of Norway (2001). Bibliometric study in support of Norway's strategy for international research collaboration. St. Hanshaugen.

OECD Countries (Australia, Austria, Belgium, Canada, Chile, Czech Republic, Denmark, Estonia, Finland, France, Germany, Greece, Hungary, Iceland, Ireland, Israel, Italy, Japan, Korea, Latvia, Luxembourg, Mexico, Netherlands, New Zealand, Norway, Poland, Portugal, Slovak Republic, Slovenia, Spain, Sweden, Switzerland, Turkey, United Kindgom, United States)

524. Christensen F.M. (2010). Reflections from an OECD workshop on environmental benefits of nanotechnology. *The International Journal of Life Cycle Assessment.* **15**(2):137–138.

525. Gruère G.P. (2012). Implications of nanotechnology growth in food and agriculture in OECD countries. *Food Policy.* **37**(2):191–198.

526. Murashov V. et al. (2009). Occupational safety and health in nanotechnology and organisation for economic cooperation and development. *Journal of Nanoparticle Research.* **11**(7):1587–1591.

527. Visser R. (2009). A sustainable development for nanotechnologies: an OECD perspective. In Hodge G. et al. (eds.), *New Global Frontiers in Regulation: The Age of Nanotechnology.* Cheltenham, UK: Edward Elgar Publishing Ltd., pp. 320–332.

OIC States (Organization of Islamic Cooperation) [Iran, Turkey, Egypt, Malaysia, Saudi Arabia, and Pakistan]

528. Bajwa R.S. et al. (2012). Nanotechnology research among some leading OIC member states. *Journal of Nanoparticle Research.* **14**(9):1–10.

Pakistan

529. Ali S.A. and Tariq M. (2014). Nanotechnology and its implication in medical science. *The Journal of the Pakistan Medical Association.* **64**(9):984–986.

530. Bajwa R.S. et al. (2013). A scientometric assessment of research output in nanoscience and nanotechnology: Pakistan perspective. *Scientometrics*. **94**(1):333–342.

Peru

531. Rojas Tapia J. and Landuro Saenz C.V. (2011). El reto de la divulgación y la formación en nanociencia y nanotecnología en Perú. *Mundo Nano. Revista Interdisciplinaria en Nanociencia y Nanotecnología*. **4**(2).
532. Zárate Vásquez J.S. (2014). Between uncertainty and individualism: scientific ethos of adversity and nanotechnology in Peru. A paper presented at the *XVIII ISA World Congress of Sociology*. Yokohama, Japan.

Philippines

533. Saloma-Akpedonu C. (2006). Studying science and technology in the Philippines as culture and practice. *Philippine Sociological Review*. **54**:30–33.

Poland

534. Bakalarczyk S. (2012). Financial aspects of nanotechnology innovation of the polish company ABC. A paper presented at *Nanocon' 2012*. Brno, Czech Republic, October 23–25.
535. Lemańczyk S. (2012). Between national pride and the scientific success of "others": the case of Polish press coverage of nanotechnology, 2004–2009. *NanoEthics*. **6**(2):101–115.
536. Mazurkiewciz A. et al. (2006). Trends in education and training development in the field of nanotechnology. *Problemy Eksploatacji*. **4**:37–47.
537. Popławska M. et al. (2015). New sector of employment: a review of data on nanoproduction, research and development in the field of nanotechnology in Poland. *Medycyna Pracy*. **66**(3):317–326.

538. Poteralska B. et al. (2011). The development of education and training systems in the field of nanotechnology. *Journal of College Teaching & Learning.* **4**(6). doi:https://doi.org/10.19030/tlc.v4i6.1570

539. Siemaszko A. (2004). Structural funds in Poland and Polish research potential in nanotechnology. *Diffusion & Defect Data Part B Solid State Phenomena.* **99**:283–286.

Portugal

540. Carvalho A. and Nunes J.A. (2013). Technology, methodology and intervention: performing nanoethics in Portugal. *NanoEthics.* **7**(2):149–160.

Qatar

541. Chouchane L. et al. (2011). Medical education and research environment in Qatar: a new epoch for translational research in the Middle East. *Journal of Translational Medicine.* **9**:16.

542. Ibala Bs. et al. (2014). Preservation of cultural heritage artifacts in Qatar using nanotechnology. In *Qatar Foundation Annual Research Conference* (No. 1, p. SSSP0165).

Romania

543. Bălăşoiu M. and Arzumanyan G.M. (eds.) (2010). *Modern Trends in Nanoscience.* Bucharest: Romanian Academy.

544. Dascalu D. et al. (2001). Nanoscale science and engineering in Romania. *Journal of Nanoparticle Research.* **3**(5–6):343–352.

545. Gorghiu L.M. and Gorghiu G. (2012). Teachers' perception related to the promotion of nanotechnology concepts in Romanian science education. *Procedia - Social and Behavioral Sciences.* **46**:4174–4180.

546. Gorghiu L.M. and Gorghiu G. (2014). Related aspects on using digital tools in the process of introducing nanotechnology in science lessons. *Acta Physica Polonica A.* **125**(2):544–547.

547. Gorghiu L.M. et al. (2013). Promoting the nano-technology concepts in secondary science education through ICT tools the Romanian and Turkish teacher's perception. *Global Journal on Technology.* **3**:577–583.

548. Morozan A. and Stamatin I. (2006). Nanotechnology in Romania. *Nanotechnology Law and Business.* **3**:533.

Russia

549. Alferov J. (2005). Nanotechnologies: prospects of R&D in Russia. *Nano Technologies and Materials Journal.* **1**:6–12 (Russian).

550. Ananyan M. (2005). Nanotechnology in Russia: from laboratory towards industry. *Nanotechnology Law and Business.* **2**:194.

551. Andrievski R.A. (2003). Modern nanoparticle research in Russia. *Journal of Nanoparticle Research.* **5**(5–6):415–418.

552. Antsiferova I.V. and Esaulova I.A. (2013). Nanotechnology research and education centers as an intellectual basis of nanotechnology in Russia. *Middle East Journal of Scientific Research.* (13):127–131.

553. Belokrylova E.A. (2001). Legal problems of nanotechnology environmental safety provision in the Russian federation: the foreign country's experience. *Nanotechnology Law and Business.* **8**:203.

554. Connolly R. (2013). State industrial policy in Russia: the nanotechnology industry. *Post-Soviet Affairs.* **29**(1):1–30.

555. Frolov D.P. et al. (2015). Building an institutional framework for nanotechnology industry in Russia. *Mediterranean Journal of Social Sciences.* **6**(3 S6):81–86.

556. Galyavieva M.S. (2014). Bibliometric analysis of the document flow of informetrics based on the Russian science citation index. *Scientific and Technical Information Processing.* **41**(4):220–229.

557. Gavrilenko V.P. et al. (2009). First Russian standards in nanotechnology. *Bulletin of the Russian Academy of Sciences: Physics.* **73**(4):433–440.

558. Josephson P.R. (2009). Russia's nanotechnology revolution. *Georgetown Journal of International Affairs.* **10**(1):149–157.

559. Karasev O. et al. (2011). Emerging technology-related markets in Russia: the case of nanotechnology. *Journal of East-West Business.* **17**(2–3):101–119.

560. Karaulova M. et al. (2014). Nanotechnology research and innovation in Russia: a bibliometric analysis. Available at SSRN: http://dx.doi.org/10.2139/ssrn.2521012

561. Lavrik O.L. et al. (2015). Nanoscience and nanotechnology in the Siberian branch of the Russian academy of sciences: bibliometric analysis and evaluation. *Journal of Nanoparticle Research.* **17**(2):1–11.

562. Maebius S. and Jamison D. (2009). Russian dominance in nanotechnology. *Nanotechnology Law and Business.* **6**:1.

563. Maltsev P. (2005). Terminology of MEMS/NST in Russia. *Nano and Microsystem Techniques.* **9**:2–6 (Russian).

564. Melkonyan M. and Kozyrev S. (2009). The current state-of-the art in the area of nanotechnology risk assessment in Russia. In *Nanomaterials: Risks and Benefits.* Netherlands: Springer, pp. 309–315.

565. Pavlov A.J. and Batova V.N. (2014). Ensuring economic security of the innovative development of nanotechnology in the Russian Federation and foreign countries. *Life Science Journal.* **11**(6s):322–325.

566. Putilov A.V. (2004). Russian program on nanoscience and technology. *Journal of Structural Chemistry.* **45**:S1–S2.

567. Reiss T. and Thielmann A. (2010). Nanotechnology research in Russia-an analysis of scientific publications and patent applications. *Nanotechnology Law and Business.* **7**:387.

568. Terekhov A.I. (2011). Scientometric approach to nanotechnology. *Applied Econometrics.* **23**(3):3–12.

569. Terekhov A.I. (2012). Evaluating the performance of Russia in the research in nanotechnology. *Journal of Nanoparticle Research.* **14**(11):1–17.

570. Tret'yakov Y.D. (2007). Challenges of nanotechnological development in Russia and abroad. *Herald of the Russian Academy of Sciences.* **77**(1):15–21.

571. Tret'yakov Y.D. and Gudilin E.A. (2009). Lessons from the foreign nanohype. *Herald of the Russian Academy of Sciences.* **79**(1):1–6.

572. Zibareva I.V. et al. (2010). Russian nanoscience: bibliometric analysis relying on the STN International databases. *Khimiya v Interesakh Ustoichivogo Razvitiya (Chem Sustain Dev).* **18**(2): 201–219.

Saudi Arabia

573. Abu-Salah K.M. et al. (2013). King Saud University: nanoscience and nanotechnology research highlights. *Green Processing and Synthesis.* **2**(2):175–177.

574. Al-Habashi N. (2007). What is nanotechnology? A brief introduction in simple lessons. Kingdom of Saudi Arabia: King Fahd National Library.

575. Rizvi S.A.H. et al. (2009). Nanotechnology within the framework of human factors engineering with special reference to developing countries like Saudi Arabia. *International Journal of Nanomanufacturing.* **4**(1–4):300–307.

Scandinavia

576. Andersen M.M. et al. (2010). Green nanotechnology in Nordic construction. *Social Capital.* **100**(134):134.

577. Enjelasi M. (2014). Diffusion of nanotechnology to businesses survey nine Nordic Universities. PhD Dissertation: Karlskrona, Sweden, Blekinge Institute of Technology.

578. Nordic Innovation Center (2010). Green nanotechnology in Nordic construction (Project # 06242). Technical University of Denmark, Chalmers University of Technology, and Advansis Ltd.

Senegal

579. Mambula C.J. (2008). Effects of factors influencing capital formation and financial management on the performance and growth of small manufacturing firms in Senegal: recommendations for policy. *International Journal of Entreprenuership.* **12**:87–106.

Serbia

580. Djurić Z. (2004). Nanoscience and nanotechnologies in Serbia. A paper presented at *Microelectronics 25th International Conference*, Belgrade, pp. 9–14.

581. Ivanović D. et al. (2015). Publications from Serbia in the science citation index expanded: a bibliometric analysis. *Scientometrics.* **105**(1):145–160.

582. Ivanović D. and Ho Y.S. (2014). Independent publications from Serbia in the science citation index expanded: a bibliometric analysis. *Scientometrics.* **101**(1):603–622.

583. Jakšić M.L. et al. (2014). Sustainable technology entrepreneurship and development–the case of Serbia. *Management.* **70**:65–73.

584. Milan B.Ž. et al. (2013). Nanotechnology and its potential applications in meat industry. *Tehnologija Mesa.* **54**(2):168–175.

585. Ribic-Zelenovic L. et al. (2009). Modern agriculture and nanotechnology. *Acta Agriculturae Serbica.* **28**:13–21.

586. Ševkušić M. and Uskoković D. (2009). State of the art in nanoscience and nanotechnology in Serbia: a preliminary bibliometric analysis. *Tehnika-Novi Materijali.* **18**(5):1–16.

587. Zibareva I.V. and Elepov B.S. (2012). Nanoscience and nanotechnology in Siberian branch of the Russian academy of sciences: bibliometric analysis based on Russian index of scientific citation. *Bibliosphere.* **4**:39–48.

Singapore

588. George S., Kaptan G., Lee J. and Frewer L. (2014). Awareness on adverse effects of nanotechnology increases negative perception among public: survey study from Singapore. *Journal of Nanoparticle Research.* **16**(12):1–11.

Slovakia

589. Braha K. et al. (2015). Innovation and economic growth: the case of Slovakia. *Visegrad Journal on Bioeconomy & Sustainable Development.* **4**(1):7–13.

Slovenia

590. Groboljsek B. and Mali F. (2012). Daily newspapers' views on nanotechnology in Slovenia. *Science Communication.* **34**(1): 30–56.

South Africa

591. Augustine B.H. and Munro O.Q. (2011). Into Africa: teaching nanoscience to undergraduates in KwaZulu-Natal, South Africa. *MRS Proceedings.* **1320**:f10.

592. Cele L.M. et al. (2009). Guest editorial: nanoscience and nanotechnology in South Africa. *South African Journal of Science.* **105**(7–8):242.

593. Claassens C.H. and Motuku M. (2006). Nanoscience and nanotechnology research and development in South Africa. *Nanotechnology Law and Business.* **3**:217.

594. Gastrow M. (2009). Thinking small: the state of nanotechnology research and development in South Africa. *Journal for New Generation Sciences.* **7**(1):1–17.

595. Kroon R.E. (2013). Nanoscience and the Scherrer equation versus the Scherrer-Gottingen equation. *South African Journal of Science.* **109**(5–6):1–2.

596. Lupton M. (2010). Regulating nanotechnology in South Africa. In Kidd M. and Hoctor S. (eds.), *Stella Iuris: Celebrating 100 years of Teaching Law in Pietermaritzburg*. Claremont, Zambia: Juta & Co Ltd., pp. 233–246.

597. Musee N. et al. (2010). A South African research agenda to investigate the potential environmental, health and safety risks of nanotechnology. *South African Journal of Science*. **106**(3–4): 1–6.

598. Pouris A. (2007). Nanoscale research in South Africa: a mapping exercise based on scientometrics. *Scientometrics*. **70**(3):541–553.

599. Pouris A. et al. (2012). Nanotechnology and biotechnology research in South Africa: technology management lessons from a developing country. In *2012 Proceedings of PICMET '12: Technology Management for Emerging Technologies*, Vancouver, BC, pp. 346–357.

600. Ward M. (2012). Overview of the DST/CSIR national innovation centre: national centre for nano-structured materials. *South African Journal of Science*. **106**(3–4):66–67.

601. Williams M.N. et al. (2010). A nanostep towards building a sustainable interdisciplinary research and educational program between the US and South Africa. *African Journal of Science*. **2**(3):133–152.

Spain

602. Arguimbau L. and Alegret S. (2010). Chemical research in the Catalan countries: a brief quantitative assessment of the agents, resources, and results. *Contributions to Science*. **6**(2):215–232.

603. Cabrer-Borrás B. and Serrano-Domingo G. (2007). Innovation and R&D spillover effects in Spanish regions: a spatial approach. *Research Policy*. **36**(9):1357–1371.

604. Chacón C., Estevao V. and Narros C. (2011). Nanotechnology in Spain: current situation and future challenges. *Convertech & e-print*. **1**(6):26–32.

605. Correia A. et al. (2006). El lento despertar de la nanotecnología en España. *Revista Sistema Madrid*. **15**:3.

606. Juanola-Feliu E. (2009). The nanotechnology revolution in Barcelona: innovation & creativity by universities. *International Management*. **13**:111–123.

607. Juanola-Feliu E. et al. (2010). Nanobiotechnologies and nanomedicine: technology transfer and commercialization in Spain. *Journal of Materials Science and Engineering*. **4**(6):71–84.

608. Minister of Industry, Tourism & Commerce (2000). Aplicaciones industriales de las nanotecnologías en España en el Horizonte 2020. Madrid: Government of Spain.

609. Oliva A. (2005). Nanoscience and nanotechnology activities in Catalonia. *Contributions to Science*. **3**(1):91–94.

610. Paez-Aviles C. et al. (2000). Spanish innovation and market on nanotechnology: an analysis within the H2020 framework (White Paper). Department of Electronics, Bioelectronics and Nanobioengineering Research Group, University of Barcelona.

611. Veltri G.A. and Crescentini A. (2011). The anchoring of nanotechnology in the Spanish national press. *International Journal of Science in Society*. **2**(2):127–138.

Sri Lanka

612. Amaradasa R.M.W. et al. (2002). Patents in a small developing economy: a case study of Sri Lanka. *Journal of Intellectual Property Rights*. **7**:395–404.

613. Dasanayaka S. (2003). Technology, poverty and the role of new technologies in eradication of poverty: the case of Sri Lanka. A paper presented at the *South Asia Conference on Technologies for Poverty Reduction*, New Delhi.

614. Ganeshan M. (2012). Public engagement in nanotechnology lessons & experiences from Sri Lanka. Colombo: National Science Foundation of Sri Lanka.

615. Karunaratne V. (2009). The Government of Sri Lanka launches nanotechnology as a priority research area. *Journal of the National Science Foundation of Sri Lanka*. **37**(2):81–82.

616. Karunaratne V. and del Alwis A. (2011). The nanotechnology and its contributions to economic development. *Economic Review*. 36–40.

617. Manamperi A. and Huzair F. (2012). Capacity building in genomics medicine and molecular diagnostics: the case of Sri Lanka. *Current Pharmacogenomics and Personalized Medicine (Formerly Current Pharmacogenomics)*. **10**(3):185–194.

618. Marikar F.M. et al. (2014). Sri Lankan medical undergraduate's awareness of nanotechnology and its risks. *Education Research International*. **2014**: Article ID 584352.

619. Rupasinghe B. and Shantha Siri J.G. (2012). Planning a regulatory framework for nanotechnology related activities in Sri Lanka. Colombo: National Science Foundation of Sri Lanka.

620. Sri Lanka Ministry of Technology & Research (2010). Science, technology & innovation strategy for Sri Lanka. Colombo.

621. Wijesekera R. (2010). Some reflections on the theme of science & education for citizenship: an essay. *Journal of the National Science Foundation of Sri Lanka*. **35**(2):63–69.

Swaziland

622. Brame J. et al. (2014). Water disinfection using nanotechnology for safer irrigation: a demonstration project in Swaziland. *Environmental Engineer and Scientist: Applied Research and Practice*. **50**(2):40–46.

Sweden

623. Boholm M. (2014). Political representations of nano in Swedish government documents. *Science and Public Policy*. **41**(5):575–596.

624. Fogerlbery H. (2008). Den svenska modellen för nanoteknik: mer effektiv än reflexiv? *Nordic Journal of Applied Ethics*. **2**:53–71.

625. Fogelberg H. and Lundqvist M.A. (2013). Integration of academic and entrepreneurial roles: the case of nanotechnology

research at Chalmers University of Technology. *Science and Public Policy.* **40**(1):127–139.

626. Fogelberg H. and Sandén B.A. (2008). Understanding reflexive systems of innovation: an analysis of Swedish nanotechnology discourse and organization. *Technology Analysis & Strategic Management.* **20**(1):65–81.

627. Meyer M. (2005). Nanotechnology in Sweden: an overview of bibliometric and patent studies. A report prepared for the Royal Swedish Academy of Engineering Sciences.

628. Perez E. and Sandgren P. (2008). Nanotechnology in Sweden: an innovation system approach to an emerging area (VINNOVA Analysis VA 2008:03). Stockholm: Swedish Governmental Agency for Innovation System.

629. Swedish Society for Nature Conservation (2012). Managing the unseen: opportunities and challenges with nanotechnology. Geneva.

630. Vico E.P. and Jacobsson S. (2012). Identifying, explaining and improving the effects of academic R&D: the case of nanotechnology in Sweden. *Science and Public Policy.* **39**(4):513–529.

Switzerland

631. Allenspach R. et al. (2010). Swiss nanotech report. Available at http://www.empa.ch/plugin/template/empa/*/91537

632. Bonfadelli H. et al. (2002). Biotechnology in Switzerland: high on the public agenda, but only moderate support. *Public Understanding of Science.* **11**(2):113–130.

633. Burri R.V. (2009). Coping with uncertainty: assessing nanotechnologies in a citizen panel in Switzerland. *Public Understanding of Science.* **18**(5):498–511.

634. Hincapié I. et al. (2015). Use of engineered nanomaterials in the construction industry with specific emphasis on paints and their flows in construction and demolition waste in Switzerland. *Waste Management.* **43**:398–406.

635. Leuenberger H. (2000). Focus on research in nanoscience and nanotechnology in Switzerland. *Journal of Nanoparticle Research.* **2**(4):391–392.

636. Schmid K. and Riediker M. (2008). Use of nanoparticles in Swiss industry: a targeted survey. *Environmental Science & Technology.* **42**(7):2253–2260.

Taiwan

637. Chen M.F. et al. (2013). Public attitudes toward nanotechnology applications in Taiwan. *Technovation.* **33**(2):88–96.
638. Cheng T.J. et al. (2009). The risk perception of nanotechnology in Taiwanese general population, workers, and experts. *Epidemiology.* **20**(6):S227.
639. Gerdsri P. (2016). National technology planning: a case study of nanotechnology for Thailand's agriculture industry. In *Hierarchical Decision Modeling.* Switzerland: Springer International Publishing, pp. 197–224.
640. Guo J.W. et al. (2014). Development of Taiwan's strategies for regulating nanotechnology-based pharmaceuticals harmonized with international considerations. *International Journal of Nanomedicine.* **9**:4773.
641. Lee C.K. et al. (2002). A catalyst to change everything: MEMS/NEMS–a paradigm of Taiwan's nanotechnology program. *Journal of Nanoparticle Research.* **4**(5):377–386.
642. Shu-Chuan L. and Tang S-M. (2006). The regulation of the potential risks to nanotechnology in Taiwan. *Asian Journal of Management and Humanity Sciences.* **1**:118–133.
643. Yen S-Y. et al. (2013). Nanotechnologies: big governance issues for the science of small (in Taiwan and beyond). *Far East Journal of Psychology and Business.* **11**(1):50–72.
644. Yueh H.P. et al. (2011). Course evaluation of the nanotechnology education and training program. *International Journal of Technology and Engineering Education.* **8**(2):1–12.

Thailand

645. Liu L. (2003). Current status of nanotech in Thailand. *Asia Pacific Nanotechnology Weekly.* **1**(19):1–4.
646. Maclurcan D.C. (2011). Southern roles in global nanotechnology innovation: perspectives from Thailand and

Australia. In *Nanotechnology and the Challenges of Equity, Equality and Development.* Netherlands: Springer, pp. 349–378.

647. Sandhu A. (2008). Thailand resorts to nanotech. *Nature Nanotechnology.* **3**(8):450–451.

648. Tanthapanichakoon W. (2005). An overview of nanotechnology in Thailand. *KONA Powder and Particle Journal.* **23**:64–68.

649. Unisearch (2004). Final report: survey for current situation of nanotechnology researchers and R&D in Thailand. Bangkok: Chulalongkorn University.

Tunisia

650. Madikizela M.N.D. (2006). The science and technology system of the Republic of Tunisia. Available online at http://portal.unesco.org/education/en/files

651. Onofri L. and Briand F. (2009). Blue biotechnology potential in Tunisia: a preliminary study of national stakeholder's involvement in setting priorities. A report prepared for the Mediterranean Science Commission.

Turkey

652. Aydogan-Duda N. (2012). Nanotechnology in Turkey. In *Making It to the Forefront.* New York: Springer, pp. 53–61.

653. Aydoğan-Duda N. and Şener İ. (2010). Entry barriers to the nanotechnology industry in Turkey. In *Nanotechnology and Microelectronics: Global Diffusion, Economics and Policy.* Hershey, PA: IGI Global, pp. 167–173.

654. Beyhan B. Nanoscience and nanotechnology research in Turkish universities: institutes, research groups and networks. *Technology Management.* **22**:185–200.

655. Beyhan B. and Teoman Pamukçu M. (2012). Nanotechnology research in Turkey: a university-driven achievement (TEKPOL Working Paper Series STPS-WP-11/07). Ankara: Middle East Technical University, Science and Technology Policies Research Center.

656. Darvish H. (2012). Assessing the diffusion of nanotechnology in Turkey: a social network analysis approach. *Collnet Journal of Scientometrics and Information Management.* **6**(1):175–183.

657. Darvish H. and Tonta Y. (2015). The diffusion of nanotechnology knowledge in Turkey. Available at http://www.issi2015.org/files/downloads/all-papers/0720.pdf

658. Kamanlıoğlu E.B. and Güzeloğlu C. (2010). Frames about nanotechnology agenda in Turkish media, 2005–2009. *International Journal of Social Behavioral Educational Economic and Management Engineering.* **4**(4):341–348.

659. Karaca F. and Öner M.A. (2015). Scenarios of nanotechnology development and usage in Turkey. *Technological Forecasting and Social Change.* **91**:327–340.

660. Kuzgun İ.K. (2011). Nanotechnology and the effects on the labor market conditions in Turkey. *China-USA Business Review.* **10**(1).

661. Öner M.A. et al. (2013). Comparison of nanotechnology acceptance in Turkey and Switzerland. *International Journal of Innovation and Technology Management.* **10**(02):1340007.

662. Senocak E. (2014). A survey on nanotechnology in the view of the Turkish public. *Science Technology & Society.* **19**(1):79–94.

Uganda

663. Ngatya K. (2009). Nanotechnology: how prepared is Uganda? *Uganda Sunday Monitor.* June 28.

Ukraine

664. Radziyevska S.O. and Chekman I.S. (2013). The principles of compiling English-Ukrainian nanoscience reference guide. *Terminolohichnyi Visnyk.* **2**(1).

665. Salihova O. (2014). State policy in the sphere of nanoscience and nanotechnologies in Ukraine in the context of EU aspirations. *Economy and Forecasting.* (3):121–136.

United Kingdom

666. Abro Q. et al. (1997). Bibliometric and patent analysis of nanotechnology: a case study from the United Kingdom. A paper presented at the *19th International Conference on Production Research*.

667. Aitken R.J. et al. (2006). Manufacture and use of nanomaterials: current status in the UK and global trends. *Occupational Medicine*. **56**(5):300–306.

668. Anderson A. et al. (2005). The framing of nanotechnologies in the British newspaper press. *Science Communication*. **27**(2):200–220.

669. Depledge M.H. et al. (2010). Nanomaterials and the environment: the views of the Royal Commission on Environmental Pollution (UK). *Environmental Toxicology and Chemistry*. **29**(1):1–4.

670. Groves C. (2013). Four scenarios for nanotechnologies in the UK, 2011–2020. *Technology Analysis & Strategic Management*. **25**(5):507–526.

671. Groves C.R. et al. (2010). CSR in the UK nanotechnology industry: attitudes and prospects (Working Paper). Cardiff: Center for Business Relationships, Accountability, Sustainability and Society. Cardiff University.

672. Groves C. et al. (2011). Is there room at the bottom for CSR? Corporate social responsibility and nanotechnology in the UK. *Journal of Business Ethics*. **101**(4):525–552.

673. Hodge G.A. and Bowman D.M. (2007). Engaging in small talk: nanotechnology policy and dialogue processes in the UK and Australia. *Australian Journal of Public Administration*. **66**(2):223–237.

674. Horton M.A. and Khan A. (2006). Medical nanotechnology in the UK: a perspective from the London Centre for Nanotechnology. *Nanomedicine: Nanotechnology, Biology and Medicine*. **2**(1):42–48.

675. Kearnes M. et al. (2006). From bio to nano: learning lessons from the UK agricultural biotechnology controversy. *Science as Culture*. **15**(4):291–307.

676. Leach R.K. et al. (2002). Nanoscience advances in the UK in support of nanotechnology. *International Journal of Nanoscience.* **1**(02):123–138.

677. Munari F. and Toschi L. (2011). Do venture capitalists have a bias against investment in academic spin-offs? Evidence from the micro-and nanotechnology sector in the UK. *Industrial and Corporate Change.* **20**(2):397–432.

678. Rogers-Hayden T. and Pidgeon N. (2006). Reflecting upon the UK's citizens' jury on nanotechnologies: NanoJury UK. *Nanotechnology Law and Business.* **3**:167.

679. Rogers-Hayden T. and Pidgeon N. (2007). Moving engagement "upstream"? Nanotechnologies and the Royal Society and Royal Academy of Engineering's inquiry. *Public Understanding of Science.* **16**(3):345–364.

680. Rogers-Hayden T. and Pidgeon N. (2008). Developments in nanotechnology public engagement in the UK: upstream towards sustainability? *Journal of Cleaner Production.* **16**(8): 1010–1013.

681. Taylor J.M. (2002). New dimensions for manufacturing: a UK strategy for nanotechnology. London: Office of Science and Technology.

682. Tolfree D. (2008). Progress in commercialising micro-nanotechnology in the UK. *International Journal of Technology Transfer and Commercialisation.* **7**(4):284–289.

683. Wienroth M. and Kearnes M. (2010). Science policy as discourse: the governance of nanotechnology in the United Kingdom'. In *Understanding Nanotechnology: Philosophy, Policy and Publics.* Heidelberg, Germany: AKA Verlag, pp. 101–120.

United States

Education

684. American Association of Community Colleges. Nanotechnology: multi-disciplinary, multi-market, multi-cultural (NSF grant 0802323). Washington, D.C.

685. Ernst J.V. (2009). Nanotechnology education: contemporary content and approaches. *The Journal of Technology Studies.* **35**(1):3.

Culture and Society

686. Bennett I. and Sarewitz D. (2006). Too little, too late? Research policies on the societal implications of nanotechnology in the United States. *Science as Culture.* **15**(4):309–325.

Ethics, Policy, and Regulation

687. Brindell J.R. (2010). Nanotechnology demands a new relationship between federal, state, and local regulatory agencies. *Nanotechnology Law and Business.* **7**:144.
688. Chittenden S.R. (2010). State and local regulation of nanotechnology: two opposing methodologies. *Nanotechnology Law and Business.* **7**:278.
689. Conti J. et al. (2011). Vulnerability and social justice as factors in emergent U.S. nanotechnology risk perceptions. *Risk Analysis.* **31**(11):1734–1748.

Informatics

690. Dudo A. et al. (2011). The emergence of nano news: tracking thematic trends and changes in US newspaper coverage of nanotechnology. *Journalism & Mass Communication Quarterly.* **88**(1):55–75.

National Nanotechnology Initiative

691. Campbell L.M. (2006). Nanotechnology and the United States national plan for research and development in support of critical infrastructure protection. *Canadian Journal of Law and Technology.* **5**(3):153.
692. Davies J.C. (2008). Nanotechnology oversight: an agenda for the new administration (PEN 13). Washington, D.C.: Project on Emerging Nanotechnologies.

693. Dunphy Guzman K.A. et al. (2006). Environmental risks of nanotechnology: national nanotechnology initiative funding, 2000–2004. *Environmental Science & Technology.* **40**(5):1401–1407.

694. Elwood T.W. (2003). Nanotechnology and the challenges it poses for the US federal government. *International Quarterly of Community Health Education.* **23**(2):89–95.

695. Gallo J. (2009). The discursive and operational foundations of the national nanotechnology initiative in the history of the national science foundation. *Perspectives on Science.* **17**(2):174–211.

696. Jung H.J. (2014). The impacts of science and technology policy interventions on university research: evidence from the US national nanotechnology initiative. *Research Policy.* **43**(1):74–91.

697. Motoyama Y. et al. (2011). The national nanotechnology initiative: federal support for science and technology, or hidden industrial policy? *Technology in Society.* **33**(1):109–118.

698. National Research Council (2002). Small wonders, endless frontiers: a review of the national nanotechnology initiative. Washington, D.C.: National Academies Press.

699. National Research Council (2013). Triennial review of the national nanotechnology initiative. Washington, D.C.: National Academies Press.

700. Roco M.C. (1999). Towards a US national nanotechnology initiative. *Journal of Nanoparticle Research.* **1**(4):435–438.

701. Roco M.C. (2001). From vision to the implementation of the US national nanotechnology initiative. *Journal of Nanoparticle Research.* **3**(1):5–11.

702. Roco M.C. (2003). National nanotechnology initiative to advance broad societal goals. *MRS Bulletin.* **28**(06):416–417.

703. Roco M.C. (2004). The US national nanotechnology initiative after 3 years (2001–2003). *Journal of Nanoparticle Research.* **6**(1):1–10.

704. Sandler R. and Kay W.D. (2006). The national nanotechnology initiative and the social good. *The Journal of Law, Medicine & Ethics.* **34**(4):675–681.

705. Sargent J.F. (2008). Nanotechnology and U.S. competitiveness: issues and options. Washington, D.C.: Congressional Research Office.

706. Sargent J.F. (2013). The national nanotechnology initiative: overview, reauthorization, and appropriations issues. Washington, D.C.: Congressional Research Office.

707. Teague C. (2005). United States national nanotechnology initiative. *Epidemiology.* **16**(5):S153.

Patents

708. Bawa R. (2004). Nanotechnology patenting in the US. *Nanotechnology Law and Business.* **1**:31.

709. Bawa R. (2007). Nanotechnology patent proliferation and the crisis at the US patent office. *Albany Law Journal of Science & Technology.* **17**:699.

710. Bennett H.S. et al. (2009). A method for assigning priorities to united states measurement system (usms) needs: nano-electrotechnologies. *Journal of Research of the National Institute of Standards and Technology.* **114**(4):237.

711. Fang Y. et al. (2014). A study of American nanotechnology development based on patent analysis. *Journal of Chemical and Pharmaceutical Research.* **6**(7):1291–1295.

Risk

712. Corley E.A. et al. (2009). Of risks and regulations: how leading U.S. nanoscientists form policy stances about nanotechnology. *Journal of Nanoparticle Research.* **11**:1573–1585.

713. Engeman C.D. et al. (2013). The hierarchy of environmental health and safety practices in the US nanotechnology workplace. *Journal of Occupational & Environmental Hygiene.* **10**(9):487–495.

714. Foster L.E. et al. (2006). Nanotechnology in the United States. Santa Monica, CA: Greenberg Traurig.

715. Freeman R. and Shukla K. (2008). Jobs in nanotech: creating a measure of job growth. *SEWP Digest*. June.

716. Granqvist N. and Laurila J. (2011). Rage against self-replicating machines: framing science and fiction in the US nanotechnology field. *Organization Studies*. **32**(2):253–280.

717. Kavetsky R. (2004). The navy's program in nanoscience and nanotechnology: a look ahead. Arlington, VA: Office of Naval Research.

718. Keller K. and Gorowara R.L. (2004). US nanotechnology directions: roadmap for nanomaterials from the chemical industry. *Chemie Ingenieur Technik*. **76**(9):1374–1374.

719. Kim K.Y. (2007). Research training and academic disciplines at the convergence of nanotechnology and biomedicine in the United States. *Nature Biotechnology*. **25**(3):359–361.

720. Kim Y. et al. (2012). Classifying US nano-scientists: of cautious innovators, regulators, and technology optimists. *Science and Public Policy*. **39**(1):30–38.

721. Manish M. (2010). 2009 NCMS study of nanotechnology in the U.S. manufacturing industry. Final report (NSF Award Number DMI-0802026). Ann Arbor, Michigan: National Center for Manufacturing Sciences.

722. Michelson E.S. (2008). Globalization at the nano frontier: the future of nanotechnology policy in the United States, China, and India. *Technology in Society*. **30**(3):405–410.

723. Miziolek A. (2002). Nanoenergetics: an emerging technology area of importance. *AMPTIAC Newsletter*. **6**(1):43–48.

724. Morris J. and Doa M.J. (2012). Transnational environmental governance of nanotechnology: a US regulatory perspective. *Nanotechnology Law and Business*. **9**:369.

725. Mouttet B. (2006). Nanotechnology and US patents: a statistical analysis. *Nanotechnology Law and Business*. **3**:309.

726. Mowery D.C. (2011). Nanotechnology and the US national innovation system: continuity and change. *The Journal of Technology Transfer*. **36**(6):697–711.

727. Murashov V. and Howard J. (2008). The US must help set international standards for nanotechnology. *Nature Nanotechnology*. **3**(11):635–636.

728. National Science & Technology Council (2000). Regional, state, and local initiatives in nanotechnology. Washington, D.C.

729. Patra D. (2011). Responsible development of nanoscience and nanotechnology: contextualizing socio-technical integration into the nanofabrication laboratories in the USA. *NanoEthics*. **5**(2):143–157.

730. Pidgeon N. et al. (2008). Deliberating the risks of nanotechnologies for energy and health applications in the united states and United Kingdom. *Nature Nanotechnology*. **4**:95–98.

731. Priest S. (2006). The North American opinion climate for nanotechnology and its products: opportunities and challenges. *Journal of Nanoparticle Research*. **8**(5):563–568.

732. Priest S. et al. (2010). Risk perceptions starting to shift? US citizens are forming opinions about nanotechnology. *Journal of Nanoparticle Research*. **12**(1):11–20.

733. Rayms-Keller A. (2012). Can nanotechnology meet emerging naval requirements? Dahlgren, VA: Naval Surface Warfare Center Dahlgren Division.

734. Roco M.C. (2002). Nanoscale science and engineering education activities in the United States. *Journal of Nanoparticle Research*. **4**:271–274.

735. Rogers J.D. et al. (2011). Assessment of fifteen nanotechnology science and engineering centers' (NSECs) outcomes and impacts: their contribution to NNI objectives and goals (NSF Award 0955089). Atlanta, GA: School of Public Policy, Georgia Institute of Technology.

736. Shapira P. and Youtie J. (2006). Measures for knowledge-based economic development: introducing data mining techniques to economic developers in the state of Georgia and the US South. *Technological Forecasting & Social Change*. **73**:950–965

737. Shapira P. and Youtie J. (2008). Emergence of nanodistricts in the United States: path dependency or new opportunities? *Economic Development Quarterly*. **22**:187–199.

738. Shapira P. and Wang J. (2007). Case study: R&D policy in the United States: the promotion of nanotechnology R&D. Atlanta: Georgia Institute of Technology, Program in Science, Technology & Innovation Policy.

739. Toumey C. et al. (2006). Dialogue on nanotech: the South Carolina citizens' school of nanotechnology. *Journal of Business Chemistry.* **3**(3):3–8.

740. U.S. EPA (2007). Proceedings of the interagency workshop on the environmental implications of nanotechnology, September 5–7. Washington, D.C. Office of Research & Development, National Center for Environmental Research.

741. U.S. Congress (2000). Nanotechnology: the state of nanoscience and its prospects for the next decade: hearing before the subcommittee on basic research of the committee on science, house of representatives. One Hundred Sixth Congress, first session, June 22, 1999, Volume 4.

742. U.S. Government Accountability Office (2010). Nanotechnology: nanomaterials are widely used in commerce, but EPA faces challenges in regulating risk (GAO-10-549). Washington, D.C.

743. U.S. Government Accountability Office (2012). Nanotechnology: improved performance information needed for environmental, health, and safety research (GAO-12-427). Washington, D.C.

744. U.S. Government Accountability Office (2014). Nanomanufacturing: emergence and implications for U.S. competitiveness, the environment, and human health (GAO-14-181SP). Washington, D.C.

745. Van Horn C.E. et al. (2009). A profile of nanotechnology degree programs in the United States. Tempe, AZ: Center for Nanotechnology in Society.

746. Voorhees Jr, T. (2014). Brief overview of current developments in nanotechnology EHS regulation in the US. *Nanotechnology Law and Business.* **11**:39.

747. Walsh J.P. and Ridge C. (2012). Knowledge production and nanotechnology: characterizing American dissertation research, 1999–2009. *Technology in Society.* **34**(2):127–137.

748. Walsh J.P. (2014). The impact of foreign-born scientists and engineers on American nanoscience research. *Science and Public Policy.* **42**(1):107–120.

749. Woodson T. (2015). Nanotechnology companies in the United States: a web-based content analysis of companies and products for poverty alleviation. *Journal of Business Chemistry.* **12**(1):3–15.

750. Yarbrough A.B. (2010). The impact of nanotechnology energetics on the department of defense by 2035. Thesis: Air War College.

751. Yawson R. (2011). Historical antecedents as precedents for nanotechnology vocational education training and workforce development. *Human Resource Development Review.* **10**(4): 417–430.

752. Youtie J. and Shapira P. (2008). Mapping the nanotechnology enterprise: a multi-indicator analysis of emerging nanodistricts in the U.S. South. *The Journal of Technology Transfer.* **33**(2):209–223.

Uruguay

753. Chiancone A. (2012). Nanociencia y nanotecnologías en Uruguay: áreas estratégicas y temáticas grupales. In Foladori G., Invernizzi N. and Záyago E. (eds.) *Perspectivas sobre el desarrollo de las nanotecnologías en América Latina.* México: Miguel Angel Porruá.

754. Chimuris R. et al. (2007). El control extranjero de las nanotecnologías mediante los derechos de propiedad. El caso de Uruguay. *Red Latinoamericana de Nanotecnología y Sociedad.* Access at www.estudiosdeldesarrollo.net/relans

Uzbekistan

755. Koneev R.I. et al. (2010). Nanomineralogy and nanogeochemistry of ores from gold deposits of Uzbekistan. *Geology of Ore Deposits.* **52**(8):755–766.

756. Madyarov S.R. (2005). Biotechnological approaches in sericultural science and technology of Uzbekistan. *International Journal of Industrial Entomology.* **11**(1):13–19.

Vatican City

757. Bibbee J.R. and Viens A.M. (2007). The inseparability of religion and politics in the neoconservative critique of biotechnology. *The American Journal of Bioethics.* **7**(10):18–20.
758. Drees W.B. (2005). Religion and science as advocacy of science and as religion versus religion. *Zygon.* **40**(3):545–554.

Venezuela

759. Cadenas M.S.L. et al. (2011). Nanoscience and nanotechnology in Venezuela. *Journal of Nanoparticle Research.* **13**(8):3101–3106.
760. Hasmy A. (2011). Formación y divulgación de la nanotecnología en Venezuela: situación y perspectiva. *En Mundo Nano.* **4**(2):72 (Spanish).
761. López M.S. et al. (2013). Socialización del conocimiento: estrategias en el campo de la nanociencia y la nanotecnología en Venezuela. *Revista de Física.* **46E**:1.

Vietnam

762. Anh T.K. et al. (2009). Nanosciences and nanotechnology of Vietnamese women: fabrication, characterization, and application of rare earth nanocompounds for photonics. *American Institute of Physics Conference Series.* **1119**:201–201.
763. Long B.T. and Toan N.D. (2015). Scientific and technological journals in Vietnam: the current state and direction of development. *Science Editing.* **2**(1):18–21.
764. Manh H.D. (2015). Scientific publications in Vietnam as seen from Scopus during 1996–2013. *Scientometrics.* **105**(1):83–95.
765. Minh P.N. et al. (2010). Applications of nanomaterials in Vietnam: opportunities and Challenges. A paper presented at the *Workshop on Nanotechnology for Sustainable Energy Solutions.* Korea.

766. Nguyen K. and Van P.H. (2014). Nanotechnology at SHTP LABS in Vietnam. *IFMBE Proceedings*. **27**:312–315.

767. Tran A.N. et al. (2010). A new platform for RFID research in Vietnam. *Advances in Natural Sciences: Nanoscience and Nanotechnology*. **1**:045015.

Yemen

768. Zaari H. et al. (2014). The investigation of electronic structure, optical and magnetic properties of MgB_2 nanosheets. In *International Renewable and Sustainable Energy Conference (IRSEC)*, Ouarzazate, pp. 931–934.

Zambia

769. Brown R. et al. (2008). POCT and rapid scale-up of centralized laboratory services in Zambia. *Point of Care*. **7**(3):200.

Zimbabwe

770. Edwardmanyarara T. and Matyanga C. (2015). Nano formulated anti-retroviral therapy drugs of potential benefit to HIV/AIDS management in Zimbabwe: a review. *Journal of International Academic Research for Multidisciplinary*. **3**(4):111–121.

771. Savers S. (2006). Zimbabwe changes attitude to biotechnology. *Biotechnology Law Report*. **25**(6):675–678.

Section 2

Economics

Nanotechnology has become a major economic player on the global stage. Nanotech companies enjoy large profits, and Wall Street and the major trading centers of the world embrace the financial risks and rewards of the very small. The literature of nanoeconomics is limited, but the following citations provide some sense of its direction and importance.

772. Abelson P.H. (2000). Funding the nanotech frontier. *Science.* **288**(5464):269–269.

773. Andersen M.M. (2011). Silent innovation: corporate strategizing in early nanotechnology evolution. *The Journal of Technology Transfer.* **36**(6):680–696.

774. APEC Center for Technology Oversight (2002). Nanotechnology: the technology for the 21st century. Bangkok.

775. Avenel E. et al. (2007). Diversification and hybridization in firm knowledge bases in nanotechnologies. *Research Policy.* **36**:864–870.

776. Beaudry C. and Allaoui S. (2012). Impact of public and private research funding on scientific production: the case of nanotechnology. *Research Policy.* **41**(9):1589–1606.

The Nanotechnology Revolution: A Global Bibliographic Perspective
Dale A. Stirling
Copyright © 2018 Pan Stanford Publishing Pte. Ltd.
ISBN 978-981-4774-19-2 (Hardcover), 978-1-315-11083-7 (eBook)
www.panstanford.com

777. Bozeman B. et al. (2008). Barriers to the diffusion of nanotechnology. *Economics* of *Innovation and New Technology.* **17**(7–8):751–763.

778. Canton J. (1999). Global future business. The strategic impact of nanoscience on the future of business and economics. San Francisco: Institute for Global Futures.

779. Chen H. et al. (2013). Nanotechnology public funding and impact analysis: a tale of two decades (1991–2010). *IEEE Nanotechnology Magazine.* **7**(1):9–14.

780. Chilcott J. et al. (2001). Nanotechnology: commercial opportunity. London: Evolution Capital Ltd.

781. Cientifica (2011). Global funding of nanotechnologies and its impact. London: Cientifica Ltd.

782. Coccia M. (2012). Evolutionary trajectories of the nanotechnology research across worldwide economic players. *Technology Analysis & Strategic Management.* **24**(10):1029–1050.

783. Cunningham S.W. and Porter A.L. (2011). Bibliometric discovery of innovation and commercialization pathways in nanotechnology. *Proceedings of PICMET '11: Technology Management in the Energy Smart World (PICMET)*, Portland, pp. 1–11.

784. Darby M.R. and Zucker L.G. (2005). Grilichesian breakthroughs: inventions of methods of inventing and firm entry in nanotechnology. *Annales d'Economie et de Statistique.* **79/80**:143–164.

785. Festel G. et al. (2010). Importance and best practice of early stage nanotechnology investments. *Nanotechnology Law and Business.* **7**:50.

786. Foladori G. and Invernizzi N. (2005). Nanotechnology in its socio-economic context. *Science Studies.* **18**(2):67–73.

787. Fonash S.J. (2009). Nanotechnology and economic resiliency. *Nano Today.* **4**(4):290–291.

788. Garrett D. (2005). Break-out in nanotech-the next potential wave of IPOs. *Nanotechnology Law and Business.* **2**:274.

789. Giordani S. (2009). Moving nanotechnology toward the market: business strategy and IP management in the value chain. *MRS Proceedings*. **1209**:1209-P05-03.

790. Graffagnini M.J. (2008). Raising venture capital for a nanoparticle therapeutics company. *Nanotechnology Law & Business*. **5**:207.

791. Graffignini M.J. (2009). Corporate strategies for nanotech companies and investors in new economic times. *Nanotechnology Law & Business*. **6**:251.

792. Granqvist N. et al. (2013). Hedging your bets: explaining executives' market labeling strategies in nanotechnology. *Organization Science*. **24**(2):395–413.

793. Hobson D.W. (2009). Commercialization of nanotechnology. *Wiley Interdisciplinary Reviews: Nanomedicine and Nanobiotechno-logy*. **1**(2):189–202.

794. Hullmann A. (2006). The economic development of nanotechnology: an indicators based analysis. Brussels: European Commission.

795. Hullmann A. (2007). Measuring and assessing the development of nanotechnology. *Scientometrics*. **70**(3):739–758.

796. James T.L. (2007). Use of reverse mergers to bypass IPOs: a new trend for nanotech companies. *Nanotechnology Law and Business*. **4**:95.

797. Jones R. (2008). The economy of promises. *Nature Nanotechnology*. **3**(2):65–66.

798. Köhler T. et al. (2003). Nanotechnology: markets & trends. Nanotechnologie: märkte und trends. *Vakuum in Forschung und Praxis*. **15**(6):292–297.

799. Liota T. and Tzitzios V. (2006). Investing in nanotechnology. *Nanotechnology Law and Business*. **3**:521.

800. Maine E. (2013). Scientist-entrepreneurs as the catalysts of nanotechnology commercialization. *Reviews in Nanoscience and Nanotechnology*. **2**(5):301–308.

801. Malanowski N. and Zweck A. (2007). Bridging the gap between foresight and market research: integrating methods to assess

the economic potential of nanotechnology. *Technological Forecasting and Social Change.* **74**(9):1805–1822.

802. Mario C. et al. (2010). Research trends in nanotechnology studies across geo-economics areas (Working Paper Ceris-Cnr, N.5/2010). Torino, Italy: Instituto di Ricera sull'Impresa e lo Sviluppo.

803. Mazzola L. (2003). Commercializing nanotechnology. *Nature.* **21**(10):1137–1143.

804. Meehan D.N. (2011). The impact of nanotechnology on oil and gas economics. *The Way Ahead.* **7**(3):18–19.

805. Milanović V. and Bučalina A. (2013). Position of the countries in nanotechnology and global competitiveness. *Management Journal for Theory and Practice Management.* 69–79.

806. Momaya K. (2011). Cooperation for competitiveness of emerging countries: learning from a case of nanotechnology. *Competitiveness Review: An International Business Journal.* **21**(2):152–170.

807. Palmberg C. et al. (2009). Nanotechnology: an overview based on indicators and statistics. Paris: OECD.

808. OECD (2013). Symposium on assessing the economic impact of nanotechnology: synthesis report. Paris.

809. Ott I. and Papilloud C. (2007). Converging institutions: shaping relationships between nanotechnologies, economy, and society. *Bulletin of Science, Technology & Society.* **27**(6):455–466.

810. Paull R. et al. (2003). Investing in nanotechnology. *Nature Biotechnology.* **21**(10):1144–1147.

811. Rasmussen B. (2007). Is the commercialization of nanotechnology different? A case study approach. *Innovation.* **9**(1):62–78.

812. Roco M.C. (2002). Government nanotechnology funding: an international outlook. *JOM.* **54**(9):22–23.

813. Roco M.C. (2005). International perspective on government nanotechnology funding in 2005. *Journal of Nanoparticle Research.* **7**(6):707–712.

814. Roco M.C. and Bainbridge W.S. (2007). Economic impacts and commercialization of nanotechnology. In *Nanotechnology:*

Societal Implications. Amsterdam, Netherlands: Springer, pp. 7–74.

815. Romig A.D. et al. (2007). An introduction to nanotechnology policy: opportunities and constraints for emerging and established economies. *Technological Forecasting and Social Change*. **74**(9):1634–1642.

816. Sakakibara M. (2014). The Role of geographic and organizational boundaries in nanotechnology collaboration. *Annals of Economics and Statistics/Annales d'Économie et de Statistique*. **115–116**:177–193.

817. Salehi M. and Niaz-Azari K. (2010). A short review on entrepreneurship in the field of evolving nanotechnology. *Middle-East Journal of Scientific Research*. **6**(4):412–417.

818. Saxl O. (2013). Nanotechnology: applications and markets, present and future. In *Ellipsometry at the Nanoscale*. Berlin Heidelberg: Springer, pp. 705–730.

819. Shapira P. and Wang J. (2010). Follow the money. *Nature*. **468**(7324): 627–628.

820. Smith D.M. et al. (2007). Reverse mergers and nanotechnology. *Nanotechnology Law and Business*. **4**:87.

821. Sparks S. (2012). *Nanotechnology: Business Applications and Commercialization*. Boca Raton, FL: CRC press.

822. Steinfeldt M. et al. (2004). Nanotechnology and sustainability (Discussion Paper 65/04). Berlin: Institute for Ecological Economy Research.

823. Stephan P. et al. (2007). The small size of the small scale market: the early-stage labor market for highly skilled nanotechnology workers. *Research Policy*. **36**(6):887–892.

824. Tolfree D. and Jackson M.J. (eds.) (2007). *Commercializing Micro-Nanotechnology Products*. Boca Raton, FL: CRC Press.

825. Walsh B. et al. (2010). A comparative methodology for estimating the economic value of innovation in nanotechnologies. Buckinghamshire, UK: Oakdene Hollins Research & Consulting.

826. Wang J. and Shapira P. (2011). Funding acknowledgement analysis: an enhanced tool to investigate research sponsorship

impacts: the case of nanotechnology. *Scientometrics.* **87**(3): 563–586.

827. Wang J. and Shapira P. (2012). Partnering with universities: a good choice for nanotechnology start-up firms? *Small Business Economics.* **38**(2):197–215.

828. Weil V. (2012). From the trenches: first-hand reports of how companies are managing nanotechnologies. *Nanotechnology Law and Business.* **9**:253.

829. Wry T. and Lounsbury M. (2013). Contextualizing the categorical imperative: category linkages, technology focus, and resource acquisition in nanotechnology entrepreneurship. *Journal of Business Venturing.* **28**(1):117–133.

830. Wry T. et al. (2014). Hybrid vigor: securing venture capital by spanning categories in nanotechnology. *Academy of Management Journal.* **57**(5):1309–1333.

831. Yadav S.K. et al. (2013). Impact of nanotechnology on socio-economic aspects: an overview. *Reviews in Nanoscience and Nanotechnology.* **2**(2):127–142.

Section 3

Education, Culture, and Society

This section of the book has two purposes. First, to highlight pertinent literature relating to nanotechnology and how it is taught, how it is learned, and how that translates to employment in the nanotechnology field. Second, to highlight literature that examines the impact of nanotechnology on culture and society, ranging from its impact on the disenfranchised to the new social media universe.

Education, Training, and Employment

832. Albe V. (2011). Nanoscience and nanotechnologies education: teachers' knowledge. Available at http://www.esera.org/media/ebook/strand13/ebook-esera2011_ALBE-13.pdf

833. Anderson A.A. et al. (2010). The changing information environment for nanotechnology: Online audiences and content. *Journal of Nanoparticle Research.* **12**(4):1083–1094.

834. Bozeman B. et al. (2007). Understanding the emergence and deployment of "nano" S&T. *Research Policy.* **36**(6):807–812.

835. Bryan L.A. et al. (2012). Facilitating teachers' development of nanoscale science, engineering, and technology content knowledge. *Nanotechnology Reviews.* **1**(1):85–95.

The Nanotechnology Revolution: A Global Bibliographic Perspective
Dale A. Stirling
Copyright © 2018 Pan Stanford Publishing Pte. Ltd.
ISBN 978-981-4774-19-2 (Hardcover), 978-1-315-11083-7 (eBook)
www.panstanford.com

836. Cavanagh S. (2009). Nanotechnology slips into schools. *Education Week.* **28**(27):1–12.

837. Chang R.P.H. (2006). A call for nanoscience education. *Nano Today.* **1**(2):6–7.

838. Chari D. et al. (2012). Disciplinary identity of nanoscience and nanotechnology research: a study of postgraduate researchers' experiences. *International Journal of Digital Society.* **3**(1):609–616.

839. Chopra N. and Reddy R.G. (2012). Undergraduate education in nanotechnology and nanoscience. *JOM.* **64**(10):1127–1129.

840. Cowan K. and Gogotsi Y. (2004). The Drexel/UPenn IGERT: creating a new model for graduate education in nanotechnology. *Journal of Materials Education.* **26**(1–3):147–152.

841. Crone W. (2010). Bringing nano to the public: a collaboration opportunity for researchers and museums. *Journal of Nano Education.* **2**(1–2):102–116.

842. Day D.A. et al. (2009). Authentic science research and the utilization of nanoscience in the non-traditional classroom setting. *MRS Proceedings.* **1233**:1233-PP04-32.

843. Drane D. et al. (2009). An evaluation of the efficacy and transferability of a nanoscience module. *Journal of Nano Education.* **1**:8–14.

844. Duncan K.A. et al. (2010). Art as an avenue to science literacy: teaching nanotechnology through stained glass. *Journal of Chemical Education.* **87**(10):1031–1038.

845. Ernst Jeremy V. (2009). Nanotechnology education: contemporary content and approaches. *Journal of Technology Studies.* **35**(1):3–8.

846. Feather J.L. and Aznar M.F. (2010). *Nanoscience Education, Workforce Training, and K-12 Resources.* Boca Raton, FL: CRC Press.

847. Foley E.T. and Hersam M.C. (2006). Assessing the need for nanotechnology education reform in the United States. *Nanotechnology Law and Business.* **3**:467.

848. Fonash S.J. (2001). Education and training of the nanotechnology workforce. *Journal of Nanoparticle Research.* **3**(1):79–82.

849. Ghattas N.I. and Carver J.S. (2012). Integrating nanotechnology into school education: a review of the literature. *Research in Science & Technological Education.* **30**(3):271–284.

850. Goodhew P. (2006). Education moves to a new scale. *Nano Today.* **1**(2):40–43.

851. Gorman M.E. et al. (2013). Integrating ethics and policy into nanotechnology education. *Journal of Nano Education.* **4**(1–2): 25–32.

852. Gottfried D.S. (2011). Review of nanotechnology in undergraduate education. *Journal of Chemical Education.* **88**(5):544–545.

853. Greenberg A. (2009). Integrating nanoscience into the classroom: perspectives on nanoscience education projects. *ACS Nano.* **3**(4):762–769.

854. Healy N. (2009). Why nano education? *Journal of Nano Education.* **1**(1):6–7.

855. Hey J.H.G. et al. (2009). Putting the discipline in interdisciplinary: using speedstorming to teach and initiate creative collaboration in nanoscience. *Journal of Nano Education.* **1**:75–85.

856. Hingant B. and Albe V. (2010). Nanosciences and nanotechnologies learning and teaching in secondary education: a review of literature. *Studies in Science Education.* **46**(2):121–152.

857. Holley S. (2009). Nano revolution – big impact: how emerging nanotechnologies will change the future of education and industry in America (and more specifically in Oklahoma). An abbreviated account. *Journal of Technology Studies.* **35**(1):9–19.

858. Hoover E. et al. (2009). Teaching small and thinking large: effects of including social and ethical implications in an interdisciplinary nanotechnology course. *Journal of Nano Education.* **1**:86–95.

859. Huffman D. et al. (2015). Integrating nanoscience and technology in the high school science classroom. *Nanotechnology Reviews.* **4**(1):81–102.

860. Itoh T. et al. (2005). Trans-disciplinary graduate and refresher programs for education, research and training in the fields of nanoscience and nanotechnology. *MRS Proceedings*. **909**:0909-PP01-06.

861. Jones M.G. et al. (2007). Differences in African-American and European-American students' engagement with nanotechnology experiences: perceptual position or assessment artifact? *Journal of Research in Science Teaching*. **44**(6):787–799.

862. Jones M.G. et al. (2015). Precollege nanotechnology education: a different kind of thinking. *Nanotechnology Reviews*. **4**(1):117–127.

863. Klimeck G. et al. (2008). Advancing education and research in nanotechnology. *Computing in Science & Engineering*. **10**(5):17–23.

864. Knobel M. et al. (2010). The perception of nanoscience and nanotechnology by children and teenagers. *Journal of Materials Education*. **32**(1–2):29–38.

865. Kurath M. and Gisler P. (2009). Informing, involving or engaging? Science communication, in the ages of atom-, bio- and nanotechnology. *Public Understanding of Science*. **18**(5):559–573.

866. Laherto A. (2010). An analysis of the educational significance of nanoscience and nanotechnology in scientific and technological literacy. *Science Education International*. **21**(3):160–175.

867. Laherto A. (2012). Nanoscience education for scientific literacy: opportunities and challenges in secondary school and in out-of-school settings (Report Series in Physics No. HU-P-D194). Helsinki: University of Helsinki, Department of Physics, Faculty of Science.

868. Lu K. (2009). A study of engineering freshmen regarding nanotechnology understanding. *Journal of STEM Education: Innovations and Research*. **10**(1/2):7.

869. Lundin Palmerius K.E. (2013). A case-based study of students' visuohaptic experiences of electric fields around molecules: shaping the development of virtual nanoscience learning

environments. *Education Research International.* **2013**: Article ID 194363.

870. Mancini-Samuelson G.J. (2013). Using nanoscience as a theme for capstone projects in an elementary education major's science course. *MRS Online Proceedings Library.* **1532**:12.

871. Meyyappan M. (2004). Nanotechnology education and training. *Journal of Materials Education.* **26**(3/4):313.

872. Mohammad A.W. et al. (2012). Elements of nanotechnology education engineering curriculum worldwide. *Procedia—Social & Behavioral Sciences.* **60**:405–412.

873. Ng W. (2009). Nanoscience and nanotechnology for the middle years. *Teaching Science.* **55**(2):16–24.

874. O'Connor C. and Hayden H. (2008). Contextualising nanotechnology in chemistry education. *Chemistry Education Research and Practice.* **9**(1):35–42.

875. Orgill M. and Wood S.A. (2014). Chemistry contributions to nanoscience and nanotechnology education: a review of the literature. *Journal of Nano Education.* **6**(2):83–108.

876. Planinsic G. et al. (2009). Themes of nanoscience for the introductory physics course. European *Journal of Physics.* **30**(4):S17.

877. Poteralska B. et al. (2011). The development of education and training systems in the field of nanotechnology. *Journal of College Teaching & Learning.* **4**(6):7–16.

878. Rasmussen A.J. and Ebbesen M. (2014). Why should nanoscience students be taught to be ethically competent? *Science and Engineering Ethics.* **20**(4):1065–1077.

879. Roco M. (2002). Nanoscale science and engineering education activities in the United States. *Journal of Nanoparticle Research.* **4**:271–274.

880. Roco M. (2003). Converging science and technology at the nanoscale: opportunities for education and training. *Nature Biotechnology.* **21**(10):1247–1249.

881. Samet C. (2009). A capstone course in nanotechnology for chemistry majors. *Journal of Nano Education.* **1**(1):15–21.

882. Schank P. et al. (2007). Can nanoscience be a catalyst for educational reform. In *Nanoethics: The Ethical and Social Implications of Nanotechnology*. New Jersey: John Wiley & Sons, pp. 277–290.

883. Schimke A. et al. (2013). Impact of local knowledge endowment on employment growth in nanotechnology. *Industrial and Corporate Change*. **22**(6):1525–1555.

884. Schwenz R.W. and Pacheco K.A. (2014). A first year experience course on nanoscience for undergraduates. *Journal of Nano Education*. **6**(2):148–151.

885. Sohlberg K. (2006). Introducing the core concepts of nanoscience and nanotechnology: two vignettes. *Journal of Chemical Education*. **83**(10):1516.

886. St A. and Sarah K. (2014). Bringing nanoscience into traditional physical and inorganic chemistry courses. *Journal of Nano Education*. **6**(2):132–138.

887. Stephan P. et al. (2007). The small size of the small scale market: the early-stage labor market for highly skilled nanotechnology workers. *Research Policy*. **36**(6):887–892.

888. Stevens S.Y. et al. (2009). *The Big Ideas of Nanoscale Science and Engineering: A Guidebook for Secondary Teachers*. Arlington, VA: NSTA Press.

889. Sweeney A.E. (2006). Social and ethical dimensions of nanoscale science and engineering research. *Science & Engineering Ethics*. **12**(3):435–464.

890. Sweeney A.E. and Seal S. (eds.) (2008). *Nanoscale Science and Engineering Education*. Stevenson Ranch, CA: American Scientific Publishers.

891. Sweeney A.E. et al. (2003). The promises and perils of nanoscience and nanotechnology: exploring emerging social and ethical issues. *Bulletin of Science, Technology & Society*. **23**(4):236–245.

892. Tarng W. et al. (2011). Development and research of web-based virtual nanotechnology laboratory for learning the basic concepts of nanoscience. *Science*. **2**(6).

893. Taylor A. et al. (2008). Bumpy, sticky and shaky: nanoscale science and the curriculum. *Science Scope.* **31**(7):28–35

894. Tomasik J. et al. (2009). Design and initial evaluation of an online nanoscience course for teachers. *Journal of Nano Education.* **1**:48–69.

895. Tretter T.R. et al. (2010). Impact of introductory nanoscience course on college freshmen's conceptions of spatial scale. *Journal of Nano Education.* **2**(1–2):53–66.

896. Trybula W. et al. (2010). The emergence of nanotechnology: establishing the new 21st century workforce. *Online Journal for Workforce Education and Development.* **3**(4):6.

897. Vogel V. et al. (2002). Education in nanotechnology: launching the first PhD program. *International Journal of Engineering Education.* **18**(5):498–505.

898. Wansom S. et al. (2009). A rubric for post-secondary degree programs in nanoscience and nanotechnology. *International Journal of Engineering Education.* **25**(3):615.

899. Weiss P.S. (2008). A conversation with Prof. Flemming Besenbacher: innovator in nanoscience and nanoscience education. *ACS Nano.* **2**(10):1979–1983.

900. Xie C. and Pallant A. (2011). The molecular workbench software: an innovative dynamic modeling tool for nanoscience education. In *Models and Modeling.* Netherlands: Springer, pp. 121–139.

901. Yawson R.M. (2012). An epistemological framework for nanoscience and nanotechnology literacy. *International Journal of Technology and Design Education.* **22**(3):297–310.

Culture and Society

902. Baccile N. and Balzerani M. (2013). Invisible et Insaisissable, quand l'art rencontre les nanosciences. Nanoscience and art: beyond the invisible and the intangible. *Nano.* **3**. http://art-science.univ-paris1.fr/plastik/document.

903. Bainbridge W.S. (2004). Sociocultural meanings of nanotechnology: research methodologies. *Journal of Nanoparticle Research.* **6**(2):285–299.

904. Bainbridge W.S. (ed.) (2007). *Nanotechnology: Societal Implications—I: Maximizing Benefits for Humanity; II: Individual Perspectives.* New York: Springer Science & Business Media.

905. Crone W.C. (2010). Bringing nano to the public: a collaboration opportunity for researchers and museums. *Journal of Nano Education.* **2**(1–2):102–116.

906. de Ridder-Vignone K.D. (2012). Public engagement and the art of nanotechnology. *Leonardo.* **45**(5):433–438.

907. de Ridder-Vignone K.D. (2012). Special section introduction: the images and art of nanotechnologies. *Leonardo.* **45**(5):431–432.

908. Gorman M.E. et al. (2004). Societal dimensions of nanotechnology. *IEEE Technology and Society Magazine.* **23**(4):55–62.

909. Griep M.H. (2010). Real life or reel life? Nanotechnology in movies. *Phi Kappa Phi Journal.* **90**(3):23.

910. Grunwald A. (2011). Ten years of research on nanotechnology and society: outcomes and achievements. In *Quantum Engagements: Social Reflections of Nanoscience and Emergent Technologies.* Heidelberg: AKA Verlag, pp. 41–58.

911. Gupta N. et al. (2012). Factors influencing societal response of nanotechnology: an expert stakeholder analysis. *Journal of Nanoparticle Research.* **14**(5):1–15.

912. Hanson R. (2006). Five nanotech social scenarios. In *Nanotechnology: Societal Implications—Individual Perspectives.* Berlin: Springer, pp. 75–130.

913. Harthorn B.H. and Mohr J.W. (eds.) (2013). *The Social Life of Nanotechnology.* New York: Routledge.

914. Hayles N.K. (ed.) (2004). *Nanoculture: Implications of the New Technoscience.* Bristol, UK: Intellect Books.

915. Kahan D.M. et al. (2009). Cultural cognition of the risks and benefits of nanotechnology. *Nature Nanotechnology.* **4**(2):87–90.

916. Kahan D.M. et al. (2011). Cultural cognition of scientific consensus. *Journal of Risk Research.* **14**(2):147–174.

917. Knipfer K. (2009). Pro or con nanotechnology? Support for critical thinking and reflective judgement at science museums. Doctoral Dissertation: Universität Tübingen.

918. Milburn C. (2010). Digital matters: video games and the cultural transcoding of nanotechnology. In *Governing Future Technologies*. Netherlands: Springer, pp. 109–127.

919. Murriello S. and Knobel M. (2008). Encountering nanotechnology in an interactive exhibition. *Journal of Museum Education*. **33**(2):221–230.

920. Murriello S. et al. (2015). Challenges of an exhibit on nanoscience and nanotechnology. *Adult Education*. **8**:05.

921. Nerlich B. (2005). From nautilus to nanobo (a) ts: the visual construction of nanoscience. *AZojono: Journal of Nanotechnology Online*. **1**.

922. Raulerson J. (2011). A poke in the eye with a sharp spike: nanoculture and the future of SF. *Science Fiction Studies*. **38**(1):183–191.

923. Roco M.C. (2003). Broader societal issues of nanotechnology. *Journal of Nanoparticle Research*. **5**(3–4):181–189.

924. Roco M.C. and Bainbridge W.S. (eds.) (2001). *Societal Implications of Nanoscience and Nanotechnology*. Washington, D.C.: National Science Foundation.

925. Roco M.C. and Bainbridge W.S. (2005). Societal implications of nanoscience and nanotechnology: maximizing human benefit. *Journal of Nanoparticle Research*. **7**(1):1–13.

926. Ronteltap A. et al. (2011). Societal response to nanotechnology: converging technologies–converging societal response research? *Journal of Nanoparticle Research*. **13**(10):4399–4410.

927. Runge K.K. et al. (2013). Tweeting nano: how public discourses about nanotechnology develop in social media environments. *Journal of Nanoparticle Research*. **15**(1):1–11.

928. Sheetz T. et al. (2005). Nanotechnology: awareness and societal concerns. *Technology in Society*. **27**(3):329–345.

929. Shih T.J. (2009). Predicting attitudes towards nanotechnology: the influence of cultural and predispositional values. PhD Dissertation: Madison, University of Wisconsin.

930. Stone J. (2006). Roots to branches: anthropology and the human dimensions of nanotechnology. *Practicing Anthropology.* **28**(2): 31–37.

931. Trujillo P.R. et al. (2011). Are (official) ethical approaches to nanotechnology affected by cultural context and tradition? A comparative analysis: Europe-USA. *Ramon Llull Journal of Applied Ethics.* **1**(2):195.

932. Weeks P. and Boyle R. (2006). What anthropology can contribute to the construction of nanotechnology policy and regulations. *Practicing Anthropology.* **28**(2):11–14.

933. Wood J. (2005). Museums bring the nano world to the masses: public outreach. *Materials Today.* **8**(12):25.

934. Wry T. et al. (2011). The cultural context of status: generating important knowledge in nanotechnology. In Pearce J.L. (ed.), *Status in Management and Organizations.* Cambridge, UK: Cambridge University Press, pp. 155–190.

935. Zülsdorf T., Coenen C., Ferrari A., Fiedeler U., Milburn C. and Wienroth M. (eds.) (2011). *Quantum Engagements: Social Reflections of Nanoscience and Emerging Technologies.* Amsterdam: IOS Press.

Section 4

Environmental and Human Health Risks

From small things, tremendous things can be created. However, in the techno-rush that is nanotechnology, its potential health and environmental effects are of real concern. Fortunately, there is a rapidly growing body of literature addressing these concerns and the following citations capture the global essence of that concern.

936. Aitken R.J. et al. (2009). EMERGNANO: a review of completed and near completed environment, health and safety research on nanomaterials and nanotechnology. Report on DEFRA project CB0409.
937. Allan S. et al. (2010). Framing risk: nanotechnologies in the news. *Journal of Risk Research.* **13**(1):29–44.
938. Anđić Z., Vujović A., Kneževoć M., Vasiljević R. and Tasić M. (2009). Nano-technologies from the aspect of human environment and safety and health at work. *Acta Facultatis Medicae Naissensis.* **27**(4).
939. Baalousha M. and Lead J.R. (2009). Overview of nanoscience in the environment. In Lead J.R. and Smith E. (eds.), *Environmental and Human Health Impacts of Nanotechnology.* New York: Blackwell Publishing, pp. 1–29.

The Nanotechnology Revolution: A Global Bibliographic Perspective
Dale A. Stirling
Copyright © 2018 Pan Stanford Publishing Pte. Ltd.
ISBN 978-981-4774-19-2 (Hardcover), 978-1-315-11083-7 (eBook)
www.panstanford.com

940. Balbus J.M. et al. (2007). Protecting workers and the environment: an environmental NGO's perspective on nanotechnology. *Journal of Nanoparticle Research.* **9**(1):11–22.

941. Bauer C. et al. (2008). Towards a framework for life cycle thinking in the assessment of nanotechnology. *Journal of Cleaner Production.* **16**(8–9):910–926.

942. Benedicta A. and Ertel J. (2008). Environmental impact of nano technology on human health. In *Standards and Thresholds for Impact Assessment.* Berlin Heidelberg: Springer, pp. 371–378.

943. Berube D.M. et al. (2010). Communicating risk in the 21st century: the case of nanotechnology. National Nanotechnology Coordination Office, Arlington (NSF Grant #0809470).

944. Berube D.M. et al. (2010). Project on emerging nanotechnologies: consumer product inventory evaluated. *Nanotechnology Law & Business.* **7**(2):152–163.

945. Besley J. et al. (2008). Expert opinion on nanotechnology: risks, benefits, and regulation. *Journal of Nanoparticle Research.* **10**(4):549–558.

946. Boholm Å. (2011). Nanotechnology, risk and communication. *Journal of Risk Research.* **14**(10):1263–1265.

947. Bomkamp S. (2010). Beyond chemicals: the lessons that toxic substance regulatory reform can learn from nanotechnology. *Indiana Law Journal.* **85**(5 Suppl):24–38.

948. Bostrom A. and Löfstedt R.E. (2010). Nanotechnology risk communication past and prologue. *Risk Analysis.* **30**(11): 1645–1662.

949. Bowman D.M. and Fitzharris M. (2007). Too small for concern? Public health and nanotechnology. *Australian and New Zealand Journal of Public Health.* **31**(4):382–384.

950. Breggin L. (2006). Harmonization of environmental, health, and safety governance approaches for nanotechnology: an overview of key themes. *Environmental Law Reporter.* **36**(12): 10909–10912

951. Breggin L.K. and Carothers L. (2006). Governing uncertainty: the nanotechnology environmental, health, and safety challenge. *Columbia Journal of Environmental Law.* **31**:285.

952. Cacciatore M.A. et al. (2009). From enabling technology to applications: the evolution of risk perceptions about nanotechnology. *Public Understanding of Science.* **20**(3):385–404.

953. Carley S. and Porter A.L. (2011). Measuring the influence of nanotechnology environmental, health and safety research. *Research Evaluation.* **20**:389–395.

954. Cattaneo A.G. et al. (2010). Nanotechnology and human health: risks and benefits. *Journal of Applied Toxicology.* **30**(8):730–744.

955. Cheng T-J. et al. (2006). Nanotechnology health risk management. *Taiwan Public Health.* **25**:169–176.

956. Cobb M. and Macoubrie J. (2004). Public perceptions about nanotechnology: risks, benefits, and trust. *Journal of Nanoparticle Research.* **6**:395–405.

957. Conti J. et al. (2011). Vulnerability and social justice as factors in emergent US nanotechnology risk perceptions. *Risk Analysis.* **31**(11):1734–1748.

958. Davies C. (2007). EPA and nanotechnology: oversight for the 21st century. Washington, D.C.: Woodrow Wilson International Center for Scholars.

959. Davis J.M. (2007). How to assess the risks of nanotechnology: learning from past experience. *Journal of Nanoscience and Nanotechnology.* **7**(2):402–409.

960. Dawson G. (2004). Health effects of nanotechnology explored. *Journal of the National Medical Association.* **96**(10):1269.

961. Dawson N. G. (2008). Sweating the small stuff: environmental risk and nanotechnology. *BioScience.* **58**(8):690–690.

962. Engeman C.D. et al. (2013). The hierarchy of environmental health and safety practices in the US nanotechnology workplace. *Journal of Occupational and Environmental Hygiene.* **10**(9):487–495.

963. Fadel T.R. et al. (2015). The challenges of nanotechnology risk management. *Nano Today.* **10**(1):6–10.

964. Fleischer T. et al. (2005). Assessing emerging technologies: methodological challenges and the case of nanotechnologies.

Technological Forecasting and Social Change. **72**(9):1112–1121.

965. Friedrichs S. and Schulte J. (2007). Environmental, health and safety aspects of nanotechnology: implications for the R&D in small company. *Science and Technology* of *Advanced Materials.* **8**:12–18.

966. Galizzi M. (2011). Firms' perceptions of health and environmental hazards and regulations: evidence from a survey of us nanotechnology companies. *Journal of Applied Business and Economics.* **12**(6):70–82.

967. Gerritzen G. et al. (2006). Review of safety practices in the nanotechnology industry phase one report: current knowledge and practices regarding environmental health and safety in the nanotechnology workplace. Prepared for the International Council on Nanotechnology. Santa Barbara, CA: University of California.

968. Glenn J.C. (2006). Nanotechnology: future military environmental health considerations. *Technological Forecasting and Social Change.* **73**(2):128–137.

969. Grassian V.H. (ed.) (2008). *Nanoscience and Nanotechnology: Environmental and Health Impacts.* Hoboken, NJ: John Wiley & Sons.

970. Hannah W. and Thompson P.B. (2008). Nanotechnology, risk and the environment: a review. *Journal of Environmental Monitoring.* **10**(3):291–300.

971. Ho S.S. et al. (2013). Factors influencing public risk–benefit considerations of nanotechnology: assessing the effects of mass media, interpersonal communication, and elaborative processing. *Public Understanding of Science.* **22**(5):606–623.

972. Hoyt V.W. and Mason E. (2008). Nanotechnology: emerging health issues. *Journal of Chemical Health and Safety.* **15**(2):10–15.

973. Hristovski K. (2009). Nanoscience and nanotechnology-environmental and health impacts. *Journal of Environmental Quality.* **38**(6):2479.

974. International Council on Nanotechnology (2006). Review of safety practices in the nanotechnology industry phase one report: current knowledge and practices regarding environmental health and safety in the nanotechnology workplace. Prepared by the University of California, Santa Barbara.

975. International Risk Governance Council (2006). White paper on nanotechnology risk governance. Geneva.

976. Kahan D.M. et al. (2007). Affect, values, and nanotechnology risk perceptions: an experimental investigation. GWU Legal Studies Research Paper No. 261.

977. Kahan D.M. et al. (2009). Cultural cognition of the risks and benefits of nanotechnology. *Nature Nanotechnology.* **4**(2):87–90.

978. Kamat P.V. et al. (2003). Nanoscience opportunities in environmental remediation. *Comptes Rendus Chimie.* **6**(8–10):999.

979. Karn B.P. and Bergeson L.L. (2009). Green nanotechnology: straddling promise and uncertainty. *Natural Resources & Environment.* 9–23.

980. Karunaratne D.G. (2015). Environmental and health impacts of nanotechnology: need for a precautionary approach. *Journal of Chemical Engineering Research Studies.* **2**(1):E1.

981. Kipen H.M. and Laskin D.L. (2005). Smaller is not always better: nanotechnology yields nanotoxicology. *American Journal of Physiology-Lung Cellular and Molecular Physiology.* **289**(5): L696–L697.

982. Knowles E.E. (2006). Nanotechnology: evolving occupational safety, health & environmental issues. *Professional Safety.* **51**(3):20.

983. Köhler A.R. and Som C. (2008). Environmental and health implications of nanotechnology: have innovators learned the lessons from past experiences? *Human & Ecological Risk Assessment.* **14**(3):512–531.

984. Lavicoli I. et al. (2014). Opportunities and challenges of nanotechnology in the green economy. *Environmental Health.* **13**:78.

985. Lee Y.L. et al. (2013). Nano-toxicology in engineering: health risk of nano-materials in built environment. *Advanced Science Letters.* **19**(9):2662–2666.

986. Li Y. et al. (2014). Nanotoxicity overview: nano-threat to susceptible populations. *International Journal of Molecular Sciences.* **15**(3):3671–3697.

987. Linkov I. et al. (2009). Nano risk governance: current developments and future perspectives. *Nanotechnology Law and Business.* **6**:203.

988. Lioy P.J. et al. (2010). Nanotechnology and exposure science what is needed to fill the research and data gaps for consumer products. *International Journal of Occupational and Environmental Health.* **16**(4):378–387.

989. Macnaghten P. (2014). Nanotechnology, risk and public perceptions. In *In Pursuit of Nanoethics.* Netherlands: Springer, pp. 167–181.

990. Marchant G.E. et al. (2000). Risk management principles for nanotechnology. Tempe: Center for the Study of Law, Science and Technology, Sandra Day O'Connor College of Law, Arizona State University.

991. Matsui Y. et al. (2009). Material scientific approach to predict nano materials risk of adverse health effects. *Journal of Physics: Conference Series.* **170**(1):012030.

992. Maynard A.D. et al. (2006). Safe handling of nanotechnology. *Nature.* **444**(7117):267–269.

993. Meder R.C. (2010). Risk management and nanotechnology: insurance concerns about small particles. *Nanotechnology Law & Business.* **7**:44.

994. Mullins M. et al. (2013). Insurability of nanomaterial production risk. *Nature Nanotechnology.* **8**:222–224.

995. Murashov V. and Howard J. (2007). Biosafety, occupational health and nanotechnology. *Applied Biosafety.* **12**(3):158.

996. Murashov V. and Howard J. (2015). Risks to health care workers from nano-enabled medical products. *Journal of Occupational & Environmental Hygiene.* **12**(6):D75–D85.

997. Myskja B.K. (2011). Trustworthy nanotechnology: risk, engagement and responsibility. *NanoEthics.* **5**(1):49–56.

998. National Science and Technology Council (2009). Human and environmental exposure assessment: report of the national nanotechnology initiative workshop. Arlington, VA.

999. Nel A. et al. (2013). Environmental health and safety considerations for nanotechnology. *Accounts of Chemical Research.* **46**(3):605–606.

1000. Nordan M.M. and Holman M.W. (2005). A prudent approach to nanotechnology environmental, health, and safety risks. *Industrial Biotechnology.* **1**(3):146–149.

1001. O'Brien N. and Cummins E. (2008). Recent developments in nanotechnology and risk assessment strategies for addressing public and environmental health concerns. *Human and Ecological Risk Assessment.* **14**(3):568–592.

1002. Petersen A. et al. (2007). Nanotechnologies, risk and society. *Health, Risk & Society.* **9**(2):117–124.

1003. Pidgeon N. and Rogers-Hayden T. (2007). Opening up nanotechnology dialogue with the publics: risk communication or 'upstream engagement'? *Health, Risk & Society.* **9**(2):191–210.

1004. Pidgeon N., Harthorn B. and Satterfield T. (2011). Nanotechnology risk perceptions and communication: emerging technologies, emerging challenges. *Risk Analysis.* **31**(11):1694–1700.

1005. Powell M.C. (2007). New risk or old risk, high risk or no risk? How scientists' standpoints shape their nanotechnology risk frames. *Health, Risk & Society.* **9**(2):173–190.

1006. Powell M.C. et al. (2008). Bottom-up risk regulation? How nanotechnology risk knowledge gaps challenge federal and state environmental agencies. *Environmental Management.* **42**(3):426–443.

1007. Ramsden J. J. (2013). Assessing the toxic risks of the nanotechnology industry. *Nanotechnology Perceptions.* **9**(2):119–134.

1008. Reisch L.A. et al. (2011). 'Better safe than sorry': consumer perceptions of and deliberations on nanotechnologies. *International Journal of Consumer Studies.* **35**(6):644–654.

1009. Reijnders L. (2008). Hazard reduction in nanotechnology. *Journal of Industrial Ecology.* **12**(3):297–306.

1010. Renn O. and Roco M.C. (2006). Nanotechnology and the need for risk governance. *Journal of Nanoparticle Research.* **8**(2):153–191.

1011. Reynolds G.H. (2003). Nanotechnology and regulatory policy: three futures. *Harvard Journal of Law and Technology.* **17**: 179.

1012. Roblegga E. et al. (2009). Health risks of nanotechnology. *EURO-NanoTox-Letters.* **1**(1):1–18.

1013. Roco M.C. (2005). Environmentally responsible development of nanotechnology. *Environmental Science & Technology.* **39**(5): 106A–112A.

1014. Rucinski T.L. (2013). Searching for the nano-needle in a green haystack: researching the environmental, health, and safety ramifications of nanotechnology. *Pace Environmental Law Review.* **30**(2):397.

1015. Satterfield T. et al. (2009). Anticipating the perceived risk of nanotechnologies. *Nature Nanotechnology.* **4**(11):752–758.

1016. Savolainen K. et al. (2010). Nanotechnologies, engineered nanomaterials and occupational health and safety: a review. *Safety Science.* **48**(8):957–963.

1017. Schmidt C.W. (2009). Nanotechnology—related environmental, health and safety research: examining the national strategy. *Environmental Health Perspectives.* **117**(4):A158–A161.

1018. Schuetz H. and Wiedemann P.M. (2008). Framing effects on risk perception of nanotechnology. *Public Understanding of Science.* **17**(3):369–379.

1019. Schug T.T. et al. (2013). ONE Nano: NIEHS's strategic initiative on the health and safety effects of engineered nanomaterials. *Environmental Health Perspectives.* **121**(4):410.

1020. Schulte P. et al. (2008). Sharpening the focus on occupational safety and health in nanotechnology. *Scandinavian Journal of Work, Environment & Health.* **34**(6):471–478.

1021. Schulte P.A. et al. (2014). Occupational safety and health criteria for responsible development of nanotechnology. *Journal of Nanoparticle Research.* **16**(1):1–17.

1022. Schultz A.C. (2007). Nanotechnology: industrial revolution or emerging hazard? *Environmental Claims Journal.* **19**(3):199–205.

1023. Senjen R. and Hansen S.F. (2011). Towards a nanorisk appraisal framework. *Comptes Rendus Physique.* **12**(7):637–647.

1024. Sequeira R. et al. (2006). The nano enterprise: a survey of health and safety concerns, considerations, and proposed improvement strategies to reduce potential adverse effects. *Human Factors & Ergonomics in Manufacturing & Service Industries.* **16**(4):343–368.

1025. Shatkin J.A. (2012). *Nanotechnology: Health and Environmental Risks.* Boca Raton, FL: CRC Press.

1026. Siegrist M. (2000). The influence of trust and perceptions of risks and benefits on the acceptance of gene technology. *Risk Analysis.* **20**(2):195–203.

1027. Simons J. et al. (2009). The slings and arrows of communication on nanotechnology. *Journal of Nanoparticle Research.* **11**(7):1555–1571.

1028. Singh S. and Nalwa H.S. (2007). Nanotechnology and health safety–toxicity and risk assessments of nanostructured materials on human health. *Journal of Nanoscience and Nanotechnology.* **7**(9):3048–3070.

1029. Smith S.E.S. et al. (2008). Americans' nanotechnology risk perception. *Journal of Industrial Ecology.* **12**(3):459–473.

1030. Sng J. and Koh D. (2008). Nanocommentary: occupational and environmental health and nanotechnology—what's new? *Occupational Medicine.* **58**(7):454–455.

1031. Springston J. (2008). Nanotechnology: understanding the occupational safety and health challenges. *Professional Safety.* **53**(10):51.

1032. Stern S.T. and McNeil S.E. (2008). Nanotechnology safety concerns revisited. *Toxicological Sciences.* **101**(1):4–21.

1033. Theodore L. and Stander L. (2012). Regulatory concerns and health/hazard risks associated with nanotechnology. *Pace Environmental Law Review.* **30**(2):469.

1034. Valverde Jr, L.J. and Linkov I. (2011). Nanotechnology: risk assessment and risk management perspective. *Nanotechnology Law & Business.* **8**:25.

1035. Von Gleich A. et al. (2008). A suggested three-tiered approach to assessing the implications of nanotechnology and influencing its development. *Journal of Cleaner Production.* **16**(8):899–909.

1036. Whatmore R.W. (2006). Nanotechnology—what is it? Should we be worried? *Occupational Medicine.* **56**(5):295–299.

1037. Williams R.A. et al. (2010). Risk characterization for nanotechnology. *Risk Analysis.* **30**(11):1671–1679.

1038. Yadav S.K. et al. (2013). Impact of nanotechnology on socio-economic aspects: an overview. *Reviews in Nanoscience & Nanotechnology.* **2**(2):127–142.

1039. Yao D. et al. (2013). Limitation and challenge faced to the researches on environmental risk of nanotechnology. *Procedia Environmental Sciences.* **18**:149–156.

1040. Youtie J. et al. (2011). The use of environmental, health and safety research in nanotechnology research. *Journal of Nanoscience & Nanotechnology.* **11**(1):158–166.

Section 5

Ethical, Legal, Policy, and Regulatory Concerns

As nanotechnology moves into the early 21st century, there is a growing body of literature that addresses concerns about ethics and the law as it relates to nanotechnology. The legislative and judicial framework that attempts to rein in nanotechnology may not keep pace with new applications and products. However, the literature cited here provides a solid framework of information.

1041. Aala M. et al. (2008). Bioethical issues of nanotechnology at a glance. *Iranian Journal of Public Health.* **37**(1):12–17.

1042. Abbott K.W. et al. (2011). Soft law oversight mechanisms for nanotechnology. *Jurimetrics.* **52**:279.

1043. Allhoff F. et al. (eds). (2007). *Nanoethics: The Ethical and Social Implications of Nanotechnology.* New York: Wiley-Interscience.

1044. Asmatulu E. et al. (2012). Recent progress in Nanoethics and its possible effects on engineering education. *International Journal of Mechanical Engineering Education.* **40**:1–10.

The Nanotechnology Revolution: A Global Bibliographic Perspective
Dale A. Stirling
Copyright © 2018 Pan Stanford Publishing Pte. Ltd.
ISBN 978-981-4774-19-2 (Hardcover), 978-1-315-11083-7 (eBook)
www.panstanford.com

1045. Beaudrie C.E.H. et al. (2013). Expert views on regulatory preparedness for managing the risks of nanotechnologies. *PLoS One.* **8**(11):e80250.

1046. Bell C. and Marrapese M. (2011). Nanotechnology standards and international legal considerations. In *Nanotechnology Standards.* New York: Springer, pp. 239–255.

1047. Bennett B. (2007). Regulating small things: genes, gametes and nanotechnology. *Journal of Law and Medicine.* **15**(1):153.

1048. Bennett D.J. and Schuurbiers D. (2005). Nanobiotechnology: responsible actions on issues in society and ethics. *NSTI Nanotech.* **2**:765–768.

1049. Bergeson L.L. (2004). The regulatory implications of nanotechnology. *Environmental Quality Management.* **14**(1):71–82.

1050. Berne R.W. (2004). Towards the conscientious development of ethical nanotechnology. *Science and Engineering Ethics.* **10**(4):627–638.

1051. Berube D.M. (2012). Decision ethics and emergent technologies: the case of nanotechnology. *European Journal of Law & Technology.* **2**(3):2012.

1052. Besley J. et al. (2008). Expert opinion on nanotechnology: risks, benefits, and regulation. *Journal of Nanoparticle Research.* **10**(4):549–558.

1053. Bosso C.J. (ed.) (2010). *Governing Uncertainty: Environmental Regulation in the Age of Nanotechnology.* Washington, D.C.: Earthscan.

1054. Bowman D.M. (2007). Patently obvious: intellectual property rights and nanotechnology. *Technology in Society.* **29**(3):307–315.

1055. Bowman D.M. and Hodge G.A. (2006). Nanotechnology: mapping the wild regulatory frontier. *Futures.* **38**(9):1060–1073.

1056. Bowman D.M. and Hodge G.A. (2007). A small matter of regulation: an international review of nanotechnology regulation. *Columbia Science & Technology Law Review.* **8**(1):1–36.

1057. Bowman D.M. and Hodge G.A. (2008). Governing nano-technology without government? *Science and Public Policy.* **35**(7):475–487.

1058. Breggin L. et al. (2009). Securing the promise of nano-technologies: towards transatlantic regulatory cooperation. London: Royal Institute of International Affairs.

1059. Brindell J.R. (2009). Nanotechnology and the dilemmas facing business and government. *The Florida Bar Journal.* **83**(7):73.

1060. Brownsword R. (2009). Nanoethics: old wine, new bottles? *Journal of Consumer Policy.* **32**(4):355–379.

1061. Campillo Vélez B.E. and Zuleta Salas G.L. (2014). Bioethics and nanotechnology. *Revista Lasallista de Investigación.* **11**(1):63–69 (Spanish).

1062. Caplan A.L. (2008). Deciphering nanoethics. *Chemical & Engineering News.* **86**(13):42–43.

1063. Castro F. (2004). Legal and regulatory concerns facing nanotechnology. *Chicago-Kent Journal of Intellectual Property.* **4**(1):140–146.

1064. Chatterjee R. (2008). The challenge of regulating nanomaterials. *Environmental Science & Technology.* **42**(2):339–343.

1065. Chen H., Roco M.C., Li X. and Lin Y. (2008). Trends in nanotechnology patents. *Nature Nanotechnology.* **3**(3):123–125.

1066. Corely E.A. et al. (2009). Of risks and regulations: how leading U.S. nano-scientists form policy stances about nanotechnology. *Journal of Nanoparticle Research.* **11**:1573–1585.

1067. Corley E.A. et al. (2013). The current status and future direction of nanotechnology regulations: a view from nano-scientists. *Review of Policy Research.* **30**(5):488–511.

1068. Corley E.A. et al. (2015). Scientists' ethical obligations and social responsibility for nanotechnology research. *Science and Engineering Ethics.* 1–22.

1069. Dang Y. et al. (2010). Trends in worldwide nanotechnology patent applications: 1991 to 2008. *Journal of Nanoparticle Research.* **12**(3):687–706.

1070. Delemarle A. and Throne-Holst H. (2013). The role of standardization in the shaping of a vision for nanotechnology. *International Journal of Innovation and Technology Management.* **10**(02):1340005.

1071. Delgato-Ramos G.C. (2006). Nano-conceptions: a sociological insight of nanotechnology concepts. *The Journal of Philosophy, Science & Law.* **6**. Available at http://jpsl.org/archives/nano-conceptions-sociological-insight/

1072. Demissie H.T. (2010). Is beneficent regulation the new better regulation? Nano-regulation in the wake of the 'new better regulation' movement. *Law, Innovation & Technology.* **2**(1):115–152.

1073. Drexler K.E. and Wejnert J. (2004). Nanotechnology and policy. *Jurimetrics.* **45**(1):1–22.

1074. Du Mont J.J. (2008). Trademarking nanotechnology: nano-lies & federal trademark registration. *AIPLA Quarterly Journal.* **36**:147.

1075. Dupuy J.P. (2007). Some pitfalls in the philosophical foundations of nanoethics. *Journal of Medicine & Philosophy.* **32**(3):237–261.

1076. Ebbesen M. (2008). The role of the humanities and social sciences in nanotechnology research and development. *NanoEthics.* **2**(1):1–13.

1077. Ebbesen M. et al. (1999). Ethics in nanotechnology: starting from scratch. *Bulletin of Science, Technology & Society.* **26**(6):451–462.

1078. Ebbesen M. et al. (2006). Ethics in nanotechnology: starting from scratch? *Bulletin of Science Technology & Society.* **26**(6):451–462.

1079. Fairbrother A. and Fairbrother J. (2009). Are environmental regulations keeping up with innovation? A case study of the nanotechnology industry. *Ecotoxicology & Environmental Safety.* **72**:1327–1330.

1080. Falkner R. and Jasper N. (2012). Regulating nanotechnologies: risk, uncertainty and the global governance gap. *Global Environmental Politics.* **12**(1):30–55.

1081. Faunce T.A. (2007). Nanotechnology in global medicine and human biosecurity: private interests, policy dilemmas and the calibration of public health law. *Journal of Law, Medicine and Ethics.* **35**(4):629–642.

1082. Ferrari A. (2010). Developments in the debate on nanoethics: traditional approaches and the need for new kinds of analysis. *NanoEthics.* **4**(1):27–52.

1083. Fiedler F.A. and Reynolds G.H. (1994). Legal problems of nanotechnology: an overview. *Southern California Interdisciplinary Law Journal.* **3**(2):593.

1084. Fink M. et al. (2012). Nanotechnology and ethics: the role of regulation versus self-commitment in shaping researchers' behavior. *Journal of Business Ethics.* **109**(4):569–581.

1085. Fisher E. (2007). The convergence of nanotechnology, policy, and ethics. *Advances in Computers.* **71**:273–296.

1086. Forrest D.R. (1989). Regulating nanotechnology development. Palo Alto: Foresight Institute.

1087. Forsberg E.M. (2012). Standardisation in the field of nanotechnology: some issues of legitimacy. *Science & Engineering Ethics.* **18**(4):719–739.

1088. Forsberg E-M. (2010). The role of ISO in the governance of nanotechnology. Oslo: Norwegian Work Research Institute & Standards Norway.

1089. Galiay P. (2011). Situation in Europe and the World: a code of conduct for responsible European research in nanoscience and nanotechnology. In *Nanoethics and Nanotoxicology.* Berlin Heidelberg: Springer, pp. 497–509.

1090. Gazsó A. et al. (2012). Regulating nanotechnologies by dialogue. *European Journal of Risk & Regulation.* **3**:103.

1091. Getto E.J. et al. (2009). Nanotechnology: will tiny particles create large legal issues? *SciTech Lawyer.* **6**(1):6–9, 15.

1092. Godman M. (2009). Philosophical and empirical investigations in nanoethics. Thesis: KTH School of Architecture & the Built Environment, Stockholm.

1093. Gordijn B. and Cutter A.M. (eds.) (2000). *In Pursuit of Nanoethics (The International Library of Ethics, Law & Technology)*. Netherlands: Springer Science.

1094. Gouvea R. et al. (2012). Emerging technologies and ethics: a race-to-the-bottom or the top? *Journal of Business Ethics.* **109**(4):553–567.

1095. Grunwald A. (2005). Nanotechnology: a new field of ethical inquiry? *Science and Engineering Ethics.* **11**(2):187–201.

1096. Grunwald A. (2010). From speculative nanoethics to explorative philosophy of nanotechnology. *NanoEthics.* **4**(2):91–101.

1097. Gupta N. et al. (2012). Factors influencing society response of nanotechnology: an expert stakeholder analysis. *Journal of Nanoparticle Research.* **14**:857.

1098. Hansen S.F. (2010). Multicriteria mapping of stakeholder preferences in regulating nanotechnology. *Journal of Nanoparticle Research.* **12**(6):1959–1970.

1099. Hartman B.M. and Naidu B.D. (2007). Nanotechnology: an update on business opportunities and regulatory challenges. *Journal of Biolaw & Business.* **10**(1):25.

1100. Hermerén G. (2011). Ethics and nanotechnologies. *Colecção Bioética.* **12**:35–50.

1101. Hester K. et al. (2015). Anticipatory ethics and governance (AEG): towards a future care orientation around nanotechnology. *NanoEthics.* **9**(2):123–136.

1102. Hodge G.A. et al. (eds.) (2009). *New Global Frontiers in Regulation: The Age of Nanotechnology*. Cheltenham: Edward Elgar Publishing.

1103. Hodge G.A. et al. (eds.) (2010). *International Handbook on Regulating Nanotechnologies*. Cheltenham: Edward Elgar Publishing.

1104. Hogle L.F. (2009). Science, ethics, and the "problems" of governing nanotechnologies. *The Journal of Law, Medicine and Ethics.* **37**(4):749–758.

1105. Hongladarom S. (2009). Nanotechnology, development and Buddhist values. *NanoEthics.* **3**(2):97–107.

1106. Huang Z. et al. (2003). Longitudinal patent analysis for nanoscale science and engineering: country, institution and technology field. *Journal of Nanoparticle Research.* **5**:333–363.

1107. Huang Z. et al. (2004). International nanotechnology development in 2003: country, institution, and technology field analysis based on USPTO patent database. *Journal of Nanoparticle Research.* **6**:325–354.

1108. Huang Z. et al. (2005). Longitudinal nanotechnology development (1991–2002): National Science Foundation funding and its impact on patents. *Journal of Nanoparticle Research.* **7**:343–376.

1109. Huang Z. et al. (2006). Connecting NSF funding to patent innovation in nanotechnology (2001–2004). *Journal of Nanoparticle Research.* **8**:859–879.

1110. Hunt G. and Mehta M. (eds.) (2013). *Nanotechnology: Risk, Ethics and Law.* London: Earthscan.

1111. Insall L. (2010). Legal aspects of the use of nanotechnology. *European Food & Feed Law Review.* **5**(1):28–32.

1112. Jamier V. et al. (2013). The social context of nanotechnology and regulating its uncertainty: a nanotechnologist approach. *Journal of Physics: Conference Series.* **429**(1):012059.

1113. Jamison A. (2009). Can nanotechnology be just? On nanotechnology and the emerging movement for global justice. *NanoEthics.* **3**(2):129–136.

1114. Jordan C.C. et al. (2012). Nanotechnology patent survey: who will be the leaders in the fifth technology revolution? *Nanotechnology Law & Business.* **9**:122–132.

1115. Johnson S. (2009). The era of nanomedicine and nanoethics: has it come, is it still coming, or will it pass us by? *American Journal of Bioethics.* **9**(10):1–2.

1116. Kahan D.M. et al. (2009). Cultural cognition of the risks and benefits of nanotechnology. *Nature Nanotechnology.* **4**(2):87–90.

1117. Kaiser M. (2006). Drawing the boundaries of nanoscience—rationalizing the concerns? *Journal of Law Medicine & Ethics.* **34**(4):667–674.

1118. Kanama D. (2006). Patent application trends in the field of nanotechnology. *Science & Technology Trends.* **21**:77–88.

1119. Karim Md. E. (2014). Nanotechnology within the legal and regulatory framework: an introductory overview. *Malayan Law Journal.* **3**. Available at http://eprints.um. edu.my/13180/1/Nanotechnology_within_the_Legal_and_ Regulato.pdf

1120. Karim Md. E. (2014). Nanotechnology in Asia: a preliminary assessment of the existing legal framework. *KLRI Journal of Law and Legislation.* **4**(2):169–223.

1121. Kato Y. (2009). Elaborating the list of nanotech-related ethical issues. *Journal of International Biotechnology Law.* **6**(4):150–155.

1122. Kermisch C. (2012). Do new ethical issues arise at each stage of nanotechnological development? *NanoEthics.* **6**(1):29–37.

1123. Kica E. and Bowman D.M. (2013). Transnational governance arrangements: legitimate alternatives to regulating nanotechnologies? *NanoEthics.* **7**(1):69–82.

1124. Kjølberg K.L. and Strand R. (2011). Conversations about responsible nanoresearch. *NanoEthics.* **5**(1):99–113.

1125. Kuzma J. and Besley J.C. (2008). Ethics of risk analysis and regulatory review: from bio-to nanotechnology. *NanoEthics.* **2**(2):149–162.

1126. Lee R. and Stokes E. (2009). Twenty-first century novel: regulating nanotechnologies. *Journal of Environmental Law.* **21**(3):469–482.

1127. Lewenstein B.V. (2011). What counts as a 'social and ethical issue' in nanotechnology? *HYLE—International Journal for Philosophy of Chemistry.* **11**(1):5–18.

1128. Li X. et al. (2007). Patent citation network in nanotechnology (1976–2004). *Journal of Nanoparticle Research.* **9**(3):337–352.

1129. Li X. et al. (2007). Worldwide nanotechnology development: a comparative study of USPTO, EPO, and JPO patents (1976–2004). *Journal of Nanoparticle Research.* **9**(6):977–1002.

1130. Lin A.C. (2007). Size matters: regulating nanotechnology. *Harvard Environmental Law Review.* **31**:349–408.

1131. Lin P. (2007). Nanotechnology bound: evaluating the case for more regulation. *NanoEthics.* **1**(2):105–122.

1132. Lin-Easton P.C. (2001). It's time for environmentalists to think small—real small. A call for the involvement of environmental lawyers in developing precautionary principles for molecular nanotechnology. *Georgetown International Environmental Law Review.* **14**(1):107.

1133. Linton J.D. and Walsh S.T. (2012). Introduction to the field of nanotechnology ethics and policy. *Journal of Business Ethics.* **109**(4):547–549.

1134. Litton P. (2007). Nanoethics: what's new? *The Hastings Center Report.* **37**(1):22–25.

1135. Lu L.Y. et al. (2012). Ethics in nanotechnology: what's being done? What's missing? *Journal of Business Ethics.* **109**(4):583–598.

1136. Ludlow K. et al. (2009). Hitting the mark or falling short with nanotechnology regulation? *Trends in Biotechnology.* **27**(11):615–620.

1137. Lupton M.L. (2011). The social, moral and ethical issues raised by nanotechnology in the field of medicine. *Medicine and Law: An International Journal.* **30**(2):187–200.

1138. Macnaghten P. et al. (2005). Nanotechnology, governance, and public deliberation: what role for the social sciences? *Science Communication.* **27**(2):268–291.

1139. Malfatti L. and Carboni D. (2013). From nanoscience to nanoethics: the viewpoint of a scientist. *Etica & Politica.* **15**:303–309.

1140. Malloy T.F. (2011). Nanotechnology regulation: a study in claims making. *ACS Nano.* **5**(1):5–12.

1141. Malloy T.F. (2012). Soft Law and nanotechnology: a functional perspective. *Jurimetrics.* **52**(3):347–358.

1142. Mandel G.N. (2008). Nanotechnology governance. *Alabama Law Review*. **59**:1–44.

1143. Mansoori G.A. and Soelaiman T.F. (2005). Nanotechnology: an introduction for the standards community. *Journal of ASTM International*. **2**(6):1–21.

1144. Marchant G.E. and Sylvester D.J. (2006). Transnational models for regulation of nanotechnology. *Journal of Law Medicine & Ethics*. **34**:714–725.

1145. Marchant G.E. and White A. (2011). An international nanoscience advisory board to improve and harmonize nanotechnology oversight. *Journal of Nanoparticle Research*. **13**(4):1489–1498.

1146. Marchant G.E. et al. (2006). Transnational models for regulation of nanotechnology. *The Journal of Law, Medicine & Ethics*. **34**(4):714–725.

1147. Marchant G.E. et al. (2009). What does the history of technology regulation teach us about nano oversight? *The Journal of Law, Medicine & Ethics*. **37**(4):724–731.

1148. Marchant G.E. et al. (2010). New soft law approach to nanotechnology oversight: a voluntary product certification scheme. *UCLA Journal of Environmental Law & Policy*. **28**:123.

1149. Maria de Souza Antunes A. et al. (2012). Trends in nanotechnology patents applied to the health sector. *Recent Patents on Nanotechnology*. **6**(1):29–43.

1150. Marinova D. And Mcaleer M. (2003). Nanotechnology strength indicators: international rankings based on US patents. *Nanotechnology*. **14**:R1–R7.

1151. Martin-Lomas M. (2005). [Bio-nanoscience and bio-nanotechnology: social and ethical aspects]. *Revista de Derecho y Genoma Humano*. **25**:91–107 (Spanish).

1152. Menaa F. (2013). Policy implications for global pervasive nanotechnology innovation. *Pharmaceutica Analytica Acta*. **5**:1.

1153. Meyer M.S. (2001). Patent citation analysis in a novel field of technology: an exploration of nano-science and nano-technology. *Scientometrics*. **51**(1):163–183.

1154. Mills K. and Fleddermann C. (2005). Getting the best from nanotechnology: approaching social and ethical implications openly and proactively. *Technology & Society Magazine.* **24**(4):18–26.

1155. Mnyusiwalla A. et al. (2003). Mind the gap: science and ethics in nanotechnology. *Nanotechnology.* **14**(3):R9.

1156. Morris J. et al. (2010). Science policy considerations for responsible nanotechnology decisions. *Nature Nanotechnology.* **6**(2):73–77.

1157. Mouttet B. (2006). Nanotechnology and US patents: a statistical analysis. *Nanotechnology Law & Business.* **3**:309.

1158. Neamatullah I. (2009). Nanotechnology: brave new world for civil tort plaintiffs. *SciTech Lawyer.* **6**(1):10–11, 21.

1159. Núñez-Mujica G.D. (2006). Employing geoethics to avoid negative nanotechnology scenarios in undeveloped countries. *Future Takes.* **5**(3).

1160. Paddock L. (2010). Integrated approach to nanotechnology governance. *UCLA Journal of Environmental Law & Policy.* **28**:251.

1161. Patra D. et al. (2009). Nanoscience and nanotechnology: ethical, legal, social and environmental issues. *Current Science.* **96**(5):651–657.

1162. Perez O. (2010). Precautionary governance and the limits of scientific knowledge: a democratic framework for regulating nanotechnology. *UCLA Journal of Environmental Law & Policy.* **28**:29.

1163. Pinson R.D. (2004). Is nanotechnology prohibited by the biological and chemical weapons convention? *Berkeley Journal of International Law.* **22**(2):279–309.

1164. Powell M.C. et al. (2008). Bottom-up risk regulation? How nanotechnology risk knowledge gaps challenge federal and state environmental agencies. *Environmental Management.* **42**(3):426–443.

1165. Rakhlin M. (2008). Regulating nanotechnology: a private-public insurance solution. *Duke Law & Technology Review.* **7**:1–20.

1166. Randles S. (2008). From nano-ethics to real-time regulation. *Journal of Industrial Ecology.* **12**(3):270–274.

1167. Rashba E. and Gamota D. (2003). Anticipatory standards and the commercialization of nanotechnology. *Journal of Nanoparticle Research.* **5**(3–4):401–407.

1168. Rashba E., Gamota D., Jamison D. and Miller J. (2004). Standards in nanotechnology. *Nanotechnology Law and Business.* **1**:185.

1169. Reese M. (2013). Nanotechnology: using co-regulation to bring regulation of modern technologies into the 21st century. *Health Matrix.* **23**:537.

1170. Renn O. and Roco M.C. (2006). Nanotechnology and the need for risk governance. *Journal of Nanoparticle Research.* **8**(2):153–191.

1171. Reynolds G.H. (2003). Nanotechnology and regulatory policy: three futures. *Harvard Journal of Law and Technology.* **17**(1):179–210.

1172. Roco M.C. (2003). Broader societal issues of nanotechnology. *Journal of Nanoparticle Research.* **5**(3–4):181–189.

1173. Roco M.C. (2005). The emergence and policy implications of converging new technologies integrated from the nanoscale. *Journal of Nanoparticle Research.* **7**(2–3):129–143.

1174. Roco M.C. and Bainbridge W.S. (2005). Societal implications of nanoscience and nanotechnology: maximizing human benefit. *Journal of Nanoparticle Research.* **7**(1):1–13.

1175. Romig A.D. (2004). Nanotechnology: scientific challenges and society benefits and risks. *Metallurgical & Materials Transactions.* **35**(12):3641–3648.

1176. Romig A.D. et al. (2007). An introduction to nanotechnology policy: opportunities and constraints for emerging and established economies. *Technological Forecasting and Social Change.* **74**(9):1634–1642.

1177. Russell A.W. (2013). Improving legitimacy in nanotechnology policy development through stakeholder and community engagement: forging new pathways. *Review of Policy Research.* **30**(5):566–587.

1178. Sandler R. (2009). Nanomedicine and nanomedical ethics. *American Journal of Bioethics* **9**(10):14–15.

1179. Santo M. et al. (2006). Text mining as a valuable tool in foresight exercises: a study on nanotechnology. *Technological Forecasting & Social Change.* **73**:1013–1027.

1180. Sargent Jr, J.F. (2012). Nanotechnology: a policy primer. Washington, D.C.: Congressional Research Service.

1181. Scheufele D.A. and Brossard D. (2008). Nanotechnology as a Moral Issue? Religion and Science in the US. *Nanotechnology.* **21**(1).

1182. Schulte P. and Salamanca-Buentello F. (2007). Ethical and scientific issues of nanotechnology in the workplace. *Environmental Health Perspectives.* **115**(1):5–12.

1183. Schummer J. (2004). Societal and ethical implications of nanotechnology. *Techné: Research in Philosophy and Technology.* **8**(2):56–87.

1184. Schummer J. (2006). Cultural diversity in nanotechnology ethics. *Interdisciplinary Science Reviews.* **31**(3):217–230.

1185. Schummer J. (2009). The popularization of emerging technologies through ethics: from nanotechnology to synthetic biology. *Spontaneous Generations: A Journal for the History and Philosophy of Science.* **2**(1):56.

1186. Segal S.H. (2004). Environmental regulation of nanotechnology: avoiding big mistakes for small machines. *Nanotechnology Law & Business.* **1**:290.

1187. Shea C.M. (2005). Future management research directions in nanotechnology: a case study. *Journal of Engineering and Technology Management.* **22**(3):185–200.

1188. Sheetz T. et al. (2005). Nanotechnology: awareness and societal concerns. *Technology in Society.* **27**(3):329–345.

1189. Sheremeta L. and Daar A.S. (2004). The case for publicly funded research on the ethical, environmental, economic, legal and social issues raised by nanoscience and nanotechnology. *Health Law Review.* **12**(3):74–77.

1190. Smadja E. (2006). Four scenarios towards more ethical futures: a case study in nanoscale science and technology. *Foresight.* **8**(6):37–47.

1191. Soltani A.M. et al. (2011). An evaluation scheme for nanotechnology policies. *Journal of Nanoparticle Research.* **13**(12):7303–7312.

1192. Spagnolo A.G. and Daloiso V. (2009). Outlining ethical issues in nanotechnologies. *Bioethics.* **23**(7):394–402.

1193. Stokes E. (2009). Regulating nanotechnologies: sizing up the options. *Legal Studies.* **29**(2):281–304.

1194. Stokes E. (2012). Nanotechnology and the products of inherited regulation. *Journal of Law & Society.* **39**(1):93–112.

1195. Stokes E. and Bowman D.M. (2012). Looking back to the future of regulating new technologies: the cases of nanotechnologies and synthetic biology. *European Journal of Risk & Regulation.* **3**:235.

1196. Susanne C. et al. (2004). What challenges offers nanotechnology to bioethics? *Revista de Derecho y Genoma Humano (Law and the Human Genome Review/Catedra de Derecho y Genoma Humano/Fundacion BBV-Diputacion Foral de Bizkaia).* **22**:27–45.

1197. Sweeney A.E. et al. (2003). The promises and perils of nanoscience and nanotechnology: exploring emerging social and ethical issues. *Bulletin of Science Technology & Society.* **23**:236–245.

1198. Swierstra T. and Rip A. (2007). Nano-ethics as NEST-ethics: patterns of moral argumentation about new and emerging science and technology. *NanoEthics.* **1**(1):3–20.

1199. Sylvester D.J. et al. (2009). Not again! Public perception, regulation, and nanotechnology. *Regulation & Governance.* **3**(2):165–185.

1200. Schummer J. and Pariotti E. (2008). Regulating nanotechnologies: risk management models and nanomedicine. *NanoEthics.* **2**(1):39–42.

1201. Thompson D.K. (2011). Small size, big dilemma: the challenge of regulating nanotechnology. *Tennessee Law Review.* **79**:621.

1202. Toumey C. and Cobb M. (2012). Nano in sight: epistemology, aesthetics, comparisons and public perceptions of images of nanoscale objects. *Leonardo*. **45**(5):409–410, 461–465.

1203. Trujillo P.R. et al. (2014). Ethical reflection on nanotechnology; but what does "being nanotechnological" mean? A contribution from an epistemically realist point of view. *Revista Catalana de Filosofia*. **16**(1):105–122.

1204. Tucker J L. (2008). Is nanotechnology the next gold rush? Not without standards! *IEEE Nanotechnology Magazine*. **2**(3):6–11.

1205. Turk V. et al. (2005). Nanologue background paper: on selected nanotechnology applications and their ethical, legal and social implications (Working Paper). Wuppertal: Wuppertal Institute for Climate, Environment and Energy.

1206. UNESCO (2006). The ethics and politics of nanotechnology. Paris.

1207. Van de Poel I. (2008). How should we do nanoethics? A network approach for discerning ethical issues in nanotechnology. *NanoEthics*. **2**(1):25–38.

1208. Violet F. (2010). Analysis of technical standards in the field of nanotechnology. *Nanotechnology Law and Business*. **7**:299.

1209. Visciano S. (2011). Nanotechnologies, bioethics and human dignity. *Journal International de Bioéthique*. **22**(1/2):17.

1210. Walsh S. and Medley T.A. (2006). Framework for responsible nanotechnology standards. *Technical Proceedings of the 2005 NSTI Nanotechnology Conference and Trade Show*, Volume 1.

1211. Wang G. and Guan J. (2010). The role of patenting activity for scientific research: a study of academic inventors from China's nanotechnology. *Journal of Informetrics*. **4**(3):338–350.

1212. Wardak A. and Gorman M.E. (2006). Using trading zones and life cycle analysis to understand nanotechnology regulation. *The Journal of Law, Medicine & Ethics*. **34**(4):695–703.

1213. Wasson A. (2004). Protecting the next small thing: nanotechnology and the reverse doctrine of equivalents. *Duke Law & Technology Review*. **3**:1–12.

1214. Weil V. (2003). Zeroing in on ethical issues in nanotechnology. *Proceedings of the IEEE.* **91**(11):1976–1979.

1215. Weil V. (2006). Introducing standards of care in the commercialization of nanotechnology. *International Journal of Applied Philosophy.* **20**(2):205–213.

1216. Wejnert J. (2004). Regulatory mechanisms for molecular nanotechnology. *Jurimetrics.* **44**(3):323–350.

1217. White G.B. (2009). Missing the boat on nanoethics. *American Journal of Bioethics* **9**(10):18–19.

1218. Wickson F. et al. (2010). Nature and nanotechnology: science, ideology and policy. *International Journal of Emerging Technologies and Society.* **8**(1):5.

1219. Wilson J. (2004). The politics of small things: nanotechnology, risk, and uncertainty. *Technology & Society Magazine.* **23**(4): 16–21.

1220. Wilson R.F. (2006). Nanotechnology: the challenge of regulating known unknowns. *The Journal of Law, Medicine & Ethics.* **34**(4):704–713.

1221. Wolfson J.R. (2003). Social and ethical issues in nanotechnology: lessons from biotechnology and other high technologies. *Biotechnology Law Report.* **4**:376–396.

1222. Wolinsky H. (2006). Nanoregulation. *EMBO Reports.* **7**(9): 858–861.

1223. Zheng J. (2014). Layout of nanotechnology patents in global market. *Advanced Materials Research.* **889**:1578–1584.

1224. Zheng J. and Cui W. (2013). International Collaboration in Nanotechnology from 1991 to 2010 Based on Patent Analysis. *Advanced Materials Research.* **771**:119–124.

1225. Zitt M. and Bassecoulard E. (2006). Delineating complex scientific fields by a hybrid lexical-citation method: an application to nanosciences. *Information Processing and Management.* **42**:1513–1531.

Section 6

History, Trends, and Future Directions

How did we get here, what are we doing here, and where will we go? These are all pertinent questions as we try to understand the past, present, and future of nanotechnology. This section of the book is for those who need to know where we've been and how that has manifested itself in the early 21st century. But it's also for those who like to look to the future and imagine where nanotechnology may take us—whether good, bad, or indifferent.

1226. Avery T. (2006). Nanoscience and literature: bridging the two cultures. *New Solutions: A Journal of Environmental and Occupational Health Policy.* **15**(4):289–307.

1227. Bai C. (2005). Global voices of science. Ascent of nanoscience in China. *Science.* **309**(5731):61–63.

1228. Balogh L. (2014). Nanoscience, nanoengineering and nanotechnology. *Journal of Geoethical Nanotechnology.* **9**(1):76–85.

1229. Behari J. (2010). Principles of nanoscience: an overview. *Indian Journal of Experimental Biology.* **48**(10):1008–1019.

1230. Bennett I. and Sarewitz D. (2006). Too little, too late? Research policies on the societal implications of nanotechnology in the United States. *Science as Culture.* **15**(4):309–325.

The Nanotechnology Revolution: A Global Bibliographic Perspective
Dale A. Stirling
Copyright © 2018 Pan Stanford Publishing Pte. Ltd.
ISBN 978-981-4774-19-2 (Hardcover), 978-1-315-11083-7 (eBook)
www.panstanford.com

1231. Berube D.M. (2006). *Nano-Hype: The Truth Behind the Nanotechnology Buzz.* Amherst, NY: Prometheus Books.

1232. Bowman D.M., Hodge G.A. and Binks P. (2007). Are we really the prey? Nanotechnology as science and science fiction. *Bulletin of Science, Technology & Society.* **27**(6):435–445.

1233. Cameron N.M.D.S. (2006). Nanotechnology and the human future. *Annals of the New York Academy of Sciences.* **1093**(1):280–300.

1234. Carroll D. (2004). The future lies in nanotechnology. *Drug Discovery & Development.* **7**(7):15.

1235. Chen H. et al. (2013). Global nanotechnology development from 1991 to 2012: patents, scientific publications, and effect of NSF funding. *Journal of Nanoparticle Research.* **15**:1951.

1236. Cherry E.H. et al. (2008). Nanoscale in perspective. *Science Scope.* **31**(8):48.

1237. Choi H. and Mody C.C. (2009). The long history of molecular electronics microelectronics origins of nanotechnology. *Social Studies of Science.* **39**(1):11–50.

1238. Chong K.P. (2004). Nanoscience and engineering in mechanics and materials. *Journal of Physics & Chemistry of Solids.* **65**(8–9):1501.

1239. Cohen M.L. (2005). Nanoscience: the quantum frontier. *Physica E: Low-Dimensional Systems and Nanostructures.* **29**(3–4):447–453.

1240. Compano R. and Hullmann A. (2002). Forecasting the development of nanotechnology with the help of science and technology indicators. *Nanotechnology.* **13**:243–247.

1241. Corbett J. et al. (2000). Nanotechnology: international developments and emerging products. *CIRP Annals-Manufacturing Technology.* **49**(2):523–545.

1242. Crandall B.C. (1996). *Nanotechnology: Molecular Speculations on Global Abundance.* Cambridge: MIT Press.

1243. Debasmita P. (2013). Nanoscience, nanotechnology, or nanotechnoscience: perceptions of Indian nanoresearchers. *Public Understanding of Science.* **22**:590–605.

1244. Delemarle A. et al. (2009). Geography of knowledge production in nanotechnologies: a flat world with many hills and mountains. *Nanotechnology Law and Business.* **6**:103.

1245. Domingo P.A.S. (2011). Mini-revolution: half a century of nanotechnology. *Mètode Annual Review.* **1**:48–55.

1246. Drexler E.K. (1986). *Engines of Creation: The Coming Era of Nanotechnology.* New York: Anchor.

1247. Drexler E.K. (2004). Nanotechnology: from Feynman to funding. *Bulletin of Science, Technology & Society.* **24**(1):21–27.

1248. Drexler K.E. (2013). *Radical Abundance: How A Revolution in Nanotechnology Will Change Civilization.* New York: Public Affairs.

1249. Drexler K.E. and Peterson C. (1991). *Unbounding the Future: The Nanotechnology Revolution.* New York: William Morrow & Company Inc.

1250. Dupas C. and Lahmani M. (2007). *Nanoscience: Nanotechnologies and Nanophysics.* New York: Springer Science & Business Media.

1251. Faber B. (2006). Popularizing nanoscience: the public rhetoric of nanotechnology, 1986–1999. *Technical Communication Quarterly.* **15**(2):141–169.

1252. Fages V. and Albe V. (2015). Social issues in nanoscience and nanotechnology master's degrees: the socio-political stakes of curricular choices. *Cultural Studies of Science Education.* **10**(2):419–435.

1253. Fanfair D. et al. (2007). The early history of nanotechnology (NSF Grant No. EEC-0407237). Rice University: Nanotechnology – Content and Context Class.

1254. Fernandes M.F.M. (2011). Nanotecnologia e historiografia da ciência do tempo presente. *Revista Brasileira de Ciência, Tecnologia e Sociedade.* **2**(1):99–108.

1255. Georgalis E.E. and Aifantis E.C. (2013). Forecasting the evolution of nanotechnology. *Nano Bulletin.* **2**(2):130215.

1256. Gerber C. and Lang H.P. (2006). How the doors to the nanoworld were opened. *Nature Nanotechnology.* **1**(1):3–5.

1257. Giergiel J. (2006). Nanoknowledge: nanotechnology yesterday, today, tomorrow. *Mechanics & Mechanical Engineering.* **10**(1):21–32.

1258. Goluchowicz K. et al. (2013). supporting successful standardization processes in complex emerging fields through quantitative analysis—the case of nanotechnology. *International Journal of Innovation & Technology Management.* **10**(2).

1259. Grieneisen M.L. and Zhang M. (2011). Nanoscience and nanotechnology: evolving definitions and growing footprint on the scientific landscape. *Small.* **7**(20):2836–2839.

1260. Guston D.H. (ed.) (2010). *Encyclopedia of Nanoscience and Society.* Thousand Oaks, CA: Sage Publications, Inc.

1261. Hingant B. and Albe V. (2010). Nanosciences and nanotechnologies learning and teaching n secondary education: a review of literature. *Studies in Science Education.* **46**(2):121–152.

1262. Hochella M.F., Jr. (2002). Nanoscience and technology: the next revolution in the earth sciences. *Earth & Planetary Science Letters.* **203**(2):593.

1263. Homaeigohar S. and Elbahri M. (2012). Nano galaxy. *Materials Today.* **15**(12):591.

1264. Hu G. et al. (2012). Visualizing nanotechnology research in Canada: evidence from publication activities, 1990–2009. *Journal of Technology Transfer.* **37**(4):550–562.

1265. Hullmann A. (2006). Who is winning the global nanorace? *Nature Nanotechnology.* **1**(2):81–83.

1266. Hullmann A. (2007). Measuring and assessing the development of nanotechnology. *Scientometrics.* **70**(3):739–758.

1267. Islam N. and Miyazaki K. (2009). Nanotechnology innovation system: understanding hidden dynamics of nanoscience fusion trajectories. *Technological Forecasting and Social Change.* **76**(1):128–140.

1268. Islam N. and Miyazaki K. (2010). An empirical analysis of nanotechnology research domains. *Technovation.* **30**(4):229–237.

1269. Jin R. (2013). Nanoscience and nanotechnology: where are we heading? *Nanotechnology Reviews.* **2**(1):3–4.

1270. Joachim C. and Plévert L. (2009). *Nanosciences: The Invisible Revolution.* Singapore: World Scientific Press.

1271. Johansson M. (2003). 'Plenty of room at the bottom': towards an anthropology of nanoscience. *Anthropology Today.* **19**(6):3–6.

1272. Jotterand F. (2006). The politicization of science and technology: its implications for nanotechnology. *The Journal of Law, Medicine & Ethics.* **34**(4):658–666.

1273. Jotwani N. and Pawar P. (2011). Future of nanotechnology. *A Quarterly Journal of KPIT Cummins Infosystems Limited.* **4**:34–39.

1274. Karunaratne V. et al. (2012). Nanotechnology in a world out of balance. *Journal of the National Science Foundation of Sri Lanka.* **40**(1):3–8.

1275. Keiper A. (2003). The nanotechnology revolution. *The New Atlantis.* 17–34.

1276. Kr T. et al. (2012). Nanotechnology: from imagination to reality. *Journal of Pearldent.* **3**(3):26–30.

1277. Kroto H. (2013). Nanoscience and nanotechnology in the 21st century. *Drug Delivery and Translational Research.* **3**(4):297–298.

1278. Kurath M. and Gisler P. (2009). Informing, involving or engaging? Science communication, in the age of atom-, bio-and nanotechnology. *Public Understanding of Science.* **18**(5):550–573.

1279. Laurent L. and Petit J-C. (2005). Nanosciences and its convergence with other technologies. *International Journal for Philosophy of Chemistry.* **11**(1):45–76.

1280. Lee J.S. (2014). Moving from convergence to divergence: the future of nanotechnology. *Nanotechnology Reviews.* **3**(5):411–412.

1281. Leinonen A. and Kivisaari S. (2009). Nanotechnology perceptions: literature review on media coverage, public opinion and ngo perspectives (Research Notes 2559). Vuorimiehentie, Finland: VTT Technical Research Centre of Finland.

1282. Lokhande A.S. (2013). Nanotechnology literature: a bibliometric study. *International Journal of Information Dissemination and Technology.* **3**(4):288–291.

1283. Lösch A. (2006). Anticipating the futures of nanotechnology: visionary images as means of communication. *Technology Analysis & Strategic Management.* **18**(3–4):393–409.

1284. Lösch A. (2006). Means of communicating innovations. A case study for the analysis and assessment of Nanotechnology's futuristic visions. *Science, Technology & Innovation Studies.* **2**(2):103.

1285. McCray W.P. (2005). Will small be beautiful? Making policies for our nanotech future. *History & Technology.* **21**(2):177–203.

1286. McGrail S. (2010). Nano dreams and nightmares: emerging technoscience and the framing and (re) interpreting of the future, present and past. *Journal of Futures Studies.* **14**(4):23–48.

1287. Mehta M.D. (2002). Nanoscience and nanotechnology: assessing the nature of innovation in these fields. *Bulletin of Science, Technology & Society.* **22**(4):269.

1288. Mehta M.D. (2004). From biotechnology to nanotechnology: what can we learn from earlier technologies? *Bulletin of Science, Technology & Society.* **24**(1):34–39.

1289. Michelson E.S. (2011). The interdisciplinary impacts of nanotechnology: a look into the future. *Science & Public Policy.* **38**(4):334–335.

1290. Milburn C. (2002). Nanotechnology in the age of posthuman engineering: science fiction as science. *Configurations.* **10**(2):261–295.

1291. Mirkovic T. and Scholes G.D. (2011). Advances in bio-nanotechnology. *The Journal of Physical Chemistry Letters.* **2**(20):2678–2679.

1292. Mody C.C.M. (2004). Small, but determined: technological determinism in nanoscience. *HYLE–International Journal for Philosophy of Chemistry.* **10**(2):101–130.

1293. Monaghan P. (2000). Nanotechnology moves from wishful thinking to real research. *Chronicle of Higher Education.* **47**(16):A21.

1294. Mulhall D. (2002). *Our Molecular Future: How Nanotechnology, Robotics, Genetics and Artificial Intelligence Will Transform Our World.* Amherst, MA: Prometheus Books.

1295. Nerlich B. (2005). From nautilus to nanobo (a) ts: the visual construction of nanoscience. *AZojono: Journal of Nanotechnology Online.* **1**.

1296. Nordmann A. (2009). Invisible origins of nanotechnology: herbert gleiter, materials science, and questions of prestige. *Perspectives on Science.* **17**(2):123–143.

1297. Palmer R.E. (2002). New directions in nanoscience: new challenges for surface analysis. *Surface and Interface Analysis.* **34**(1):3–9.

1298. Parr D. (2005). Will nanotechnology make the world a better place? *Trends in Biotechnology.* **23**(8):395–398.

1299. Partridge J.G. (2009). Trends in nanotechnology 2008. *Small.* **5**(1):20–21.

1300. Pautrat J.L. (2011). Nanosciences: evolution or revolution? *Comptes Rendus Physique.* **12**(7):605–613.

1301. Peterson C.L. (1992). Nanotechnology: evolution of the concept. *Journal of the British Interplanetary Society.* **45**:395–400.

1302. Pitkethly M. (2008). Nanotechnology: past, present, and future. *Nano Today.* **3**(3):6.

1303. Porter A.L. and Youtie J. (2009). Where does nanotechnology belong in the map of science? *Nature Nanotechnology.* **4**(9):534–536.

1304. Rocco M.C. and Bainbridge W.S. (2005). Societal implications of nanoscience and nanotechnology: maximizing human benefit. *Journal of Nanoparticle Research.* **7**:1–13.

1305. Salerno M. et al. (2008). Designing foresight studies for nanoscience and nanotechnology (NST) future developments. *Technological Forecasting & Social Change.* **75**(8):1202–1223.

1306. Sandler R. (2007). Nanotechnology and social context. *Technology & Society.* **27**(6):446–454.

1307. Schaefer H.E. (2010). *Nanoscience: The Science of the Small in Physics, Engineering, Chemistry, Biology and Medicine.* New York: Springer Science & Business Media.

1308. Schaper-Rinkel P. (2013). The role of future-oriented technology analysis in the governance of emerging technologies: the example of nanotechnology. *Technological Forecasting & Social Change.* **80**(3):444–452.

1309. Scheufele D.A. and Lewenstein B.V. (2005). The public and nanotechnology: how citizens make sense of emerging technologies. *Journal of Nanoparticle Research.* **7**(6):659–667.

1310. Schmidt K.F. (2007). Nanofrontiers: visions for the future of nanotechnology. Washington, D.C.: Woodrow Wilson International Center for Scholars, Project on Emerging Nanotechnologies.

1311. Schulte J. (ed.) (2005). *Nanotechnology: Global Strategies, Industry Trends and Applications.* New York: John Wiley & Sons.

1312. Schummer J. (2004). Bibliography of studies on nanoscience and nanotechnology. *HYLE–International Journal for Philosophy of Chemistry.* **10**(2).

1313. Schwarz J.A., Contescu C.I. and Petyera K. (eds.) (2004). *Dekker Encyclopedia of Nanoscience and Nanotechnology.* New York: Marcel Dekker, Inc.

1314. Seear K. et al. (2009). The social and economic impacts of nanotechnologies: a literature review, final report. Victoria, Australia: Monash University.

1315. Selin C. (2007). Expectations and the emergence of nanotechnology. *Science, Technology & Human Values.* **32**(2):196–220.

1316. Selin C. (2011). Negotiating plausibility: intervening in the future of nanotechnology. *Science and Engineering Ethics.* **17**(4):723–737.

1317. Shelley T. (2006). *Nanotechnology: New Promises, New Dangers.* New York: Zed Books.

1318. Shew A. (2008). Nanotech's history: an interesting, interdisciplinary, ideological split. *Bulletin of Science, Technology & Society.* **28**(5):390–399.

1319. Stebbing M. (2009). Avoiding the trust deficit: public engagement, values, the precautionary principle and the future of nanotechnology. *Journal of Bioethical Inquiry.* **6**(1):37–48.

1320. Thompson D. (2007). Michael Faraday's recognition of ruby gold: the birth of modern nanotechnology. *Gold Bulletin.* **40**(4):267–269.

1321. Tomalia D.A. (2009). In quest of a systematic framework for unifying and defining nanoscience. *Journal of Nanoparticle Research.* **11**(6):1251–1310.

1322. Toumey C.P. (2008). Reading Feynman into nanotechnology. *Techné: Research in Philosophy & Technology.* **12**(3):133–168.

1323. Toumey C. (2012). Probing the history of nanotechnology. *Nature Nanotechnology.* **7**(4):205–206.

1324. Walsh J.P. (2015). The impact of foreign-born scientists and engineers on American nanoscience research. *Science & Public Policy.* **42**:107–120.

1325. Weiss P.S. (2010). Nanoscience and nanotechnology: present and future. *ACS Nano.* **4**(4):1771–1772.

1326. Werth hammer N.R. et al. (1992). Nanotechnology: the past and the future. *Science, New Series.* **255**(5042):268–269.

1327. Whitesides G.M. (2005). Nanoscience, nanotechnology, and chemistry. *Small.* **1**(2):172–179.

1328. Wickson F. (2008). Narratives of nature and nanotechnology. *Nature Nanotechnology.* **3**(6):313–315.

1329. Wilson J. et al. (2002). Nanotechnology in bloom. *Popular Mechanics.* **179**(6):22.

1330. Wood S. et al. (2008). Crystallizing the nanotechnology debate. *Technology Analysis & Strategic Management.* **20**(1):13–27.

1331. Woyke A. (2007). Nanotechnology as a new key technology? An attempt of a historical and systematical comparison with other technologies. *Journal for General Philosophy of Science.* **38**(2):329–346.

1332. Youtie J. et al. (2008). Assessing the nature of nanotechnology: can we uncover an emerging general purpose technology? *Journal of Technology Transfer.* **33**:315–329.

1333. Zhenxing L. and Xuebiao Z. (2005). The latest progresses in nano-biological technology. *World Sci-tech R & D.* **1**:010.

1334. Zucker L.G. and Darby M.R. (2006). Evolution of nanotechnology from science to firm. *Journal of Technology Transfer.* **32**(6):123–130.

Section 7

Nanoinformation

One gauge of the progress and maturation of a discipline, school of thought, or new technologies is to study the related body of literature. This section of the book cites essential literature that examines the writings of nanotechnology. It consists mostly of journal articles and conference papers. Articles pertaining to nanoinformatics in a specific country or region can be found in Section 1 of the bibliography.

1335. Alander J.T. (1994). An indexed bibliography of genetic algorithms in nanotechnology (Report Series No. 94-1-NANO). Vaasa, Finland: University of Vaasa, Department of Electrical & Energy Engineering: Automation.

1336. Alencar M.S.M. et al. (2007). Nanopatenting patterns in relation to product life cycle. *Technological Forecasting & Social Change.* **74**:1661–1680.

1337. Andrievski R.A. and Klyuchareva S.V. (2011). Journal information flow in nanotechnology. *Journal of Nanoparticle Research.* **13**(12):6221–6230.

1338. Arora S.K. et al. (2013). Capturing new developments in an emerging technology: an updated search strategy for identifying nanotechnology research outputs. *Scientometrics.* **95**(1):351–370.

The Nanotechnology Revolution: A Global Bibliographic Perspective
Dale A. Stirling
Copyright © 2018 Pan Stanford Publishing Pte. Ltd.
ISBN 978-981-4774-19-2 (Hardcover), 978-1-315-11083-7 (eBook)
www.panstanford.com

1339. Arora S.K. et al. (2014). Measuring the development of a common scientific lexicon in nanotechnology. *Journal of Nanoparticle Research.* **16**(1):1–11.

1340. Avery T. (2005). Nanoscience and literature: bridging the two cultures. *New Solutions.* **15**(4):289–307.

1341. Ávila-Robinson A. and Miyazaki K. (2014). Assessing nanotechnology potentials: interplay between the paths of knowledge evolution and the patterns of competence building. *International Journal of Technology Intelligence and Planning.* **10**(1):1–28.

1342. Barpujari I. (2010). The patent regime and nanotechnology: issues and challenges. *Journal of Intellectual Property Rights.* **15**(3):206–213.

1343. Bartol T. and Stopar K. (2015). Nano language and distribution of article title terms according to power laws. *Scientometrics.* **103**(2):435–451.

1344. Bass S.D. and Kurgan L.A. (2010). Discovery of factors influencing patent value based on machine learning in patents in the field of nanotechnology. *Scientometrics.* **82**(2):217–241.

1345. Bassecoulard E. et al. (2007). Mapping nanosciences by citation flows: a preliminary analysis. *Scientometrics.* **70**(3):859–880.

1346. Bates M.E. et al. (2015). How decision analysis can further nanoinformatics. *Beilstein Journal of Nanotechnology.* **6**(1):1594–1600.

1347. Bawa R. et al. (2005). The nanotechnology patent "gold rush". *Journal of Intellectual Property Rights.* **10**(5):426–433.

1348. Bei L. and Xiangdong C. (2015). Identification of emerging technologies in nanotechnology based on citing coupling clustering of patents. *Journal of Intelligence.* **5**:7.

1349. Biglu M.H. et al. (2011). Scientometrics analysis of nanotechnology in MEDLINE. *BioImpacts.* **1**(3):193–198.

1350. Boholm M. and Boholm Å. (2012). The many faces of nano in newspaper reporting. *Journal of Nanoparticle Research.* **14**(2):1–18.

1351. Borisova L.F. et al. (2007). Bionanotechnology: a bibliometric analysis using science citation index database (1995–2006). *Scientific and Technical Information Processing.* **34**(4):212–218.

1352. Bowman D.M. (2007). Patently obvious: intellectual property rights and nanotechnology. *Technology in Society.* **29**:307–315.

1353. Braun T. et al. (1997). Nanoscience and nanotechnology on the balance. *Scientometrics.* **38**(2):321–325.

1354. Braun T. et al. (2007). Gatekeeping patterns in nano-titled journals. *Scientometrics.* **70**(3):651–667.

1355. British Standards Institute (2007). Terminology for nanomaterials (PAS 136). London: British Standards Institute.

1356. Calero C. et al. (2006). How to identify research groups using publication analysis: an example in the field of nanotechnology. *Scientometrics.* **66**(2):365–376.

1357. Chau M. et al. (2006). Building a scientific knowledge web portal: the NanoPort experience. *Decision Support Systems.* **42**:1216–1238.

1358. Chen H. and Roco M.C. (2008). *Mapping Nanotechnology Innovations and Knowledge: Global and Longitudinal Patent and Literature Analysis.* New York: Springer Science+Business Media, LLC.

1359. Chen H. et al. (2009). Trends in nanotechnology patents. *Nature Nanotechnology.* **3**:124.

1360. Chiesa S. et al. (2008). Building an index of nanomedical resources: an automatic approach based on text mining. In *Knowledge-Based Intelligent Information and Engineering Systems.* Berlin Heidelberg: Springer, pp. 50–57.

1361. Coffrin T. and MacDonald C. (2004). Ethical and social issues in nanotechnology. Annotated bibliography. Available online at http://www.ethicsweb.ca/nanotechnology/bibliography.html.

1362. Compañó R. and Hullmann A. (2002). Forecasting the development of nanotechnology with the help of science and technology indicators. *Nanotechnology.* **13**(3):243.

1363. Cunningham S.W. and Porter A.L. (2011). Bibliometric discovery of innovation and commercialization pathways in nanotechnology. In *2011 Proceedings of PICMET '11: Technology Management in the Energy Smart World (PICMET)*, Portland, OR, 2011, pp. 1–11.

1364. De la Iglesia D. et al. (2011). International efforts in nanoinformatics research applied to nanomedicine. *Methods of Information in Medicine.* **50**(1):84.

1365. Didegah F. and Thelwall M. (2013). Determinants of research citation impact in nanoscience and nanotechnology. *Journal of the American Society for Information Science and Technology.* **64**(5):1055–1064.

1366. Doyle M.E. (2006). Nanotechnology: a brief literature review. Madison: University of Wisconsin, Food Research Institute.

1367. Dudo A. et al. (2011). The emergence of nano news: tracking thematic trends and changes in US newspaper coverage of nanotechnology. *Journalism & Mass Communication Quarterly.* **88**(1):55–75.

1368. Eggleson K. (2013). Dual-use nanoresearch of concern: recognizing threat and safeguarding the power of nanobiomedical research advances in the wake of the H5N1 controversy. *Nanomedicine: Nanotechnology, Biology and Medicine.* **9**(3):316–321.

1369. Eto H. (2003). Interdisciplinary information input and output of a nanotechnology project. *Scientometrics.* **58**(1):5–33

1370. European Science Foundation (2011). Nanoscience and the long-term future of information technology. Strasbourg, France.

1371. Faber B. (2006). Popularizing nanoscience: the public rhetoric of nanotechnology, 1986–1999. *Technical Communication Quarterly.* **15**(2):141–169.

1372. Fleischer T. et al. (2005). Assessing emerging technologies: Methodological challenges and the case of nanotechnologies. *Technological Forecasting & Social Change.* **72**:1112–1121

1373. Forsberg E.M. (2012). Standardization in the field of nano-technology: some issues of legitimacy. *Science and Engineering Ethics.* **18**(4):719–739.

1374. Friedman S.M. and Egolf B.P. (2005). Nanotechnology: risks and the media. *IEEE Technology and Society Magazine.* **24**:5–11.

1375. Friedman S.M. and Egolf B.P. (2011). A longitudinal study of newspaper and wire service coverage of nanotechnology risks. *Risk Analysis.* **31**(11):1701–1717.

1376. García-Remesal M. et al. (2012). Using nanoinformatics methods for automatically identifying relevant nanotoxicology entities from the literature. *BioMed Research International.* **2013**: Article ID 410294.

1377. Garfield E. and Pudovkin A.I. (2003). From materials science to nano-ceramics: citation analysis identifies the key journals and players. *Journal of Ceramic Processing Research.* **4**(4):155–167.

1378. González-Nilo F. et al. (2011). Nanoinformatics: an emerging area of information technology at the intersection of bioinformatics, computational chemistry and nanobiotechnology. *Biological Research.* 44(1):43–51.

1379. Graham S.J.H. and Iacopetta M. (2014). Nanotechnology and the emergence of a general purpose technology. *Annals of Economics and Statistics.* **115/116**:5–35.

1380. Grieneisen M.L. and Zhang M. (2011). Nanoscience and nanotechnology: evolving definitions and growing footprint on the scientific landscape. *Small.* **7**:2836–2839.

1381. Hassanzadeh M. and Khodadust R. (2011). Co-authorship and co-citation in nanotechnology: a social network approach. Istanbul: Bilgi University.

1382. Heinze T. (2004). Nanoscience and nanotechnology in analysis of publications and patent applications including comparisons with the United States. *Nanotechnology Law & Business.* **1**(4): Article 10.

1383. Heinze T. et al. (2007). Identifying creative research accomplishments: methodology and results for nanotechnology and human genetics. *Scientometrics.* **70**(1):125–152.

1384. Huang C. et al. (2011). Nanoscience and technology publications and patents: a review of social science studies and search strategies. *The Journal of Technology Transfer.* **36**(2):145–172.

1385. Hullmann A. and Meyer M. (2003). Publications and patents in nanotechnology: an overview of previous studies and the state of the art. *Scientometrics.* **58**(3):507–527.

1386. Islam N. and Miyazaki K. (2010). An empirical analysis of nanotechnology research domains. *Technovation.* **30**(4):229–237.

1387. Joksimović D. et al. (2014). The analysis of the publications in the most active countries in nanotechnology. *Marketing.* **45**(3):201–212.

1388. Karpagam R. (2014). Literature in nanotechnology among G20 countries: a scientometrics study based on scopus database. PhD Dissertation: Chromepet, Chennai, Anna University.

1389. Karpagam R. et al. (2011). Publication trend on nanotechnology among G15 countries: a bibliometric study. COLLNET *Journal of Scientometrics and Information Management.* **5**(1):61–80.

1390. Kim J.H. (2012). A hyperlink and semantic network analysis of the triple helix (university-government-industry): the interorganizational communication structure of nanotechnology. *Journal of Computer-Mediated Communication.* **17**:152–170.

1391. Koppikar V. et al. (2004). Current trends in nanotech patents: a view from inside the patent office. *Nanotechnology Law & Business.* **1**(1):1–7

1392. Kostoff R.N. et al. (2006a). The growth of nanotechnology literature. *Nanotechnology Perspectives.* **2**:229–247.

1393. Kostoff R.N. et al. (2006b). The seminal literature of global nanotechnology research. *Journal of Nanoparticle Research.* **8**:193–213.

1394. Kostoff R.N. et al. (2006c). The structure and infrastructure of the global nanotechnology literature. *Journal of Nanoparticle Research.* **8**:301–321.

1395. Kostoff R.N. et al. (2007a). Global nanotechnology research literature overview. *Current Science.* **92**(11):1492–1498.

1396. Kostoff R.N. et al. (2007b). Global nanotechnology research metrics. *Scientometrics.* **70**(3):565–601.

1397. Kostoff R.N. et al. (2007c). Structure of the nanoscience and nanotechnology instrumentation literature. *Current Nanoscience.* **3**(2):135–154.

1398. Kostoff R.N. et al. (2007d). Technical structure of the global nanoscience and nanotechnology literature. *Journal of Nanoparticle Research.* **9**(5):701–724.

1399. Kostoff R.N. et al. (2009). Seminal nanotechnology literature: a review. *Journal of Nanoscience and Nanotechnology.* **9**(11):6239–6270.

1400. Kricka L.J. and Fortina P. (2002). Nanotechnology and applications: an all-language survey including books and patents. *Clinical Chemistry.* **48**(4):662–665.

1401. Kuusi O. and Meyer M. (2007). Anticipating technological breakthroughs: using bibliographic coupling to explore the nanotubes paradigm. *Scientometrics.* **70**(3):759–777

1402. Leydesdorff L. (2008). The delineation of nanoscience and nanotechnology in terms of journals and patents: a most recent update. *Scientometrics.* **76**(1):159–167.

1403. Leydesdorff L. (2013). An evaluation of impacts in "nanoscience & nanotechnology": steps towards standards for citation analysis. *Scientometrics.* **94**(1):35–55.

1404. Leydesdorff L. and Rafols I. (2011). Indicators of the interdisciplinarity of journals: diversity, centrality, and citations. *Journal of Informetrics.* **5**(1):87–100.

1405. Leydesdorff L. and Zhou P. (2007). Nanotechnology as a field of science: its delineation in terms of journals and patents. *Scientometrics*. **70**(3):693–713.

1406. Li X. et al. (2008). A longitudinal analysis of nanotechnology literature. *Journal of Nanoparticle Research*. **10**:3–22.

1407. Lin M. and Zhang J. (2007). Language trends in nanoscience and technology: the case of Chinese-language publications. *Scientometrics*. **70**(3):555–564.

1408. Listerman T. (2008). Framing of science in opinion-leading news: international comparison of biotechnology issue coverage. *Public Understanding of Science*. doi:10. 117710963662508089539

1409. Liu X. and Webster T.J. (2013). Nanoinformatics for biomedicine: emerging approaches and applications. *International Journal of Nanomedicine*. **8**(Suppl 1):1.

1410. Lokhande A.S. (2013). Nanotechnology literature: a bibliometric study. *International Journal of Information Dissemination & Technology*. **3**(4):28.

1411. Lv P.H. et al. (2011). Bibliometric trend analysis on global graphene research. *Scientometrics*. **88**(2):399–419.

1412. Maghrebi M. et al. (2011). A collective and abridged lexical query for delineation of nanotechnology publications. *Scientometrics*. **86**:15–25.

1413. Maimon O. and Browarnik A. (2010). NHECD-Nano health and environmental commented database. In *Data Mining and Knowledge Discovery Handbook*. New York: Springer, pp. 1221–1241.

1414. Menéndez-Manjón A. et al. (2011). Nano-energy research trends: bibliometrical analysis of nanotechnology research in the energy sector. *Journal of Nanoparticle Research*. **13**(9):3911–3922.

1415. Meyer M. (1998). Nanotechnology: interdisciplinary, patterns of collaboration and differences in application. *Scientometrics*. **42**:195–205.

1416. Meyer M. (2000). Does science push technology? Patents citing scientific literature. *Research Policy*. **29**(3):409–434.

1417. Meyer M. (2000). Patent citations in a novel field of technology: what can they tell about interactions of emerging communities of science and technology? *Scientometrics.* **48**:151–178.

1418. Meyer M. (2001). Patent citations in a novel field of technology: an exploration of nanoscience and nano-technology. *Scientometrics.* **51**:163–183.

1419. Miyazaki K. and Islam N. (2007). Nanotechnology systems of innovation—an analysis of industry and academia research activities. *Technovation.* **27**(11):661–675.

1420. Mogoutov A. and Kahane B. (2007). Data search strategy for science and technology emergence: a scalable and evolutionary query for nanotechnology tracking. *Research Policy.* **36**:893–903.

1421. Motoyama Y. and Eisler M.N. (2011). Bibliometry and nanotechnology: a meta-analysis. *Technological Forecasting & Social Change.* **78**(7):1174–1182.

1422. Munoz-Sandoval E. (2014). Trends in nanoscience, nanotechnology, and carbon nanotubes: a bibliometric approach. *Journal of Nanoparticle Research.* **16**(1):1–22.

1423. Nazim M. and Ahmed M. (2008). A bibliometric analysis on nanotechnology research. *Annals of Library & Information Studies.* **55**(4):292–299.

1424. Nicolau D. and Esposto A.S. (2012). Challenges in measuring and capturing scientific knowledge in the emerging nanosciences. *International Journal of Innovation* and *Management and Technology.* **3**(6):784.

1425. O'Neill S. et al. (2007). Broad claiming in nanotechnology patents: is litigation inevitable. *Nanotechnology Law & Business.* **4**:29.

1426. O'Niell S. et al. (2011). The structure and analysis of nanotechnology co-author and citation networks. *Scientometrics.* **89**(1):119–138.

1427. Ouellette L.L. (2012). Nanotechnology patents are useful but could be improved. *Nature Nanotechnology.* **7**(12):770–771.

1428. Pei R. and Porter A.L. (2007). Profiling leading scientists in nano-biomedical science: interdisciplinary and potential leading indicators of research directions. *R&D Management.* **41**(3):288–306.

1429. Porter A.L. et al. (2008). Refining search terms for nanotechnology. *Journal of Nanoparticle Research.* **10**(5):715–728.

1430. Robinson D.K.R. et al. (2007). Tracking the evolution of new and emerging S&T via statement-linkages: vision assessment in molecular machines. *Scientometrics.* **70**(3):831–858.

1431. Roco M.C. (1999). Nanoparticles and nanotechnology research. *Journal of Nanoparticle Research.* **1**(1):1–6.

1432. Roco M.C. (2001). International strategy for nanotechnology research. *Journal of Nanoparticle Research.* **3**(5–6):353–360.

1433. Runge K.K. et al. (2013). Tweeting nano: how public discourses about nanotechnology develop in social media environments. *Journal of Nanoparticle Research.* **15**(1):1–11.

1434. Santo M.d.M. et al. (2006). Text mining as a valuable tool in foresight exercises: a study on nanotechnology. *Technological Forecasting & Social Change.* **73**:1013–1027.

1435. Schummer J. (2004a). Bibliography of studies on nanoscience and nanotechnology. *HYLE–International Journal for Philosophy of Chemistry.* **10**(2):1–8.

1436. Schummer J. (2004b). Multidisciplinary, interdisciplinary, and patterns of research collaboration in nanoscience and nanotechnology. *Scientometrics.* **59**:425–465.

1437. Schummer J. (2005). Reading nano: the public interest in nanotechnology as reflected in book purchase patterns. *Public Understanding of Science.* **14**(2).

1438. Schummer J. (2007). The global institutionalization of nanotechnology research: a bibliometric approach to the assessment of science policy. *Scientometrics.* **70**(3):669–692.

1439. Sekhon B.S. (2014). Nanoinformatics: an overview. *Research & Reviews: A Journal of Pharmaceutical Science.* **5**(1):21–25.

1440. Shapira P. et al. (2010). The emergence of social science research on nanotechnology. *Scientometrics.* **85**(2):595–611.

1441. Sharrott D. and Gupta S. (2011). How to cope with the expiration of early nanotechnology patents. *Nanotechnology Law & Business.* **8**:159.

1442. Simons J. et al. (2009). The slings and arrows of communication on nanotechnology. *Journal of Nanoparticle Research.* **11**(7):1555–1571.

1443. Sotudeh H. and Khoshian N. (2014). Gender, web presence and scientific productivity in nanoscience and nanotechnology. *Scientometrics.* **99**(3):717–736.

1444. Stanishevskaya I. (2004). Nanotechnology and the information era. *Mississippi Libraries.* **4**:103–106.

1445. Stephens L.F. (2005). News narratives about nano S&T in major U.S. and non-U.S. newspapers. *Science Communication.* **27**:175–199.

1446. Stopar K. et al. (2016). Citation analysis and mapping of nanoscience and nanotechnology: identifying the scope and interdisciplinarity of research. *Scientometrics.* **106**(2):563–581.

1447. Surulinathi M. and Pu M. (2011). A citationist perspective on the work of IEEE transactions on nanotechnology: historiographic citation analysis.

1448. Takeda Y. et al. (2009). Nanobiotechnology as an emerging research domain from nanotechnology: a bibliometric approach. *Scientometrics.* **80**(1):23–38.

1449. Veltri G.A. (2012). Microblogging and nanotweets: nanotechnology on Twitter. *Public Understanding of Science.* **22**(7):832–849.

1450. Walsh J.P. and Ridge C. (2012). Knowledge production and nanotechnology: characterizing American dissertation research, 1999–2009. *Technology in Society.* **34**(2):127–137.

1451. Wang J. and Shapira P. (2011). Funding acknowledgement analysis: an enhanced tool to investigate research sponsorship impacts: the case of nanotechnology. *Scientometrics.* **87**(3):563–586.

1452. Winarski T. and Stoker-Townsend E. (2005). Nanotechnology thriving on patents. *Intellectual Property Today.* 26–30.

1453. Youtie J. et al. (2008). Nanotechnology publications and citations by leading countries and blocs. *Journal of Nanoparticle Research.* **10**:981–986.

1454. Zhao Q. and Guan J. (2013). Love dynamics between science and technology: some evidences in nanoscience and nanotechnology. *Scientometrics.* **94**(1):113–132.

1455. Zheng J. (2014). Layout of nanotechnology patents in global market. *Advanced Materials Research.* **889**:1578–1584.

1456. Zibareva I.V. (2015). A review of information resources on nanoscience, nanotechnology, and nanomaterials. *Scientific and Technical Information Processing.* **42**(2):93–111.

1457. Zibareva I.V. et al. (2014). Nanocatalysis: a bibliometric analysis. *Kinetics and Catalysis.* **55**(1):1–11.

1458. Zitt M. and Bassecoulard E. (2006). Delineating complex scientific fields by a hybrid lexical-citation method: an application to nanosciences. *Information Processing & Management.* **42**(6):1513–1531.

Section 8

Nanodevices, Materials, Structures, and Systems

From documents cited in this section of the book, the reader will come to understand the nuts and bolts of nanotechnology through its many devices, materials, structures, and systems. These "architectural" building blocks of the very small allow for the creation of nanotechnology products and materials which are being introduced to the world at a rapid clip.

General Overviews

1459. Atwood J.L. and Steed J.W. (eds.) (2008). *Organic Nanostructures*. New York: John Wiley & Sons.

1460. Bisquert J. (2008). Physical electrochemistry of nanostructured devices. *Physical Chemistry Chemical Physics.* **10**(1):49–72.

1461. Flory F. et al. (2011). Optical properties of nanostructured materials: a review. *Journal of Nanophotonics.* **5**(1):052502.

1462. Gleiter H. (2000). Nanostructured materials: basic concepts and microstructure. *Acta Materialia.* **48**(1):1–29.

The Nanotechnology Revolution: A Global Bibliographic Perspective
Dale A. Stirling
Copyright © 2018 Pan Stanford Publishing Pte. Ltd.
ISBN 978-981-4774-19-2 (Hardcover), 978-1-315-11083-7 (eBook)
www.panstanford.com

1463. Kharisov B.I. et al. (2012). *Handbook of Less Common Nanostructures.* Boca Raton, FL: CRC Press.

1464. Koch C.C. (2006). *Nanostructured Materials: Processing, Properties and Applications.* William Andrew.

1465. Moriarty P. (2001). Nanostructured materials. *Reports on Progress in Physics.* **64**(3):297.

1466. Muhammed M. and Tsakalakos T. (2003). Nanostructured materials and nanotechnology: overview. *Journal of the Korean Ceramic Society.* **40**(11):1027–1046.

1467. Nalwa H.S. (ed.) (2001). *Nanostructured Materials and Nanotechnology: Concise Edition.* Houston: Gulf Professional Publishing.

1468. Reithmaier J.P. (2009). *Nanostructured Materials for Advanced Technological Applications.* New York: Springer Science & Business Media.

1469. Rittner M.N. and Abraham T. (1998). Nanostructured materials: an overview and commercial analysis. *JOM Journal of the Minerals, Metals and Materials Society.* **50**(1):37–38.

1470. Rosei F. (2004). Nanostructured surfaces: challenges and frontiers in nanotechnology. *Journal of Physics: Condensed Matter.* **16**(17):S1373.

1471. Siegel R.W. (1993). Nanostructured materials-mind over matter. *Nanostructured Materials.* **3**(1):1–18.

1472. Siegel R.W. (1999). *Nanostructure Science and Technology: R & D Status and Trends in Nanoparticles, Nanostructured Materials and Nanodevices.* New York: Springer Science & Business Media.

1473. Ying J.Y. (ed.) (2001). *Nanostructured Materials.* San Diego: Academic Press.

Dendrimers

1474. Abbasai E. et al. (2014). Dendrimers: synthesis, applications and properties. *Nanoscale Research Letters.* **9**(1):247.

1475. Aulenta F. et al. (2003). Dendrimers: a new class of nanoscopic containers and delivery devices. *European Polymer Journal.* **39**(9):1741–1771.

1476. Balzani V. et al. (2003). Luminescent dendrimers. Recent advances. *Dendrimers.* **5**:159–191.

1477. Bosman A.W. et al. (1999). About dendrimers: structure, physical properties, and applications. *Chemical Reviews.* **99**(7):1665–1688.

1478. Caminade A.M. et al. (2005). Characterization of dendrimers. *Advanced Drug Delivery Reviews.* **57**(15):2130–2146.

1479. Cloninger M.J. (2002). Biological applications of dendrimers. *Current Opinion in Chemical Biology.* **6**(6):742–748.

1480. Donnio B. et al. (2007). Liquid crystalline dendrimers. *Chemical Society Reviews.* **36**(9):1495–1513.

1481. Dykes G.M. (2001). Dendrimers: a review of their appeal and applications. *Journal of Chemical Technology and Biotechnology.* **76**(9):903–918.

1482. Fischer M. and Vögtle F. (1999). Dendrimers: from design to application: a progress report. *Angewandte Chemie International Edition.* **38**(7):884–905.

1483. Froehling P.E. (2001). Dendrimers and dyes: a review. *Dyes and Pigments.* **48**:187–195.

1484. Grayson S.M. and Frechet J.M. (2001). Convergent dendrons and dendrimers: from synthesis to applications. *Chemical Reviews.* **101**(12):3819–3868.

1485. Inoue K. (2000). Functional dendrimers, hyperbranched and star polymers. *Progress in Polymer Science.* **25**(4):453–571.

1486. Klajnert B. and Bryszewska M. (2000). Dendrimers: properties and applications. *Acta Biochimica Polonica.* **48**(1):199–208.

1487. Kofoed J. and Reymond J.L. (2005). Dendrimers as artificial enzymes. *Current Opinion in Chemical Biology.* **9**(6):656–664.

1488. Niederhafner P. et al. (2005). Peptide dendrimers. *Journal of Peptide Science.* **11**(12):757–788.

1489. Patel H.N. and Patel P.M. (2013). Dendrimer applications: a review. *International Journal of Pharma and Bio Sciences.* **4**(2):454–463.

1490. Ponomarenko S.A. et al. (1996). Liquid crystalline carbosilane dendrimers: first generation. *Liquid Crystals.* **21**(1):1–12.

1491. Service R.F. (1995). Denrimer: dream molecules approach real applications. *Science.* **267**:458–459.

1492. Tomalia D.A. and Fréchet J.M. (2002). Discovery of dendrimers and dendritic polymers: a brief historical perspective. *Journal of Polymer Science Part A: Polymer Chemistry.* **40**(16):2719–2728.

1493. Tomalia D.A. et al. (1986). Dendrimers II: architecture, nanostructure and supramolecular chemistry. *Macromolecules.* **19**:2466.

1494. Tully D.C. and Frechet J.M.J. (2001). Dendrimers at surfaces and interfaces: chemistry and applications. *Chemical Communications.* **6**(14):1229–1239.

1495. Vögtle F. (2000). *Dendrimers II: Architecture, Nanostructure and Supramolecular Chemistry*, vol. 2. New York: Springer Science & Business Media.

1496. Vögtle F. et al. (2000). Functional dendrimers. *Progress in Polymer Science.* **25**(7):987–1041.

MEMS (Microelectromechanical Systems)

1497. Angell J.B. et al. (1983). Silicon micromechanical devices. *Scientific American.* **248**:44–55.

1498. Bao M. and Wang W. (1996). Future of microelectromechanical systems (MEMS). *Sensors and Actuators A: Physical.* **56**(1):135–141.

1499. Bishop D. et al. (2001). Microelectro–mechanical systems: technology and applications. *MRS Bulletin.* **26**(4):282–288.

1500. Bogue R. (2007). MEMS sensors: past, present and future. *Sensor Review.* **27**(1):7–13.

1501. Bryzek J. (1996). Impact of MEMS technology on society. *Sensors and Actuators A: Physical.* **56**(1):1–9.

1502. Fujita H. (1997). A decade of MEMS and its future. In *Proceedings IEEE The Tenth Annual International Workshop on Micro Electro Mechanical Systems. An Investigation of Micro*

Structures, Sensors, Actuators, Machines and Robots, Nagoya, pp. 1–7.

1503. Gad-el-Hak M. (ed.) (2001). *The MEMS Handbook*. Boca Raton, FL: CRC Press.

1504. Hesketh P.J. (ed.) (2012). *BioNanoFluidic MEMS*. New York: Springer Science+Business Media, LLC.

1505. Ho C.M. and Tai Y.C. (1996). Review: MEMS and its applications for flow control. *Journal of Fluids Engineering.* **118**(3):437–447.

1506. Ho C.M. and Tai Y.C. (1998). Micro-electro-mechanical-systems (MEMS) and fluid flows. *Annual Review of Fluid Mechanics.* **30**(1):579–612.

1507. Judy J.W. (2001). Microelectromechanical systems (MEMS): fabrication, design and applications. *Smart Materials and Structures.* **10**(6):1115.

1508. Jung E. (2003). Packaging options for MEMS devices. *MRS Bulletin.* **28**(1):51–54.

1509. Lavu S. et al. (2005). Avoiding MEMS failures. *Electronics Systems and Software.* **3**(5):22–25.

1510. Kim S.H. et al. (2007). Nanotribology and MEMS. *Nano Today.* **2**(5):22–29.

1511. Ko W.H. (2007). Trends and frontiers of MEMS. *Sensors and Actuators A: Physical.* **136**(1):62–67.

1512. Liu C. (2007). Recent developments in polymer MEMS. *Advanced Materials.* **19**(22):3783–3790.

1513. Maboudian R. (1998). Surface processes in MEMS technology. *Surface Science Reports.* **30**(6):207–269.

1514. Marinis T.F. (2009). *The Future of MEMS*. Cambridge, MA: Draper Laboratory.

1515. Nguyen N. et al. (2002). MEMS-micropumps: a review. *Journal of fluids Engineering.* **124**(2):384–392.

1516. Niarchos D. (2003). Magnetic MEMS: key issues and some applications. *Sensors and Actuators A: Physical.* **109**(1):166–173.

1517. Rebeiz G.M. and Muldavin J.B. (2001). RF MEMS switches and switch circuits. IEEE Microwave Magazine. **2**(4):59–71.

1518. Rossi C. et al. (2007). Nanoenergetic materials for MEMS: a review. *Journal of Microelectromechanical Systems.* **16**(4):919–931.

1519. Shi F. et al. (1996). dynamic analysis of micro-electro-mechanical systems. *International Journal for Numerical Methods in Engineering.* **39**(24):4119–4139.

1520. Spearing S.M. (2000). Materials issues in microelectromechanical systems (MEMS). *Acta Materialia.* **48**(1):179–196.

1521. Tanner D.M. (2009). MEMS reliability: where are we now? *Microelectronics Reliability.* 49(9):937–940.

1522. van Spengen W.M. (2003). MEMS reliability from a failure mechanisms perspective. *Microelectronics Reliability.* **43**(7):1049–1060.

1523. Walker J.A. (2000). The future of MEMS in telecommunications networks. *Journal of Micromechanics & Microengineering.* **10**(3):R1–R7.

1524. Zhao Y.P. et al. (2003). Mechanics of adhesion in MEMS: a review. *Journal of Adhesion Science and Technology.* **17**(4):519–546.

Nanobearings

1525. Coffey T. (2006). C60 molecular bearings and the phenomenon of nanomapping. *Physical Review Letters.* **96**(18): 186104.

1526. Galvao D.S. (2004). Molecular dynamics simulations of C60 nanobearings. *Chemical Physic Letters.* **386**:425–429.

1527. Shenhai P.M. and Zhao Y. (2010). Schematic construction of flanged nanobearings from double-walled carbon nanotubes. *Nanoscale.* **8**:1500–1504.

1528. Shenhai P.M. et al. (2010). Sustained smooth dynamics in short-sleeved nanobearings based on double-walled carbon nanotubes. *Nanotechnology.* **21**(49).

1529. Shintani K. et al. (2009). Molecular dynamics of carbon nanobearings. *MRS Proceedings.* **1204**:1204-K05-79.

Nanobelts

1530. Hughes W. L. and Wang Z.L. (2003). Nanobelts as nanocantilevers. *Applied Physics Letters.* **82**(17):2886–2888.

1531. Ma C. et al. (2003). Nanobelts, nanocombs, and nanowindmills of wurtzite ZnS. *Advanced Materials.* **15**(3):228–231.

1532. Ni S. et al. (2011). Fabrication of VO2(B) nanobelts and their application in lithium ion batteries. *Journal of Nanomaterials.* **2011**: Article ID 961389.

1533. Pan Z.W. et al. (2001). Nanobelts of semiconducting oxides. *Science.* **291**(5510):1947–1949.

1534. Payne C.M. et al. (2014). Synthesis and crystal structure of gold nanobelts. *Chemistry of Materials.* **26**(6):1999–2004.

1535. Wang Z.L. (2003). Nanobelts, nanowires, and nanodiskettes of semiconducting oxides from materials to nanodevices. *Advanced Materials.* **15**(5):4.

1536. Wang Z.L. (2004). Functional oxide nanobelts: materials, properties and potential applications in Nanosystems and biotechnology. *Annual Review: Physical Chemistry.* **55**:159–196.

1537. Wang Z.L. (ed.) (2013). *Nanowires and Nanobelts: Materials, Properties and Devices. Metal and Semiconductor Nanowires,* vol. 1. New York: Springer Science & Business Media.

Nanobows

1538. Hughes W.L. and Wang Z.L. (2004). Formation of piezoelectric single-crystal nanorings and nanobows. *Journal of the American Chemical Society.* **126**(21):6703–6709.

1539. Hughes W.L. and Wang Z.L. (2005). Controlled synthesis and manipulation of ZnO nanorings and nanobows. *Applied Physics Letters.* **86**(4):043106.

Nanocantilevers

1540. Fukushima K. et al. (2001). Characterization of silicon nanocantilevers. 生産研究, **53**(2):139–142.

1541. Hwang K.S. et al. (2009). Micro-and nanocantilever devices and systems for biomolecule detection. *Annual Review of Analytical Chemistry.* **2**:77–98.

1542. Joshi C. et al. (2010). Quantum entanglement of nanocantilevers. *Physical Review A.* **82**(4):043846.

1543. Nilsson S.G. et al. (2003). Fabrication and mechanical characterization of ultrashort nanocantilevers. *Applied Physics Letters.* **83**(5):990–992.

1544. Perisanu S. et al. (2010). Beyond the linear and duffing regimes in nanomechanics: circularly polarized mechanical resonances of nanocantilevers. *Physical Review B.* 81(16): 165440.

Nanocapsules

1545. Couvreur P. et al. (1977). Nanocapsules: a new type of lysosomotropic carrier. *FEBS Letters.* **84**(2):323–326.

1546. Couvreur P. et al. (2002). Nanocapsule technology: a review. *Critical Reviews in Therapeutic Drug Carrier Systems.* **19**(12).

1547. Dergunov S.A. et al. (2010). Nanocapsules with "invisible" walls. *Chemical Communications.* **46**(9):1485–1487.

1548. Jin P. et al. (2010). Mixed metal-organic nanocapsules. *Coordination Chemistry Reviews.* **254**(15):1760–1768.

1549. Meier W. (2000). Polymer nanocapsules. *Chemical Society Reviews.* **29**(5):295–303.

1550. Ruysschaert T. et al. (2004). Liposome-based nanocapsules. *IEEE Transactions on Nanobioscience.* **3**(1):49–55.

1551. Saito Y. (1995). Nanoparticles and filled nanocapsules. *Carbon.* **33**(7):979–988.

1552. Sauer M. and Meier W. (2001). Responsive nanocapsules. *Chemical Communications.* (1):55–56.

1553. Sukhorukov G. et al. (2005). Intelligent micro-and nanocapsules. *Progress in Polymer Science.* **30**(8):885–897.

1554. Taylor T.M. et al. (2000). Liposomal nanocapsules in food science and agriculture. *Critical Reviews in Food Science & Nutrition.* **45**(7–8):587–605.

Nanoclays

1555. Patel H.A. et al. (2006). Nanoclays for polymer nanocomposites, paints, inks, greases and cosmetics formulations, drug delivery vehicle and waste water treatment. *Bulletin of Materials Science.* **29**(2):133–145.

1556. Patel K. (2012). The use of nanoclays as a construction material. *International Journal of Engineering Research & Applications.* **2**(4):1382–1386.

1557. Posati T. et al. (2010). Synthesis and characterization of luminescent nanoclays. *Crystal Growth & Design.* **10**(7):2847–2850.

1558. Uddin F. (2008). Clays, nanoclays, and montmorillonite minerals. *Metallurgical & Materials Transactions A.* **39**(12):2804–2814.

Nanoclusters

1559. Baletto F. and Ferrando R. (2005). Structural properties of nanoclusters: energetic, thermodynamic, and kinetic effects. *Reviews of Modern Physics.* **77**(1):371.

1560. Catlow C.R.A. et al. (2010). Modelling nano-clusters and nucleation. *Physical Chemistry Chemical Physics.* **12**(4):786–811.

1561. Díez I. and Ras R.H. (2011). Fluorescent silver nanoclusters. *Nanoscale.* **3**(5):1963–1970.

1562. Schmid G. et al. (1999). Current and future applications of nanoclusters. *Chemical Society Reviews.* **28**(3):179–185.

Nanocombs

1563. Huang Y. et al. (2006). Bicrystalline zinc oxide nanocombs. *Journal of Nanoscience and Nanotechnology.* **6**(8):2566–2570.

1564. Kaya S. and Atar E. (2011). Electrochemically grown metallic nanocomb structures on nanoporous alumina templates. *Applied Physics Letters.* **98**(22):223105.

1565. Singh S. et al. (2010). Microstructural study of assorted zno nanostructures: nanocombs, nanocones and microspheres. *Journal of Nanoscience & Nanotechnology.* **10**(4):2458–2462.

1566. Wang A.H., Lu L.X., Song H.Z., Song J.F., Bao T. and Lu C. (2012). Preparation and properties of the ZnO nanocombs. *Acta Photonica Sinica.* **41**(6):728–731.

1567. Xu T. et al. (2012). Growth and structure of pure ZnO micro/nanocombs. *Journal of Nanomaterials.* 2012: Article ID 797935.

1568. Zang C.H. et al. (2010). The synthesis and optical properties of ZnO nanocombs. *Journal of Nanoscience and Nanotechnology.* **10**(4):2370–2374.

Nanocomposites

1569. Advani S.G. (2007). *Processing and Properties of Nanocomposites.* Singapore: World Scientific Publishers.

1570. Alexandre M. and Dubois P. (2000). Polymer-layered silicate nanocomposites: preparation, properties and uses of a new class of materials. *Materials Science and Engineering: R: Reports.* **28**(1–2):1–63.

1571. Arora A. and Padua G.W. (2010). Review: nanocomposites in food packaging. *Journal of Food Science.* **75**(1):R43–R49.

1572. Bai S. and Shen X. (2012). Graphene–inorganic nanocomposites. *RSC Advances.* **2**(1):64–98.

1573. Balogh L. et al. (1999). Formation of silver and gold dendrimer nanocomposites. *Journal of Nanoparticle Research.* **1**(3):353–368.

1574. Beyer G. (2002). Nanocomposites: a new class of flame retardants for polymers. *Plastics, Additives and Compounding.* **4**(10):22–28.

1575. Bhattacharya S. et al. (2008). *Polymeric Nanocomposites Theory and Practice.* Munich: Carl Hanser Publishers.

1576. Bryan C. (2005). Metal nanoparticle—conjugated polymer nanocomposites. *Chemical Communications.* (27):3375–3384.

1577. Burke N.A. et al. (2002). Magnetic nanocomposites: preparation and characterization of polymer-coated iron nanoparticles. *Chemistry of Materials.* **14**(11):4752–4761.

1578. Burnside S.D. and Giannelis E.P. (2000). Nanostructure and properties of polysiloxane-layered silicate nanocomposites. *Journal of Polymer Science Part B: Polymer Physics.* **38**(12):1595–1604.

1579. Cai D. and Song M. (2010). Recent advance in functionalized graphene/polymer nanocomposites. *Journal of Materials Chemistry.* **20**(37):7906–7915.

1580. Cao X. et al. (2005). Polyurethane/clay nanocomposites foams: processing, structure and properties. *Polymer.* **46**(3): 775–783.

1581. Caseri W.R. (2006). Nanocomposites of polymers and inorganic particles: preparation, structure and properties. *Materials Science and Technology.* **22**(7):807–817.

1582. Chan C.M. et al. (2002). Polypropylene/calcium carbonate nanocomposites. *Polymer.* **43**(10):2981–2992.

1583. Chen B. and Evans J.R. (2005). Thermoplastic starch-clay nanocomposites and their characteristics. *Carbohydrate Polymers.* **61**(4):455–463.

1584. Chen X.M. et al. (2002). Novel electrically conductive polypropylene/graphite nanocomposites. *Journal of Materials Science Letters.* **21**(3):213–214.

1585. Choudalakis G. and Gotsis A.D. (2009). Permeability of polymer/clay nanocomposites: a review. *European Polymer Journal.* **45**(4):967–984.

1586. Chronakis I.S. (2005). Novel nanocomposites and nanoceramics based on polymer nanofibers using electrospinning process: a review. *Journal of Materials Processing Technology.* **167**(2):283–293.

1587. Crosby A.J. and Lee J.Y. (2007). Polymer nanocomposites: the "nano" effect on mechanical properties. *Polymer Reviews.* **47**(2):217–229.

1588. Dallas P. et al. (2011). Silver polymeric nanocomposites as advanced antimicrobial agents: classification, synthetic paths, applications, and perspectives. *Advances in Colloid and Interface Science.* **166**(1):119–135.

1589. De Azeredo H.M. (2009). Nanocomposites for food packaging applications. *Food Research International.* **42**(9):1240–1253.

1590. Delozier D.M. et al. (2002). Preparation and characterization of polyimide/organoclay nanocomposites. *Polymer.* **43**(3):813–822.

1591. Du H. et al. (2002). Synthesis, characterization, and nonlinear optical properties of hybridized CdS-polystyrene nanocomposites. *Chemistry of Materials.* **14**(10):4473–4479.

1592. Dufresne A. (2006). Comparing the mechanical properties of high performances polymer nanocomposites from biological sources. *Journal of Nanoscience and Nanotechnology.* **6**(2):322–330.

1593. Evora V.M. and Shukla A. (2003). Fabrication, characterization, and dynamic behavior of polyester/TiO2 nanocomposites. *Materials Science and Engineering: A.* **361**(1):358–366.

1594. Fischer H. (2003). Polymer nanocomposites: from fundamental research to specific applications. *Materials Science and Engineering: C.* **23**(6):763–772.

1595. Fischer H.R. et al. (1999). Nanocomposites from polymers and layered minerals. *Acta Polymerica.* **50**(4):122–126.

1596. Gangopadhyay R. and De A. (2000). Conducting polymer nanocomposites: a brief overview. *Chemistry of Materials.* **12**(3):608–622.

1597. Garces J.M. et al. (2000). Polymeric nanocomposites for automotive applications. *Advanced Materials.* **12**(23):1835–1839.

1598. Gacitua W. et al. (2005). Polymer nanocomposites: synthetic and natural fillers a review. *Maderas. Ciencia y Tecnología.* **7**(3):159–178.

1599. Giannelis E.P. (1998). Polymer-layered silicate nanocomposites: synthesis, properties and applications. *Applied Organometallic Chemistry.* **12**:675–680.

1600. Gilman J.W. et al. (1997). Nanocomposites: a revolutionary new flame retardant approach. *Sampe Journal.* **33**:40–46.

1601. Godovsky D.Y. (2000). Device applications of polymer-nanocomposites. In *Biopolymers PVA Hydrogels, Anionic Polymerisation Nanocomposites*. Berlin Heidelberg: Springer, pp. 163–205.

1602. Guangming C. et al. (1999). Advance in polymer layered silicate nanocomposites. *Polymer Bulletin-Beijing*. (4):1–10.

1603. Hsueh H.B. and Chen C.Y. (2003). Preparation and properties of LDHs/epoxy nanocomposites. *Polymer*. **44**(18):5275–5283.

1604. Hubbe M.A., Rojas O.J., Lucia L.A. and Sain M. (2008). Cellulosic nanocomposites: a review. *BioResources*. **3**(3):929–980.

1605. Huang Z.M. et al. (2003). A review on polymer nanofibers by electrospinning and their applications in nanocomposites. *Composites Science and Technology*. **63**(15):2223–2253.

1606. Hubbe M.A. et al. (2008). Cellulosic nanocomposites: a review. *BioResources*. **3**(3):929–980.

1607. Hussain F. et al. (2006). Review article: polymer-matrix nanocomposites, processing, manufacturing, and application: an overview. *Journal of Composite Materials*. **40**(17):1511–1575.

1608. Jancar J. et al. (2010). Current issues in research on structure–property relationships in polymer nanocomposites. *Polymer*. **51**(15):3321–3343.

1609. Jordan J., Jacob K.I., Tannenbaum R., Sharaf M.A. and Jasiuk I. (2005). Experimental trends in polymer nanocomposites: a review. *Materials Science and Engineering: A*. **393**(1):1–11.

1610. Kalia S. et al. (2011). Cellulose-based bio-and nanocomposites: a review. *International Journal of Polymer Science*. **2011**: Article ID 837875.

1611. Kannan R.Y. et al. (2005). Polyhedral oligomeric silsesquioxane nanocomposites: the next generation material for biomedical applications. *Accounts of Chemical Research*. **38**(11):879–884.

1612. Karger-Kocsis J. and Wu C.M. (2004). Thermoset rubber/layered silicate nanocomposites. Status and future trends. *Polymer Engineering and Science*. **44**(6):1083–1093.

1613. Kim B.K. et al. (2003). Morphology and properties of waterborne polyurethane/clay nanocomposites. *European Polymer Journal.* **39**(1):85–91.

1614. Kim H. et al. (2010). Graphene/polymer nanocomposites. *Macromolecules.* **43**(16):6515–6530.

1615. Komarneni S. (1992). Nanocomposites. *Journal of Materials Chemistry.* **2**(12):1219–1230.

1616. Kornmann X. et al. (2002). High performance epoxy-layered silicate nanocomposites. *Polymer Engineering & Science.* **42**(9):1815–1826.

1617. Krishnamoorti R. and Vaia R.A. (2007). Polymer nanocomposites. *Journal of Polymer Science Part B: Polymer Physics.* **45**(24):3252–3256.

1618. Kuo S.W. and Chang F.C. (2011). POSS related polymer nanocomposites. *Progress in Polymer Science.* **36**(12):1649–1696.

1619. Lagaron J.M. et al. (2005). Improving packaged food quality and safety. Part 2: Nanocomposites. *Food Additives and Contaminants.* **22**(10):994–998.

1620. Lambert T.N. et al. (2009). Synthesis and characterization of titania−graphene nanocomposites. *The Journal of Physical Chemistry C.* **113**(46):19812–19823.

1621. Lan T. and Pinnavaia T.J. (1994). Clay-reinforced epoxy nanocomposites. *Chemistry of Materials.* **6**(12):2216–2219.

1622. Laus M. et al. (1997). New hybrid nanocomposites based on an organophilic clay and poly (styrene-b-butadiene) copolymers. *Journal of Materials Research.* **12**(11):3134–3139.

1623. LeBaron P.C., Wang Z. and Pinnavaia T.J. (1999). Polymer-layered silicate nanocomposites: an overview. *Applied Clay Science.* **15**(1):11–29.

1624. Lee H.S. et al. (2005). TPO based nanocomposites. Part 1. Morphology and mechanical properties. *Polymer.* **46**(25):11673–11689.

1625. Li X. et al. (2001). Preparation and characterization of poly (butyleneterephthalate)/organoclay nanocomposites. *Macromolecular Rapid Communications.* **22**(16):1306–1312.

1626. Liu J. et al. (2011). Magnetic nanocomposites with mesoporous structures: synthesis and applications. *Small.* **7**(4): 425–443.

1627. Liu T. X. et al. (2003). Morphology, thermal and mechanical behavior of polyamide 6/layered-silicate nanocomposites. *Composites Science and Technology.* **63**(3):331–337.

1628. Luo J.J. and Daniel I.M. (2003). Characterization and modeling of mechanical behavior of polymer/clay nanocomposites. *Composites Science and Technology.* **63**(11):1607–1616.

1629. Magaraphan R. et al. (2001). Preparation, structure, properties and thermal behavior of rigid-rod polyimide/ montmorillonite nanocomposites. *Composites Science and Technology.* **61**(9): 1253–1264.

1630. Manias E. (2007). Nanocomposites: stiffer by design. *Nature Materials.* **6**(1):9–11.

1631. Manias E. et al. (2001). Polypropylene/montmorillonite nanocomposites. Review of the synthetic routes and materials properties. *Chemistry of Materials.* **13**(10):3516–3523.

1632. Mehta S. et al. (2004). Thermoplastic olefin/clay nanocomposites: morphology and mechanical properties. *Journal of Applied Polymer Science.* **92**(2):928–936.

1633. Meldrum A. et al. (2001). Nanocomposites formed by ion implantation: recent developments and future opportunities. *Nuclear Instruments and Methods in Physics Research Section B: Beam Interactions with Materials and Atoms.* **178**(1):7–16.

1634. Meneghetti P. and Qutubuddin S. (2006). Synthesis, thermal properties and applications of polymer-clay nanocomposites. *Thermochimica Acta.* **442**(1):74–77.

1635. Moon R.J. et al. (2011). Cellulose nanomaterials review: structure, properties and nanocomposites. *Chemical Society Reviews.* **40**(7):3941–3994.

1636. Morgan A.B. (2006). Flame retarded polymer layered silicate nanocomposites: a review of commercial and open literature systems. *Polymers for Advanced Technologies.* **17**(4):206–217.

1637. Murugan R. and Ramakrishna S. (2005). Development of nanocomposites for bone grafting. *Composites Science and Technology.* **65**(15):2385–2406.

1638. Nguyen Q.T. and Baird D.G. (2006). Preparation of polymer-clay nanocomposites and their properties. *Advances in Polymer Technology.* **25**(4):270–285.

1639. Niihara K. (1991). New design concept of structural ceramics–ceramic nanocomposites. *Nippon Seramikkusu Kyokai Gakujutsu Ronbunshi.* **99**(10):974–982.

1640. Okada A. and Usuki A. (2006). Twenty years of polymer-clay nanocomposites. *Macromolecular Materials and Engineering.* **291**(12):1449–1476.

1641. Pandey J.K. et al. (2005). An overview on the degradability of polymer nanocomposites. *Polymer Degradation and Stability.* **88**(2):234–250.

1642. Paul D.R. and Robeson L.M. (2008). Polymer nanotechnology: nanocomposites. *Polymer.* **49**(15):3187–3204.

1643. Pavlidou S. and Papaspyrides C.D. (2008). A review on polymer–layered silicate nanocomposites. *Progress in Polymer Science.* **33**(12):1119–1198.

1644. Percy M.J. et al. (2000). Synthesis and characterization of vinyl polymer-silica colloidal nanocomposites. *Langmuir.* **16**(17):6913–6920.

1645. Petrović Z.S. et al. (2000). Structure and properties of polyurethane–silica nanocomposites. *Journal of Applied Polymer Science.* **76**(2):133–151.

1646. Podsiadlo P. et al. (2005). Molecularly engineered nanocomposites: layer-by-layer assembly of cellulose nanocrystals. *Biomacromolecules.* **6**(6):2914–2918.

1647. Podsiadlo P. et al. (2007). Ultrastrong and stiff layered polymer nanocomposites. *Science.* **318**(5847):80–83.

1648. Pokropivnyi V.V. (2002). Two-Dimensional nanocomposites: photonic crystals and nanomembranes (review). Part 1.

Types and preparation. *Powder Metallurgy and Metal Ceramics.* **41**(5):264–272.

1649. Pokropivnyi V.V. (2002). Two-dimensional nanocomposites: photonic crystals and nanomembranes (review). Part 2. Properties and applications. *Powder Metallurgy and Metal Ceramics.* **41**(7):369–381.

1650. Porter D. et al. (2000). Nanocomposite fire retardants: a review. *Fire and Materials.* **24**(1):45–52.

1651. Provenzano V. and Holtz R.L. (1995). Nanocomposites for high temperature applications. *Materials Science and Engineering: A.* **204**(1):125–134.

1652. Ragosta G. et al. (2005). Epoxy-silica particulate nanocomposites: chemical interactions, reinforcement and fracture toughness. *Polymer.* **46**(23):10506–10516.

1653. Ray S.S. and Okamoto M. (2003). Polymer/layered silicate nanocomposites: a review from preparation to processing. *Progress in Polymer Science.* **28**(11):1539–1641.

1654. Ray S.S. et al. (2003). New polylactide/layered silicate nanocomposites. 5. Designing of materials with desired properties. *Polymer.* **44**(21):6633–6646.

1655. Sadhu S. and Bhowmick A.K. (2004). Preparation and properties of styrene–butadiene rubber based nanocomposites: the influence of the structural and processing parameters. *Journal of Applied Polymer Science.* **92**(2):698–709.

1656. Satarkar N.S. et al. (2010). Hydrogel nanocomposites: a review of applications as remote controlled biomaterials. *Soft Matter.* **6**(11):2364–2371.

1657. Schaefer D.W. and Justice R.S. (2007). How nano are nanocomposites? *Macromolecules.* **40**(24):8501–8517.

1658. Schmidt D. et al. (2002). New advances in polymer/layered silicate nanocomposites. *Current Opinion in Solid State and Materials Science.* **6**(3):205–212.

1659. Schultz P.A. (2000). Nanomaterials and the interface between nanotechnology and environment. *Revista Visa em Debate Sociedade, Ciencia Tecnologia.* **1**(14):52–56.

1660. Seydibeyoğlu M.Ö. and Oksman K. (2008). Novel nanocomposites based on polyurethane and micro fibrillated cellulose. *Composites Science and Technology.* **68**(3):908–914.

1661. Shennan D. (2010). Nanomaterials in advanced composites. Stamford, CT: Hexcel.

1662. Shi J.L. and Zhang L.X. (2004). Nanocomposites from ordered mesoporous materials. *Journal of Materials Chemistry.* **14**(5):795–806.

1663. Sternitzke M. (1997). Structural ceramic nanocomposites. *Journal of the European Ceramic Society.* **17**(9):1061–1082.

1664. Sun L. et al. (2009). Energy absorption capability of nanocomposites: a review. *Composites Science and Technology.* **69**(14):2392–2409.

1665. Tate J.S. et al. (2009). Bio-based nanocomposites: an alternative to traditional composites. *Journal of Technology Studies.* **35**(1):25–32.

1666. Thostenson E.T. et al. (2005). Nanocomposites in context. *Composites Science and Technology.* **65**(3):491–516.

1667. Tjong S.C. (2006). Structural and mechanical properties of polymer nanocomposites. *Materials Science and Engineering: R: Reports.* 53(3):73–197.

1668. Ton-That M.T. et al. (2004). Polyolefin nanocomposites: formulation and development. *Polymer Engineering and Science.* **44**(7):1212.

1669. Tsujimoto T. et al. (2003). Green nanocomposites from renewable resources: biodegradable plant oil-silica hybrid coatings. *Macromolecular Rapid Communications.* **24**(12): 711–714.

1670. Uyama H. et al. (2003). Green nanocomposites from renewable resources: plant oil-clay hybrid materials. *Chemistry of Materials.* **15**(13):2492–2494.

1671. Vaia R.A. and Giannelis E.P. (2001). Polymer nanocomposites: status and opportunities. *MRS Bulletin.* **26**(05):394–401.

1672. Vaia R.A. and Wagner H.D. (2004). Framework for nanocomposites. *Materials Today.* **7**(11):33–7.

1673. Vaia R.A. et al. (1995). New polymer electrolyte nanocomposites: melt intercalation of poly (ethylene oxide) in mica-type silicates. *Advanced Materials.* **7**(2):154–156.

1674. Veprek S. and Argon A.S. (2001). Mechanical properties of superhard nanocomposites. *Surface and Coatings Technology.* **146**:175–182.

1675. Vollath D. and Szabó D.V. (2004). Synthesis and properties of nanocomposites. *Advanced Engineering Materials.* **6**(3):117–127.

1676. Wagner H.D. (2007). Nanocomposites: paving the way to stronger materials. *Nature Nanotechnology.* **2**(12):742–744.

1677. Wagner H.D. and Vaia R.A. (2004). Nanocomposites: issues at the interface. *Materials Today.* **7**(11):38–42.

1678. Wang K.H. et al. (2001). Synthesis and characterization of maleated polyethylene/clay nanocomposites. *Polymer.* **42**(24):9819–9826.

1679. Wang X. et al. (2006). Preparation, characterization and antimicrobial activity of chitosan/layered silicate nanocomposites. *Polymer.* **47**(19):6738–6744.

1680. Wang Y. et al. (2000). Preparation and characterization of rubber–clay nanocomposites. *Journal of Applied Polymer Science.* **78**(11):1879–1883.

1681. Wang Y. et al. (2005). Preparation and properties of natural rubber/rectorite nanocomposites. *European Polymer Journal.* **41**(11):2776–2783.

1682. Wetzel B. et al. (2006). Epoxy nanocomposites–fracture and toughening mechanisms. *Engineering Fracture Mechanics.* **73**(16): 2375–2398.

1683. Winey K.I. and Vaia R.A. (2007). Polymer nanocomposites. *MRS Bulletin.* **32**(04):314–322.

1684. Wohlleben W. et al. (2011). On the lifecycle of nanocomposites: comparing released fragments and their in-vivo hazards from three release mechanisms and four nanocomposites. *Small.* **7**(16):2384–2395.

1685. Young R.J. et al. (2012). The mechanics of graphene nanocomposites: a review. *Composites Science and Technology*. **72**(12):1459–1476.

1686. Zanetti M. et al. (2000). Polymer layered silicate nanocomposites. *Macromolecular Materials and Engineering*. **279**(1):1–9.

1687. Zeng Q.H. et al. (2005). Clay-based polymer nanocomposites: research and commercial development. *Journal of Nanoscience and Nanotechnology*. **5**(10):1574–1592.

1688. Zhang Z. et al. (2004). Creep resistant polymeric nanocomposites. *Polymer*. **45**(10):3481–3485.

1689. Zheng L. et al. (2001). Novel polyolefin nanocomposites: synthesis and characterizations of metallocene-catalyzed polyolefin polyhedral oligomeric silsesquioxane copolymers. *Macromolecules*. **34**(23):8034–8039.

1690. Zhu J. et al. (2001). Fire properties of polystyrene-clay nanocomposites. *Chemistry of Materials*. **13**(10):3774–3780.

1691. Zhu Y. et al. (2010). Magnetic nanocomposites: a new perspective in catalysis. *ChemCatChem*. **2**(4):365–374.

Nanocones

1692. Brinkmann G. and Van Cleemput N. (2011). Classification and generation of nanocones. *Discrete Applied Mathematics*. **159**(15):1528–1539.

1693. Charlier J.C. and Rignanese G.M. (2001). Electronic structure of carbon nanocones. *Physical Review Letters*. **86**(26):5970.

1694. Heiberg-Andersen H. et al. (2008). Carbon nanocones: a variety of non-crystalline graphite. *Journal of Non-Crystalline Solids*. **354**(47):5247–5249.

1695. Klein D.J. and Balaban A.T. (2006). The eight classes of positive-curvature graphitic nanocones. *Journal of Chemical Information & Modeling*. **46**(1):307–320.

1696. Machado M. et al. (2003). Electronic properties of selected BN nanocones. *Materials Characterization*. **50**(2):179–182.

1697. Mota R. et al. (2003). Structural and electronic properties of 240 nanocones. *Physica Status Solidi*. **2**:799–802.

1698. Trzaskowski B. et al. (2007). Functionalization of carbon nanocones by free radicals: a theoretical study. *Chemical Physics Letters.* **444**(4):314–318.

1699. Wei J.X. et al. (2007). Mechanical properties of carbon nanocones. *Applied Physics Letters.* **91**(26):261906.

Nanocrystals

1700. Alivisatos A.P. (1996). Perspectives on the physical chemistry of semiconductor nanocrystals. *The Journal of Physical Chemistry.* **100**(31):13226–13239.

1701. Alivisatos A.P. (2000). Naturally aligned nanocrystals. *Science.* **289**(5480):736–737.

1702. Alivisatos P. (2004). The use of nanocrystals in biological detection. *Nature Biotechnology.* **22**(1):47–52.

1703. Buhro W.E. and Colvin V.L. (2003). Semiconductor nanocrystals: shape matters. *Nature Materials.* **2**(3):138–139.

1704. Dalpian G.M. and Chelikowsky J.R. (2006). Self-purification in semi-conductor nanocrystals. *Physical Review Letters.* **96**(22):226802.

1705. Efros A.L. and Rosen M. (2000). The Electronic Structure of Semiconductor Nanocrystals. *Annual Review of Materials Science.* **30**(1):475–521.

1706. El-Sayed M.A. (2004). Small is different: shape-, size-, and composition-dependent properties of some colloidal semiconductor nanocrystals. *Accounts of Chemical Research.* **37**(5):326–333.

1707. Erb U. (1995). Electrodeposited nanocrystals: synthesis, properties and industrial applications. *Nanostructured Materials.* **6**(5):533–538.

1708. Eychmüller A. (2000). Structure and photophysics of semiconductor nanocrystals. *The Journal of Physical Chemistry B.* **104**(28):6514–6528.

1709. Habibi Y. et al. (2010). Cellulose nanocrystals: chemistry, self-assembly, and applications. *Chemical Reviews.* **110**(6):3479–3500.

1710. Hao E. et al. (2004). Synthesis and optical properties of "branched" gold nanocrystals. *Nano Letters.* **4**(2):327–330.

1711. Heitmann J. et al. (2005). Silicon nanocrystals: size matters. *Advanced Materials.* **17**(7):795–803.

1712. Junghanns J.U.A. and Müller R.H. (2008). Nanocrystal technology, drug delivery and clinical applications. *International Journal of Nanomedicine.* **3**(3):295.

1713. Kim F. et al. (2004). Platonic gold nanocrystals. *Angewandte Chemie.* **116**(28):3759–3763.

1714. Leff D.V. et al. (1996). Synthesis and characterization of hydrophobic, organically-soluble gold nanocrystals functionalized with primary amines. *Langmuir.* **12**(20):4723–4730.

1715. Link S. and El-Sayed M.A. (2003). Optical properties and ultrafast dynamics of metallic nanocrystals. *Annual Review of Physical Chemistry.* **54**(1):331–366.

1716. Liu Z. et al. (2002). Metal nanocrystal memories. I. Device design and fabrication. *IEEE Transactions on Electron Devices.* **49**(9):1606–1613.

1717. Murray C.B. et al. (2000). Synthesis and characterization of monodisperse nanocrystals and close-packed nanocrystal assemblies. *Annual Review of Materials Science.* **30**(1):545–610.

1718. Norris D.J. et al. (2008). Doped nanocrystals. *Science.* **319**(5871): 1776–1779.

1719. Parak W.J. et al. (2003). Biological applications of colloidal nanocrystals. *Nanotechnology.* **14**(7):R15.

1720. Rao C.N.R. et al. (2002). Size-dependent chemistry: properties of nanocrystals. *Chemistry-A European Journal.* **8**(1):28–35.

1721. Reiss P. et al. (2009). Core/shell semiconductor nanocrystals. *Small.* **5**(2):154–168.

1722. Tao A.R. et al. (2008). Langmuir–Blodgettry of nanocrystals and nanowires. *Accounts of Chemical Research.* **41**(12):1662–1673.

1723. Tao A.R. et al. (2008). Shape control of colloidal metal nanocrystals. *Small.* **4**(3):310–325.

1724. Tiwari S. et al. (1996). A silicon nanocrystals based memory. *Applied Physics Letters.* **68**(10):1377–1379.

1725. Wang D. and Lieber C.M. (2003). Inorganic materials: nanocrystals branch out. *Nature Materials.* **2**(6):355–356.

1726. Wang D. and Li Y. (2011). Bimetallic nanocrystals: liquid-phase synthesis and catalytic applications. *Advanced Materials.* **23**(9):1044–1060.

1727. Wang G. et al. (2011). Lanthanide-doped nanocrystals: synthesis, optical-magnetic properties, and applications. *Accounts of Chemical Research.* **44**(5):322–332.

1728. Wang X. et al. (2005). A general strategy for nanocrystal synthesis. *Nature.* **437**(7055):121–124.

1729. Xia Y. et al. (2009). Shape-controlled synthesis of metal nanocrystals: simple chemistry meets complex physics? *Angewandte Chemie International Edition.* **48**(1):60–103.

1730. Xiong Y. et al. (2007). Nanocrystals with unconventional shapes: a class of promising catalysts. *Angewandte Chemie International Edition.* **46**(38):7157–7159.

1731. Yin M. et al. (2005). Copper oxide nanocrystals. *Journal of the American Chemical Society.* **127**(26):9506–9511.

1732. Yip S. (1998). Nanocrystals: the strongest size. *Nature.* **391**(6667):532–533.

1733. Yu S.H. et al. (2012). Nanocrystals. *CrystEngComm.* **14**(22): 7531–7534.

Nanodevices

1734. Demming A. (2011). Nanodevices come to life. *Nanotechnology.* **22**(9):090201.

1735. Heddle J.G. (2008). Protein cages, rings and tubes: useful components of future nanodevices? *Nanotechnology, Science and Applications.* **1**:67.

1736. Hess H. et al. (2004). Powering nanodevices with biomolecular motors. *Chemistry-A European Journal.* **10**(9):2110–2116.

1737. Liedl T. et al. (2007). DNA-based nanodevices. *Nano Today.* **2**(2):36–41.

1738. Kumar C.S. (ed.) (2006). *Nanodevices for the Life Sciences.* New York: Wiley-VCH.

1739. Oda S. (2003). NeoSilicon materials and silicon nanodevices. *Materials Science & Engineering,B.* **101**(1):19–23.

1740. Simmel F.C. and Dittmer W.U. (2005). DNA nanodevices. *Small.* **1**(3):284–299.

1741. Sussman H.E. (2003). Nanodevices hold promise for gene therapy. *Drug Discovery Today.* **8**(13):564–565.

1742. Volkov A.G. et al. (2007). Nanodevices in nature: electrochemical aspects. *Electrochimica Acta.* **52**(8):2905–2912.

1743. Yang J.J. et al. (2009). A family of electronically reconfigurable nanodevices. *Advanced Materials.* **21**(37):3754–3758.

1744. Zhirnov V.V. and Cavin R.K. (2008). Nanodevices: charge of the heavy brigade. *Nature Nanotechnology.* **3**(7):377–378.

Nanodiamonds

1745. Bondar V.S. and Puzyr A.P. (2004). Nanodiamonds for biological investigations. *Physics of the Solid State.* **46**:716–719.

1746. Chao J.I. et al. (2007). Toxicity and detection of carboxylated nanodiamonds on human lung epithelial cells. *FASEB Journal.* **21**(5):A267.

1747. Chukhaeva S.I. (2004). Synthesis, properties, and applications of fractionated nanodiamonds. *Physics of the Solid State.* **46**(4):625–628.

1748. Danilenko V.V. (2004). On the history of the discovery of Nano diamond synthesis. *Physics of the Solid State.* **46**:595–599.

1749. Danilenko V.V. (2010). Nanodiamonds: problems and prospects. *Journal of Superhard Materials.* **32**(5):301–310.

1750. Dolmatov V.Y. (2007). Detonation-synthesis nanodiamonds: synthesis, structure, properties and applications. *Russian Chemical Reviews.* **76**(4):339.

1751. Greentree A.D. et al. (2010). 21st-century applications of nanodiamonds. *Optics and Photonics News.* **21**(9):20–25.

1752. Ho D. (ed.) (2009). *Nanodiamonds: Applications in Biology and Nanoscale Medicine.* New York: Springer Science & Business Media.

1753. Hui Y.Y. et al. (2010). Nanodiamonds for optical bioimaging. *Journal of Physics D: Applied Physics.* **43**(37):374021.

1754. Kazi S. (2014). A review article on nanodiamonds discussing their properties and applications. *International Journal of Pharmaceutical Science Invention.* **3**(7):40–45.

1755. Kulakova I.I. (2004). Surface chemistry of nanodiamonds. *Physics of the Solid State.* **46**(4):636–643.

1756. Kurmashev V.I. et al. (2004). Nanodiamonds in magnetic recording system technologies. *Physics of the Solid State.* **46**(4):696–702.

1757. Li H. (2006). Developments and applications of nanodiamonds. *Diamond & Abrasives Engineering.* **3**:021.

1758. Mattson W.D. et al. (2009). Exploiting unique features of nanodiamonds as an advanced energy source (No. ARL-TR-4783). Aberdeen Proving Ground, MD: Army Research Lab, Weapons & Materials Research Directorate.

1759. Mochalin V.N. et al. (2012). The properties and applications of nanodiamonds. *Nature Nanotechnology.* **7**(1):11–23.

1760. Munke A. et al. Why are some nanodiamonds toxic? Uppsala, Sweden: Uppsala University. Available online at http://www.smss.se/symposium/wp-content/uploads/2015/10/SMSS2015_Total_Final.pdf#page=13

1761. Paci J.T. et al. (2013). Understanding the surfaces of nanodiamonds. *The Journal of Physical Chemistry C.* **117**(33):17256–17267.

1762. Panich A.M. (2012). Nuclear magnetic resonance studies of nanodiamonds. *Critical Reviews in Solid State and Materials Sciences.* **37**(4):276–303.

1763. Pozdnyakova I.O. (2004). Applications of nanodiamonds for separation and purification of proteins. *Physics of the Solid State.* **46**(4):758–760.

1764. Puzyr A.P. et al. (2007). Nanodiamonds with novel properties: a biological study. *Diamond and Related Materials.* **16**(12):2124–2128.

1765. Raty J.Y. and Galli G. (2005). Optical properties and structure of nanodiamonds. *Journal of Electroanalytical Chemistry.* **584**(1):9–12.

1766. Schrand A.M. et al. (2007). Are diamond nanoparticles cytotoxic? *Journal of Physical Chemistry B.* **111**:2–7.

1767. Schrand A.M. et al. (2009). Nano diamond particles: properties and perspectives for bioapplications. *Critical Reviews in Solid State and Materials Sciences.* **34**:18–74.

1768. Shakun A. et al. (2014). Hard nanodiamonds in soft rubbers—past, present and future: a review. *Composites Part A: Applied Science and Manufacturing.* **64**:49–69.

1769. Song Q. et al. (2013). The surface functionalization and applications of nanodiamonds. *Materials Review.* **5**:004.

1770. Tamburri E. et al. (2015). Nanodiamonds: the ways forward. *AIP Conference Proceedings.* **1667**:020001.

1771. Terranova M.L. et al. (2015). Nanodiamonds for field emission: state of the art. *Nanoscale.* **7**(12):5094–5114.

1772. Vogt B. (2004). Nanodiamonds increase the life of automotive paints. IDR. *Industrial Diamond Review.* (3):30–31.

1773. Yuan Y. et al. (2010). Pulmonary toxicity and translocation of nanodiamonds in mice. *Diamond and Related Materials.* **19**(4):291–299.

Nanodiscs

1774. Alhmoud H. et al. (2015). Porous silicon nanodiscs for targeted drug delivery. *Advanced Functional Materials.* **25**(7): 1137–1145.

1775. Bayburt T.H. and Sligar S.G. (2010). Membrane protein assembly into nanodiscs. *FEBS Letters.* **584**(9):1721–1727.

1776. Min J.H. et al. (2014). Magnetic nanodiscs fabricated from multilayered nanowires. *Journal of Nanoscience & Nanotechnology.* **14**(10):7923–7928.

1777. Schuler M.A., Denisov I.G. and Sligar S.G. (2013). Nanodiscs as a new tool to examine lipid–protein interactions. *Lipid-Protein Interactions: Methods and Protocols.* 415–433.

Nanoemulsions

1778. Balamohan P. et al. (2013). Nanoemulsion: synthesis, characterization and its applications. *Journal of Bionanoscience.* **7**(4):323–333.

1779. Boonme P. et al. (2009). Microemulsions and nanoemulsions: novel vehicles for whitening cosmeceuticals. *Journal of Biomedical Nanotechnology.* **5**(4):373–383.

1780. Chouksey R.K. et al. (2011). Nanoemulsion: a review. *Inventi Impact: Pharm Tech.* **2**(1).

1781. Costa J.A. et al. (2012). Evaluation of nanoemulsions in the cleaning of polymeric resins. *Colloids & Surfaces.* **415**:112–118.

1782. de Campos V.E. et al. (2012). Nanoemulsions as delivery systems for lipophilic drugs. *Journal of Nanoscience and Nanotechnology.* **12**(3):2881–2890.

1783. Fryd M.M. and Mason T.G. (2012). Advanced nanoemulsions. *Annual Review of Physical Chemistry.* **63**:493–518.

1784. Graves S. et al. (2005). Structure of concentrated nanoemulsions. *The Journal of Chemical Physics.* **122**(13):134703.

1785. Harwansh R.K. et al. (2011). Nanoemulsion as potential vehicles for transdermal delivery of pure phytopharmaceuticals and poorly soluble drug. *International Journal of Drug Delivery.* **3**(2):209–218.

1786. Kentish S. et al. (2008). The use of ultrasonics for nanoemulsion preparation. *Innovative Food Science & Emerging Technologies.* **9**(2):170–175.

1787. Khatri S. et al. (2013). Nanoemulsions in cancer therapy. *Indo Global Journal of Pharmaceutical Sciences.* **3**(2):124–133.

1788. Koroleva M.Y. and Yurtov E.V. (2012). Nanoemulsions: the properties, methods of preparation and promising applications. *Russian Chemical Reviews.* **81**(1):21.

1789. Lovelyn C. and Attama A.A. (2011). Current state of nanoemulsions in drug delivery. *Journal of Biomaterials and Nanobiotechnology*. **2**(05):626.

1790. Lu Y. et al. (2012). Absorption, disposition and pharmacokinetics of nanoemulsions. *Current Drug Metabolism*. **13**(4):396–417.

1791. Mason T.G. et al. (2006). Extreme emulsification: formation and structure of nanoemulsions. *Condensed Matter Physics*. **9**(1):193–199.

1792. Mason T.G. et al. (2006). Nanoemulsions: formation, structure, and physical properties. *Journal of Physics: Condensed Matter*. **18**(41):R635.

1793. McClements D.J. (2011). Edible nanoemulsions: fabrication, properties, and functional performance. *Soft Matter*. **7**(6):2297–2316.

1794. McClements D.J. (2012). Nanoemulsions versus microemulsions: terminology, differences, and similarities. *Soft Matter*. **8**(6):1719–1729.

1795. Nanjwade B.K. et al. (2013). Nanoemulsions formation and their potential applications. *Reviews in Nanoscience and Nanotechnology*. **2**(4):261–274.

1796. Nazarzadeh E. et al. (2013). On the growth mechanisms of nanoemulsions. *Journal of Colloid and Interface Science*. **397**:154–162.

1797. Patel R.P. and Joshi J.R. (2012). An overview on nanoemulsion: a novel approach. *International Journal of Pharmaceutical Sciences and Research*. **3**(12):4640–4650.

1798. Patel R.B. et al. (2016). Recent survey on patents of nanoemulsions. *Current Drug Delivery*. **13**(6):857–881.

1799. Setya S. et al. (2014). Nanoemulsions: formulation methods and stability aspects. *World Journal of Pharmacy* and *Pharmaceutical Sciences*. **3**:2214–2228.

1800. Shah P. et al. (2010). Nanoemulsion: a pharmaceutical review. *Systematic Reviews in Pharmacy*. **1**(1):24.

1801. Sharma S. and Sarangdevot K. (2012). Nanoemulsions for cosmetics. *International Journal of Advanced Research in Pharmaceutical & Bio Sciences.* **2**(3):408–415.

1802. Sigward E. et al. (2013). Formulation and cytotoxicity evaluation of new self-emulsifying multiple W/O/W nanoemulsions. *International Journal of Nanomedicine.* **8**:611.

1803. Silva H.D. et al. (2012). Nanoemulsions for food applications: development and characterization. *Food and Bioprocess Technology.* **5**(3):854–867.

1804. Sonneville-Aubrun O. et al. (2004). Nanoemulsions: a new vehicle for skincare products. *Advances in Colloid and Interface Science.* **108**:145–149.

1805. Tal-Figiel B. and Figiel W. (2008). Micro-and nanoemulsions in cosmetic and pharmaceutical products. *Journal of Dispersion Science and Technology.* **29**(4):611–616.

1806. Thakur N. et al. (2012). Nanoemulsions: a review on various pharmaceutical application. *Global Journal of Pharmacology.* **6**(3):222–225.

1807. Troncoso E. et al. (2012). Fabrication, characterization and lipase digestibility of food-grade nanoemulsions. *Food Hydrocolloids.* **27**(2):355–363.

1808. Wilking J.N. (2008). The structure and rheology of nanoemulsions. Doctoral Dissertation: Los Angeles, University of California.

1809. Wu Y. et al. (2013). The application of nanoemulsion in dermatology: an overview. *Journal of Drug Targeting.* **21**(4): 321–327.

Nanofibers

1810. Alemdar A. and Sain M. (2008). Biocomposites from wheat straw nanofibers: morphology, thermal and mechanical properties. *Composites Science and Technology.* **68**(2):557–565.

1811. Al-Saleh M.H. and Sundararaj U. (2009). A review of vapor grown carbon nanofiber/polymer conductive composites. *Carbon.* **47**(1):2–22.

1812. Barnes C.P. et al. (2007). Nanofiber technology: designing the next generation of tissue engineering scaffolds. *Advanced Drug Delivery Reviews.* **59**(14):1413–1433.

1813. De Jong K.P. and Geus J.W. (2000). Carbon nanofibers: catalytic synthesis and applications. *Catalysis Reviews.* **42**(4): 481–510.

1814. Endo M. and Kroto H.W. (1992). Formation of carbon nanofibers. *The Journal of Physical Chemistry.* **96**(17):6941–6944.

1815. Grafe T. and Graham K. (2003). Polymeric nanofibers and nanofiber webs: a new class of nonwovens. *International Nonwovens Journal.* **12**:51–55.

1816. Huang J. et al. (2003). Polyaniline nanofibers: facile synthesis and chemical sensors. *Journal of the American Chemical Society.* **125**(2):314–315.

1817. Jayakumar R. et al. (2010). Novel chitin and chitosan nanofibers in biomedical applications. *Biotechnology Advances.* **28**(1):142–150.

1818. Kumbar S.G. et al. (2008). Electrospun nanofiber scaffolds: engineering soft tissues. *Biomedical Materials.* **3**(3):034002.

1819. Matthews J.A. et al. (2002). Electrospinning of collagen nanofibers. *Biomacromolecules.* **3**(2):232–238.

1820. Nogi M. et al. (2009). Optically transparent nanofiber paper. *Advanced Materials.* **21**(16):1595–1598.

1821. Ramakrishna S. et al. (2006). Electrospun nanofibers: solving global issues. *Materials Today.* **9**(3):40–50.

1822. Rodriguez N.M. (1993). A review of catalytically grown carbon nanofibers. *Journal of Materials Research.* **8**(12):3233–3250.

1823. Rodriguez N.M. et al. (1994). Carbon nanofibers: a unique catalyst support medium. *The journal of Physical Chemistry.* **98**(50):13108–13111.

1824. Schiffman J.D. and Schauer C.L. (2008). A review: electrospinning of biopolymer nanofibers and their applications. *Polymer Reviews.* **48**(2):317–352.

1825. Tan E.P.S. and Lim C.T. (2006). Mechanical characterization of nanofibers: a review. *Composites Science and Technology.* **66**(9):1102–1111.

1826. Thavasi V. et al. (2008). Electrospun nanofibers in energy and environmental applications. *Energy & Environmental Science.* **1**(2):205–221.

1827. Vasita R. and Katti D.S. (2006). Nanofibers and their applications in tissue engineering. *International Journal of Nanomedicine.* **1**(1):15.

Nanofilms

1828. Haynie D.T. and Zhao W. (2009). Present and future prospects for polypeptide multilayer nanofilms in biotechnology and medicine. *Journal of Nanoscience and Nanotechnology.* **9**(6): 3562–3567.

1829. Haynie D.T. et al. (2006). Protein-inspired multilayer nanofilms: science, technology and medicine. *Nanomedicine: Nanotechnology, Biology and Medicine.* **2**(3):150–157.

1830. He J. and Kunitake T. (2006). Are ceramic nanofilms a soft matter? *Soft Matter.* **2**(2):119–125.

1831. Jiang M.Q. et al. (2011). Metallic glass nanofilms. *Journal of Non-Crystalline Solids.* **357**(7):1621–1627.

1832. Kurbatsky V.P. and Pogosov V.V. (2010). Optical conductivity of metal nanofilms and nanowires: the rectangular-box model. *Physical Review B.* **81**(15):155404.

1833. Mattoli V. et al. (2011). Freestanding functionalized nanofilms for biomedical applications. *Procedia Computer Science.* **7**:337–339.

1834. Sinibaldi E. et al. (2010). Magnetic nanofilms for biomedical applications. *Journal of Nanotechnology in Engineering and Medicine.* **1**(2):021008.

1835. Stich N. et al. (2001). Nanofilms and nanoclusters: energy sources driving fluorophores of biochip bound labels. *Journal of Nanoscience and Nanotechnology.* **1**(4):397–405.

1836. Thangaraj M. and Mills D. (2010). Design of smart nanofilms for regenerative medicine. *FASEB Journal.* **24**:181–184.

Nanoflakes

1837. Deng L.W. et al. (2002). Fabrication and microwave absorbing abilities of nanoflakes. *Journal of Functional Materials & Devices.* **8**(3):271–275.

1838. Heli H. and Yadegari H. (2010). Nanoflakes of the cobaltous oxide, CoO: synthesis and characterization. *Electrochimica Acta.* **55**(6):2139–2148.

1839. Kuc A. et al. (2010). Structural and electronic properties of graphene nanoflakes. *Physical Review B.* **81**(8):085430.

1840. Mazur M. (2004). Electrochemically prepared silver nanoflakes and nanowires. *Electrochemistry Communications.* **6**(4):400–403.

1841. Shang N.G. et al. (2002). Uniform carbon nanoflake films and their field emissions. *Chemical Physics Letters.* **358**(3):187–191.

1842. Silva A.M. et al. (2010). Graphene nanoflakes: thermal stability, infrared signatures, and potential applications in the field of spintronics and optical nanodevices. *The Journal of Physical Chemistry C.* **114**(41):17472–17485.

1843. Wohner N. et al. (2014). Energetic stability of graphene nanoflakes and nanocones. *Carbon.* **67**:721–735.

Nanofluids

1844. Abgrall P. and Nguyen N.T. (2008). Nanofluidic devices and their applications. *Analytical Chemistry.* **80**(7):2326–2341.

1845. Abgrall P. and Nguyen N.T. (2014). *Nanofluidics.* Norwood, MA: Artech House.

1846. Bai C. and Wang L. (2010). Constructual structure of nanofluids. *Journal of Applied Physics.* **108**(7):074317.

1847. Buongiorno J. et al. (2009). A benchmark study on the thermal conductivity of nanofluids. *Journal of Applied Physics.* **106**(9):094312.

1848. Chandrasekar M. and Suresh S. (2009). A review on the mechanisms of heat transport in nanofluids. *Heat Transfer Engineering.* **30**(14):1136–1150.

1849. Chaudhury M.K. (2003). Complex fluids: spread the word about nanofluids. *Nature.* **423**(6936):131–132.

1850. Chen H. et al. (2007). Rheological behaviour of nanofluids. *New Journal of Physics.* **9**(10):367.

1851. Choi S.U. (2008). Nanofluids: a new field of scientific research and innovative applications. *Heat Transfer Engineering.* **29**(5):429–431.

1852. Choi S.U. (2009). Nanofluids: from vision to reality through research. *Journal of Heat Transfer.* **131**(3):033106.

1853. Das S.K. (2006). Nanofluids: the cooling medium of the future. *Heat Transfer Engineering.* **27**(10):1–2.

1854. Das S.K. et al. (2007). *Nanofluids: Science and Technology.* New York: John Wiley & Sons.

1855. Das S.K. and Stephen U.S. (2009). A review of heat transfer in nanofluids. *Advances in Heat Transfer.* **41**:81–197.

1856. Daungthongsuk W. and Wongwises S. (2007). A critical review of convective heat transfer of nanofluids. *Renewable and Sustainable Energy Reviews.* **11**(5):797–817.

1857. Eijkel J.C. and Van Den Berg A. (2005). Nanofluidics: what is it and what can we expect from it? *Microfluidics and Nanofluidics.* **1**(3):249–267.

1858. Escher W. et al. (2011). On the cooling of electronics with nanofluids. *Journal of Heat Transfer.* **133**(5):051401.

1859. Fan J. and Wang L. (2011). Review of heat conduction in nanofluids. *Journal of Heat Transfer.* **133**(4):040801.

1860. Godson L. et al. (2010). Enhancement of heat transfer using nanofluids: an overview. *Renewable and Sustainable Energy Reviews.* **14**(2):629–641.

1861. Gupta H.K. et al. (2012). An overview of nanofluids: a new media towards green environment. *International Journal of Environmental Sciences.* **3**(1):433–440.

1862. Haddad Z. et al. (2012). A review on natural convective heat transfer of nanofluids. *Renewable and Sustainable Energy Reviews.* **16**(7):5363–5378.

1863. Haddad Z. et al. (2014). A review on how the researchers prepare their nanofluids. *International Journal of Thermal Sciences.* **76**:168–189.

1864. Hwang Y. et al. (2006). Thermal conductivity and lubrication characteristics of nanofluids. *Current Applied Physics.* **6**(Suppl 1):e67–e71.

1865. Jagannathan R. and Irvin G.C. (2005). Nanofluids: a new class of materials produced from nanoparticle assemblies. *Advanced Functional Materials.* **15**(9):1501–1510.

1866. Kaufui V. et al. (2010). Applications of nanofluids: current and future. *Advances in Mechanical Engineering.* **2010**: Article ID 519659.

1867. Keblinski P. et al. (2008). Thermal conductance of nanofluids: is the controversy over? *Journal of Nanoparticle Research.* **10**(7):1089–1097.

1868. Khanafer K. and Vafai K. (2011). A critical synthesis of thermophysical characteristics of nanofluids. *International Journal of Heat and Mass Transfer.* **54**(19):4410–4428.

1869. Mahbubul I.M. et al. (2012). Latest developments on the viscosity of nanofluids. *International Journal of Heat and Mass Transfer.* **55**(4):874–885.

1870. Mahian O. et al. (2013). A review of the applications of nanofluids in solar energy. *International Journal of Heat and Mass Transfer.* **57**(2):582–594.

1871. Michaelides E.E. (2013). Transport properties of nanofluids. A critical review. *Journal of Non-Equilibrium Thermodynamics.* **38**(1):1–79.

1872. Mijatovic D. et al. (2005). Technologies for nanofluidic systems: top-down vs. bottom-up: a review. *Lab on a Chip.* **5**(5):492–500.

1873. Murshed S.S. and de Castro C.N. (2014). Superior thermal features of carbon nanotubes-based nanofluids: a review. *Renewable and Sustainable Energy Reviews.* **37**:155–167.

1874. Murshed S.S. et al. (2011). A review of boiling and convective heat transfer with nanofluids. *Renewable and Sustainable Energy Reviews.* **15**(5):2342–2354.

1875. Paul G. et al. (2010). Techniques for measuring the thermal conductivity of nanofluids: a review. *Renewable and Sustainable Energy Reviews.* **14**(7):1913–1924.

1876. Phelan P.E. et al. (2005). Nanofluids for heat transfer applications. *Annual Review of Heat Transfer.* **14**(14).

1877. Philip J. and Shima P.D. (2012). Thermal properties of nanofluids. *Advances in Colloid and Interface Science.* **183**:30–45.

1878. Prakash S. et al. (2008). Nanofluidics: systems and applications. *IEEE Sensors Journal.* **8**(5):441–450.

1879. Puliti G. et al. (2011). Nanofluids and their properties. *Applied Mechanics Reviews.* **64**(3):030803.

1880. Rudyak V.Y. et al. (2010). On the thermal conductivity of nanofluids. *Technical Physics Letters.* **36**(7):660–662.

1881. Saidur R. et al. (2011). A review on applications and challenges of nanofluids. *Renewable and Sustainable Energy Reviews.* **15**(3):1646–1668.

1882. Sarkar J. (2011). A critical review on convective heat transfer correlations of nanofluids. *Renewable and Sustainable Energy Reviews.* 15(6):3271–3277.

1883. Senthilraja S. et al. (2010). Nanofluid applications in future automobiles: comprehensive review of existing data. *Nano-Micro Letters.* **2**(4):306–310.

1884. Sheng-shan Bi. et al. (2008). Applications of nanoparticles in domestic refrigerators. *Applied Thermal Engineering.* **28**:1834–1843.

1885. Singh A.K. (2008). Thermal conductivity of nanofluids. *Defence Science Journal.* **58**(5):600–607.

1886. Sokolov V.V. (2010). Wave propagation in magnetic nanofluids: a review. *Acoustical Physics.* **56**(6):972–988.

1887. Stephen U. and Choi S. (2009). Nanofluids: from vision to reality through research. *Journal of Heat Transfer.* **131**(4):631–648.

1888. Sureshkumar R. et al. (2013). Heat transfer characteristics of nanofluids in heat pipes: a review. *Renewable and Sustainable Energy Reviews.* **20**:397–410.

1889. Taylor R.A. and Phelan P.E. (2009). Pool boiling of nanofluids: comprehensive review of existing data and limited new data. *International Journal of Heat and Mass Transfer.* **52**(23):5339–5347.

1890. Taylor R. et al. (2013). Small particles, big impacts: a review of the diverse applications of nanofluids. *Journal of Applied Physics.* **113**(1):011301.

1891. Trisaksri V. and Wongwises S. (2007). Critical review of heat transfer characteristics of nanofluids. *Renewable and Sustainable Energy Reviews.* **11**(3):512–523.

1892. Wang L. and Fan J. (2010). Nanofluids research: key issues. *Nanoscale Research Letters.* **5**(8):1241–1252.

1893. Vekas L. (2004). Magnetic nanofluids properties and some applications. *Romanian Journal of Physics.* **49**(9–10):707–721.

1894. Wang X.Q. (2008). Review on nanofluids. II. Experiments and applications. *Brazilian Journal* of *Chemical Engineering.* **25**(4):631–648.

1895. Wang X.Q. and Mujumdar A.S. (2008). A review on nanofluids. I. Theoretical and numerical investigations. *Brazilian Journal of Chemical Engineering.* **25**(4):613–630.

1896. Wang X.Q. and Mujumdar A.S. (2007). Heat transfer characteristics of nanofluids: a review. *International Journal of Thermal Sciences.* **46**(1):1–19.

1897. Wang L. and Wei X. (2009). Nanofluids: synthesis, heat conduction, and extension. *Journal of Heat Transfer.* **131**(3): 033102.

1898. Wen D. et al. (2009). Review of nanofluids for heat transfer applications. *Particuology.* **7**:141–150.

1899. Wong K.V. and De Leon O. (2010). Applications of nanofluids: current and future. *Advances in Mechanical Engineering.* **2**:519659.

1900. Wu D. et al. (2009). Critical issues in nanofluids preparation, characterization and thermal conductivity. *Current Nanoscience.* **5**(1):103–112.

1901. Xuan Y. et al. (2003). Aggregation structure and thermal conductivity of nanofluids. *AIChE Journal.* **49**(4):1038–1043.

1902. Yu W. and Xie H. (2012). A review on nanofluids: preparation, stability mechanisms, and applications. *Journal of Nanomaterials.* 2012: Article ID 435873.

1903. Yu W. et al. (2012). Comparative review of turbulent heat transfer of nanofluids. *International Journal of Heat and Mass Transfer.* **55**(21):5380–5396.

Nanofoams

1904. Blinc R. et al. (2007). Carbon nanofoam as a potential hydrogen storage material. *Physica Status Solidi B.* **244**(11):4308–4310.

1905. Carter K.R. et al. (1995). Polyimide nanofoams for low dielectric applications. *MRS Proceedings.* **381**:79.

1906. Charlier Y. et al. (1993). High temperature polyimide nanofoams. *MRS Proceedings.* **323**:277.

1907. Detsi E. et al. (2016). Metallic muscles and beyond: nanofoams at work. *Journal of Materials Science.* **51**(1):615–634.

1908. Hedrick J.L. et al. (1995). High temperature nanofoams derived from rigid and semi-rigid polyimides. *Polymer.* **36**(14):2685–2697.

1909. Iino T. and Nakamura K. (2009). Acoustic and acousto-optic characteristics of silicon nanofoam. *Japanese Journal of Applied Physics.* **48**(7S):07GE01.

1910. Lakshmanan P. (1996). Polyimide nanofoam: materials design and development. *Polymer Preprints (American Chemical Society, Division of Polymer Chemistry).* **37**(1):136–137.

1911. Li S. et al. (2009). Magnetic carbon nanofoams. *Journal of Nanoscience and Nanotechnology.* **9**(2):1133–1136.

1912. Liqun C. and Shixin C. (2005). Nano-foam plastics. *New Chemical Materials.* **7**:018.

1913. McGrath J.E. et al. (1996). Polyimide nanofoams: materials design and development. *Polymer Preprints (USA).* **37**(1):136–137.

1914. Rode A.V. et al. (2004). Unconventional magnetism in all-carbon nanofoam. *Physical Review B.* **70**(5):054407.

1915. Sundarram S. (2013). Fabrication and characterization of open celled micro and nano foams. PhD Dissertation: Austin, TX, University of Texas.

1916. Zhang Y.H. et al. (2007). Synthesis and characterization of polymer nanofoams. *Key Engineering Materials.* **334**:821–824.

1917. Zhihuan X. (2000). Development of polyimide nanofoam films. *New Chemical Materials.* **5**:008.

1918. Zhu Z. et al. (2008). Challenges to the formation of nano-cells in foaming processes. *International Polymer Processing.* 23(3):270–276.

Nanohorns

1919. Ajima K. et al. (2005). Carbon nanohorns as anticancer drug carriers. *Molecular Pharmaceutics.* **2**(6):475–480.

1920. Berber S. et al. (2000). Electronic and structural properties of carbon nanohorns. *Physical Review B.* **62**(4):R2291.

1921. Lahiani M.H. et al. (2015). Interaction of carbon nanohorns with plants: uptake and biological effects. *Carbon.* **81**:607–619.

1922. Miyawaki J. et al. (2008). Toxicity of single-walled carbon nanohorns. *ACS Nano.* **2**(2):213–226.

1923. Pagona G. et al. (2008). Properties, applications and functionalisation of carbon nanohorns. *International Journal of Nanotechnology.* **6**(1–2):176–195.

1924. Zhu S. and Xu G. (2010). Single-walled carbon nanohorns and their applications. *Nanoscale.* **2**(12):2538–2549.

Nanolayers

1925. Akashev L.A. et al. (2013). Synthesis of aluminum nitride nanolayers. *Technical Physics Letters.* **39**(2):154–156.

1926. Bernal-Lara T.E. et al. (2005). Structure and thermal stability of polyethylene nanolayers. *Polymer.* **46**(9):3043–3055.

1927. Fernández B. et al. (2010). Glow discharge analysis of nanostructured materials and nanolayers: a review. *Analytica Chimica Acta.* **679**(1):7–16.

1928. Penn L.S. et al. (2002). Formation of tethered nanolayers: three regimes of kinetics. *Macromolecules.* **35**(18):7054–7066.

1929. Saber O. (2009). Nanostructural materials; nanolayers and nanofibers. *International Journal of Nanoparticles.* **2**(1–6): 209–215.

1930. Shakhvorostov D. et al. (2006). Structure and mechanical properties of tribologically induced nanolayers. *Wear.* **260**(4):433–437.

Nanomachines

1931. Bakalis E. (2010). Designing nanomachines: a theoretical and computational approach. *Journal of Computational & Theoretical Nanoscience.* **7**(9):1783–1799.

1932. Browne W.R. and Feringa B.L. (2006). Making molecular machines work. *Nature Nanotechnology.* **1**(1):25–35.

1933. Carey P.R. et al. (2004). Transcarboxylase: one of nature's early nanomachines. *IUBMB Life.* **56**(10):575–583.

1934. Gao W. and Wang J. (2014). The environmental impact of micro/nanomachines: a review. *ACS Nano.* **8**(4):3170–3180.

1935. Freitas R.A. (2006). Molecular manufacturing: too dangerous to allow. *Nanotechnology Perceptions.* **2**:15–24.

1936. Kay E.R. and Leigh D.A. (2006). Photochemistry: lighting up nanomachines. *Nature.* **440**(7082):286–287.

1937. Mantle T.J. (2001). Enzymes: nature's nanomachines. *Biochemical Society Transactions.* **29**(2):331–336.

1938. Montemagno C.D. (2001). Nanomachines: a roadmap for realizing the vision. *Journal of Nanoparticle Research.* **3**(1):1–3.

1939. Ozin G.A. et al. (2005). Dream nanomachines. *Advanced Materials.* **17**(24):3011–3018.

1940. Popov V.L. (2002a). Nanomachinery: a general approach to inducing directed motion at the atomic level. *Technical Physics.* **47**(11):1397–1407.

1941. Popov V.L. (2002b). The theory of quasistatic nanomachines. *Technical Physics Letters.* **28**(5):385–390.

1942. Rudall B.H. (1998). Advances in nanomachines. *Kybernetes.* **27**(8).

1943. Russell S.J. (2003). Rise of the nanomachines. *Nature Biotechnology.* **21**(8):872–873.

1944. Tyreman M.J.A. and Molloy J.E. (2003). Molecular motors: nature's nanomachines. *IEEE Proceedings - Nanobiotechnology.* **150**(3):95–102.

1945. Wang J. (2013). *Nanomachines: Fundamentals and Applications.* New York: John Wiley & Sons.

Nanomagnets

1946. Cowburn R.P. et al. (1999). Single-domain circular nanomagnets. *Physical Review Letters.* **83**(5):1042.

1947. Gatteschi D. and Sessoli R. (2004). Molecular nanomagnets: the first 10 years. *Journal of Magnetism and Magnetic Materials.* **272**:1030–1036.

1948. Gatteschi D. et al. (2006). *Molecular Nanomagnets.* New York: Oxford University Press.

1949. Singh N. et al. (2004). Fabrication of large area nanomagnets. *Nanotechnology.* **15**(11):1539.

1950. Tartaj P. (2006). Nanomagnets-from fundamental physics to biomedicine. *Current Nanoscience.* **2**(1):43–53.

1951. Wernsdorfer W. (2010). Molecular nanomagnets: towards molecular spintronics. *International Journal of Nanotechnology.* **7**(4–8): 497–522.

Nanomaterials

1952. Abbott L.C. and Maynard A.D. (2010). Exposure assessment approaches for engineered nanomaterials. *Risk Analysis.* **30**(11):1634–1644.

1953. Aguilar Z. (2013). *Nanomaterials for Medical Applications.* Waltham, MA: Elsevier, Inc.

1954. Aragay G. et al. (2012). Nanomaterials for sensing and destroying pesticides. *Chemical Reviews.* **112**(10):5317–5338.

1955. Arts J.H.E. et al. (2014). A critical appraisal of existing concepts for the grouping of nanomaterials. *Regulatory Toxicology and Pharmacology.* **70**(2):492–506.

1956. Balshaw D.M. et al. (2005). Research strategies for safety evaluation of nanomaterials, part III: nanoscale technologies for assessing risk and improving public health. *Toxicological Sciences.* **88**(2):298–306.

1957. Barceló D. and Farré M. (2012). *Analysis and Risk of Nanomaterials in Environmental and Food Samples,* Volume 59 (Comprehensive Analytical Chemistry). Oxford, UK: Elsevier.

1958. Beaudrie C.E.H. and Kandlikar M. (2011). Horses for courses: risk information and decision making in the regulation of nanomaterials. *Journal of Nanoparticle Research.* **13**(4):1477–1488.

1959. Bergamaschi E. (2009). Occupational exposure to nanomaterials: present knowledge and future development. *Nanotoxicology.* **3**:194–201.

1960. Bitounis D. et al. (2016). Detection and analysis of nanoparticles in patients: a critical review of the status quo of clinical nanotoxicology. *Biomaterials.* **76**:302–312.

1961. Blonder R. (2011). The story of nanomaterials in modern technology: an advanced course for chemistry teachers. *Journal of Chemical Education.* **88**(1):49–52.

1962. Bolt H.M. (2014). Grouping of nanomaterials for risk assessment. *Archives of Toxicology.* **88**(12):2077–2078.

1963. Borm P. et al. (2006). Research strategies for safety evaluation of nanomaterials, part V: role of dissolution in biological fate and effects of nanoscale particles. *Toxicological Sciences.* **90**(1):23–32.

1964. Borm P.J. et al. (2006). The potential risks of nanomaterials: a review carried out for ECETOC. *Particle and Fibre Toxicology.* **3**(1):11.

1965. Borm P. et al. (2009). Toxicology of Nanomaterials: permanent interactive learning. *Particle and Fibre Toxicology.* **6**:3.

1966. Boxall A. et al. (2007). Engineered nanomaterials in soils and water: how do they behave and could they pose a risk to human health? *Nanomedicine (Future Medicine).* **2**:919–927.

1967. Brayner R. et al. (eds.) (2012). *Nanomaterials: A Danger or Promise? A Chemical and Biological Perspective.* New York: Springer-Verlag.

1968. Breggin L.K. and Porter R.D. (2008). Application of the toxics release inventory to nanomaterials (PEN Brief No. 2). Washington, D.C.: Woodrow Wilson International Center for Scholars.

1969. Brunelli A. (2013). Advanced physico-chemical characterization of engineered nanomaterials in nanotoxicology. PhD Dissertation: Venice, Italy, Università Ca' Foscari Venezia.

1970. Buzea C. et al. (2007). Nanomaterials and nanoparticles: sources and toxicity. *Biointerphases.* **2**(4):MR17-MR71.

1971. Card J.W. and Magnuson B.A. (2010). A method to assess the quality of studies that examine the toxicity of engineered nanomaterials. *International Journal of Toxicology.* **29**:402–410.

1972. Caskey L. et al. (2011). Nanomaterials: the good, the bad & the ugly: a case study. *Professional Safety.* **56**(8):49–55.

1973. Cattaneo A.G. et al. (2009). Ecotoxicology of nanomaterials: the role of invertebrate testing. *Invertebrate Survival Journal.* **6**:78–97.

1974. Cayton R.H. (2001). Nanomaterials for abrasion-resistant coatings. *Paint & Coatings Industry.* **17**(10):122.

1975. Center for Environmental Law (2009). Addressing nanomaterials as an issue of global concern. Washington, D.C.

1976. Chalupka S. (2012). Management of occupational exposure to advanced nanomaterials. *Workplace Health & Safety.* **60**:556.

1977. Chen Z. et al. (2011). A review of environmental effects and management of nanomaterials. *Toxicology & Environmental Chemistry.* **93**(6):1227–1250.

1978. Choi J. et al. (2009). The impact of toxicity testing costs on nanomaterial regulation. *Environmental Science and Technology.* **43**(9):3030–3034.

1979. Cohen J. et al. (2011). Property transformations of engineered nanomaterial in different physiologic fluids: implications to nanotoxicology. Abstract No. 212 presented at the *AAAR 30th Annual Conference.* Rosen Shingle Creek Resort, Orlando, FL. October 3–7.

1980. Colvin V.L. (2003). The potential environmental impact of engineered nanomaterials. *Nature Biotechnology.* **21**:1166–1170.

1981. Conti J.A. et al. (2008). Health and safety practices in the nanomaterials workplace: results from an international survey. *Environmental Science & Technology.* **42**(9):3155–3162.

1982. Culha M. and Altunbek M. (2011). Nanotoxicology: how to test the safety of engineered nanomaterials? *Current Opinion in Biotechnology.* **22**:S28–S29.

1983. Delgado G.C. (2010). Economics and governance of nanomaterials: potential and risks. *Technology in Society.* **32**(2):137–144.

1984. de Lapuente Pérez J. (2015). Interesting nanomaterials for nanotoxicology. *Revista de Toxicología.* **31**(1).

1985. de Paula A.J. et al. (2016). Nanotoxicology of carbon-based nanomaterials. In *Bioengineering Applications of Carbon Nanostructures.* Switzerland: Springer International Publishing, pp. 105–137.

1986. Dhawan A. and Sharma V. (2010). Toxicity assessment of nanomaterials: methods and challenges. *Analytical & Bioanalytical Chemistry.* **398**(2):589–605.

1987. Dolez P. et al. (2013). Advancement in Quebec research on the prevention of risks related to occupational exposure to nanomaterials. *Relations Industrielles.* **68**(4):623–642.

1988. Edelstein A.S. and Cammaratra R.C. (eds.) (1998). *Nanomaterials: Synthesis, Properties, and Applications.* New York: Taylor & Francis Group.

1989. Fadeel B. (2014). *Handbook of Safety Assessment of Nanomaterials: From Toxicological Testing to Personalized Medicine* (Pan Stanford Series on Biomedical Nanotechnology, Volume 5). Boca Raton, FL: CRC Press.

1990. Farré M. et al. (2009). Ecotoxicity and analysis of nanomaterials in the aquatic environment. *Analytical & Bioanalytical Chemistry.* **393**(1):81–95.

1991. Farre M. et al. (2011). Analysis and assessment of the occurrence, the fate and the behavior of nanomaterials in the environment. *Trends in Analytical Chemistry.* **30**(3):517–27.

1992. Federal Institute for Occupational Safety & Health (2012). Guidance for handling and use of nanomaterials in the workplace. Berlin, Germany.

1993. Gaddam R.R. et al. (2015). Recent developments of camphor based carbon nanomaterial: their latent applications and future prospects. *Nano-Structures & Nano-Objects.* **3**:1–8.

1994. Garde K. et al. (2013). Understanding cytotoxicity of engineered nanomaterials. *Electrochemical Society Transactions.* **50**:33–39.

1995. Gazsó A. and Fries R. (2012). Nanomaterials and occupational safety: an overview. *European Journal of Risk Regulation.* **3**(4):594–601.

1996. Gendre L. et al. (2015). Nanomaterials life cycle analysis: health and safety practices, standards and regulations— past, present and future perspective. *International Research Journal of Pure & Applied Chemistry.* **5**(3):208–228.

1997. Giacobbe F. et al. (2009). Risk assessment model of occupational exposure to nanomaterials. *Human & Experimental Toxicology.* **28**:401–406.

1998. Gonzalez L. et al. (2008). Genotoxicity of engineered nanomaterials: a critical review. *Nanotoxicology.* **2**(4):252–273.

1999. Gottschalk F. and Nowack B. (2011). The release of engineered nanomaterials to the environment. *Journal of Environmental Monitoring.* **13**(5):1145–1155.

2000. Greaves-Holmes W.A. (2009). Guide for the safe handling of engineered and fabricated nanomaterials. *The Journal of Technology Studies.* **1**:33–39.

2001. Grieger K. et al. (2009). The known unknowns of nanomaterials: describing and characterizing uncertainty within environmental, health and safety risks. *Nanotoxicology.* **3**(3):222–233.

2002. Grieger K. et al. (2010). Redefining risk research priorities for nanomaterials. *Journal of Nanoparticle Research.* **12**:383–392.

2003. Grieger K.D. et al. (2012). Environmental risk analysis for nanomaterials: review and evaluation of frameworks. *Nanotoxicology.* **6**(2):196–212.

2004. Groso A. et al. (2010). Management of nanomaterials safety in research environment. *Particle and Fibre Toxicology.* **7**:40.

2005. Guidotti T.L. (2010). The regulation of occupational exposure to nanomaterials: a proposal. *Archives of Environmental & Occupational Health.* **65**(2):57–58.

2006. Gurulingappa P. and Kaul G. (2014). Health hazards associated with nanomaterials. *Toxicology & Industrial Health.* **30**:499–519.

2007. Handy R.D. et al. (2008). The ecotoxicology and chemistry of manufactured nanoparticles. *Ecotoxicology.* **17**(4):287–314.

2008. Handy R.D. et al. (2008). The ecotoxicology and chemistry of manufactured nanoparticles: current status, knowledge gaps, challenges, and future needs. *Ecotoxicology.* **17**(4):315–325.

2009. Hansen S.F. et al. (2007). Categorization framework to aid hazard identification of Nanomaterials. *Nanotoxicology.* **1**(3):243–250.

2010. Hansen S.F. et al. (2008). Categorization framework to aid exposure assessment of nanomaterials in consumer products. *Ecotoxicology.* **17**(5):438–447.

2011. Harty S.B. (2011). Regulating nanomaterials. *Ceramic Industry.* **161**(12):26–27.

2012. Holden P.A. et al. (2013). Ecological nanotoxicology: integrating nanomaterial hazard considerations across the subcellular, population. Community, and ecosystems levels. *Accounts of Chemical Research.* **46**(3):813–822.

2013. Holgate S.T. (2010). Exposure, uptake, distribution and toxicity of nanomaterials in humans. *Journal of Biomedical Nanotechnology.* **6**(1):1–19.

2014. Holsapple M. P. et al. (2005). Research strategies for safety evaluation of nanomaterials, part II: toxicological and safety evaluation of nanomaterials, current challenges and data needs. *Toxicological Sciences.* **88**(1):12–17.

2015. Holzinger M. et al. (2014). Nanomaterials for biosensing applications: a review. *Frontiers in Chemistry.* **2**:63.

2016. Hou W.C. et al. (2013). Biological accumulation of engineered nanomaterials: a review of current knowledge. *Environmental Science: Process & Impacts.* **15**(1):103–122.

2017. Hristozov D.R. et al. (2012). Risk assessment of engineered nanomaterials: a review of available data and approaches from a regulatory perspective. *Nanotoxicology.* **6**:880–898.

2018. Hubbell J.A. and Chilkoti A. (2012). Nanomaterials for drug delivery. *Science, New Series.* **337**(6092):303–305.

2019. Hunt W.H. Jr. (2004). Nanomaterials: nomenclature, novelty, and necessity. *JOM.* **56**(10):13–18.

2020. Hussain S. et al. (2012). Interactions of nanomaterials with the immune system. *Wiley Interdisciplinary Reviews: Nanomedicine and Nanobiotechnology.* **4**(2):169–183.

2021. Iavicoli I. et al. (2013). The effects of nanomaterials as endocrine disruptors. *International Journal of Molecular Sciences.* **14**(8):16732–16801.

2022. Johnson D.R. et al. (2010). Potential for occupational exposure to engineered carbon-based nanomaterials in environmental laboratory studies. *Environmental Health Perspectives.* **118**(1):49–54.

2023. Johnston H. et al. (2013). Engineered nanomaterial risk. Lessons learned from completed nanotoxicology studies:

potential solutions to current and future challenges. *Critical Reviews in Toxicology.* **43**(1):1–20.

2024. Kagan V.E. et al. (2010). Fantastic voyage and opportunities of engineered nanomaterials: what are the potential risks of occupational exposures? *Journal of Occupational & Environmental Medicine.* **52**:943–946.

2025. Karshak K. et al. (2014). Examining the cellular uptake and toxicity of engineered nanomaterials. *Electrochemical Society Transactions.* **61**:15–21.

2026. Keumpel E.D. et al. (2012). Risk assessment and risk management of nanomaterials in the workplace: translating research to practice. *Annals of Occupational Hygiene.* **56**(5):491–505.

2027. Klain S.J. et al. (2008). Nanomaterials in the environment: behavior, fate, bioavailability, and effects. *Environmental Toxicology & Chemistry.* **27**(9):1825–1851.

2028. Klein C.L. et al. (2012). Hazard identification of inhaled nanomaterials. *Archives of Toxicology.* **86**(7):1137–1151.

2029. Kreider M.L. et al. (2013). Protecting workers from risks associated with nanomaterials. Part 2. Best practices in risk management. *Occupational Health & Safety.* **82**(9):20–24.

2030. Kuempel E.D. et al. (2012). Risk assessment and risk management of nanomaterials in the workplace: translating research to practice. *Annals of Occupational Hygiene.* **56**:491–505.

2031. Kwak J.I. and Youn-Joo A. (2015). Ecotoxicological effects of nanomaterials on earthworms: a review. *Human & Ecological Risk Assessment.* **18**(21):1566–1575.

2032. Lai D.Y. (2012). Toward toxicity testing of nanomaterials in the 21st century: a paradigm for moving forward. *Wiley Interdisciplinary Reviews: Nanomedicine & Nanobiotechnology.* **4**(1):1–15.

2033. Landsiedel R. et al. (2012). Toxico/biokinetics of nanomaterials. *Archives of Toxicology.* **86**(7):1021–1060.

2034. Lee J. et al. (2010). Nanomaterials in the construction industry: a review of their applications and environmental health and safety considerations. *ACS Nanotechnology.* **4**(7):3580–3590.

2035. Lewinski N. et al. (2008). Cytotoxicity of nanoparticles. *Small.* **4**(1):26–49.

2036. Lidén G. (2011). The European Commission tries to define nanomaterials. *Annals of Occupational Hygiene.* **55**:1–5.

2037. Lin D. et al. (2010). Fate and transport of engineered nanomaterials in the environment. *Journal of Environmental Quality.* **39**(6):1896–1908.

2038. Loux N.T. et al. (2011). Issues in assessing environmental exposures to manufactured nanomaterials. *International Journal of Environmental Research & Public Health.* **8**(9):3562–3578.

2039. Lowry G.V. et al. (2010). Environmental occurrences, behavior, fate and ecological effects of nanomaterials: an introduction to the special series. *Journal of Environmental Quality.* **39**(6):1867–1874.

2040. Magnuson B.A. et al. (2011). A brief review of the occurrence, use, and safety of food-related nanomaterials. *Journal of Food Science.* **76**(6):R126–R133.

2041. Makarucha A.J. et al. (2011). Nanomaterials in biological environments: a review of computer modeling studies. *European Biophysics Journal.* **40**(2):103–115.

2042. Maniratanachote R. (2009). Nanotoxicology: a safety evaluation of nanomaterials. *Thai Journal of Toxicology.* **2552**:40–43.

2043. Manzetti S. and Andersen O. (2012). Toxicological aspects of nanomaterials used in energy harvesting consumer electronics. *Renewable and Sustainable Energy Reviews.* **16**(4):2102–2110.

2044. Marquis F.D.S. (2011). The role of nanomaterials systems in energy and environment: renewable energy. *JOM.* **63**(1):43.

2045. Masunaga T. (2014). [Nanomaterials in cosmetics: present situation and future]. *Yakugaku Zasshi.* **134**(1):39–43.

2046. Maynard A.D. and Aitken R.J. (2007). Assessing exposure to airborne nanomaterials: current abilities and future requirements. *Nanotoxicology.* **1**(1):26–41.

2047. Miller G. and Wickson F. (2015). Risk analysis of nanomaterials: exposing nanotechnology's naked emperor. *Review of Policy Research.* **32**(4):485–512.

2048. Miseljic M. and Olsen S.I. (2014). Life-cycle assessment of engineered nanomaterials: a literature review of assessment status. *Journal of Nanoparticle Research.* **16**(6):1–33.

2049. MIT (2015). Nanomaterials toxicity. Available online at http://ehs. mit.edu/site/content/nanomaterials-toxicity

2050. Mogharabi M. et al. (2014). Toxicity of nanomaterials: an undermined issue. *Daru.* **22**(8):1–5.

2051. Moon R.J. et al. (2011). Cellulose nanomaterials review: structure, properties and nanocomposites. *Chemical Society Reviews.* **40**(7):3941–3994.

2052. Morimoto Y. et al. (2010). Hazard assessment of manufactured nanomaterials. *Journal of Occupational Health.* **52**(6):325–334.

2053. Morris V.J. (2011). Emerging roles of engineered nanomaterials in the food industry. *Trends in Biotechnology.* **29**(10):509–516.

2054. Murashov V. et al. (2012). Progression of occupational risk management with advances in nanomaterials. *Journal of Occupational Health.* **9**:D12–D22.

2055. Nasterlack M. et al. (2008). Considerations in occupational medical surveillance of employees handling nanoparticles. *International Archives of Occupational & Environmental Health.* **81**:721–726.

2056. Nayar P.G. (2011). Nanomaterials and nanotoxicity in the present scenario: a short review. *Journal of the Indian Society of Toxicology.* **7**(2):11–15.

2057. Nel A. et al. (2012). Nanomaterial toxicity testing in the 21st century: use of a predictive toxicological approach and high-throughput screening. *Accounts of Chemical Research.* **46**(3):607–621.

2058. NIOSH (2009). Approaches to safe nanotechnology: managing the health and safety concerns associated with engineered nanomaterials (DHHS Publication 2009-125). Cincinnati, OH: U.S. Department of Health and Human Services, Centers for Disease Control and Prevention, National Institute for Occupational Safety and Health.

2059. NIOSH (2009). Current intelligence bulletin 60: interim guidance for medical screening and hazard surveillance for workers potentially exposed to engineered nanoparticles. Cincinnati, OH: U.S. Department of Health and Human Services, Centers for Disease Control and Prevention, National Institute for Occupational Safety and Health.

2060. NIOSH (2012). General safe practices for working with engineered nanomaterials in research laboratories (DHHS Publication No. 2012-147). Cincinnati, OH: U.S. Department of Health and Human Services, Centers for Disease Control and Prevention, National Institute for Occupational Safety and Health.

2061. NIOSH (2013). Current strategies for engineering controls in nanomaterial production and downstream handling processes (DHHS Publication No. 2014-102). Cincinnati, OH: U.S. Department of Health and Human Services, Centers for Disease Control and Prevention, National Institute for Occupational Safety and Health.

2062. Nyström A.M. and Fadeel B. (2012). Safety assessment of nanomaterials: implications for nanomedicine. *Journal of Controlled Release*. **161**(2):403–408.

2063. Oberdörster G. et al. (2005). Principles for characterizing the potential human health effects from exposure to nanomaterials: elements of a screening strategy. *Particle and Fibre Toxicology*. **2**:35.

2064. O'Brien N.J. and Cummins E.J. (2011). A risk assessment framework for assessing metallic nanomaterials of environmental concern: aquatic exposure and behavior. *Risk Analysis*. **31**(5):706–726.

2065. Oesch F. and Landsiedel R. (2012). Genotoxicity investigations on nanomaterials. *Archives of Toxicology.* **86**(7):985–994.

2066. Osman T.M. et al. (2006). The commercialization of nanomaterials: today and tomorrow. *JOM.* **58**(4):21–24.

2067. Owen R. and Handy R. (2007). Formulating the problems for environmental risk assessment of nanomaterials. *Environmental Science & Technology.* **41**(16):5582–5588.

2068. Owen R. et al. (2009). Strategic approaches for the management of environmental risk uncertainties posed by nanomaterials. *Nanomaterials: Risks and Benefits.* 369–384.

2069. Papp T. et al. (2008). Human health implications of nanomaterial exposure. *Nanotoxicology.* **2**(1):9–27.

2070. Deepa Parvathi V. and Rajagopal K. (2014). Nanotoxicology testing: potential of Drosophila in toxicity assessment of nanomaterials. *International Journal of Nanoscience and Nanotechnology.* **5**(1):25–35.

2071. Peralta-Videa J.R. et al. (2011). Nanomaterials and the environment: a review for the biennium 2008–2010. *Journal of Hazardous Materials.* **186**(1):1–15.

2072. Petkewich R. (2008). Probing hazards of nanomaterials. *Chemical & Engineering News.* **86**(42):53–56.

2073. Pietroiusti A. (2012). Health implications of engineered nanomaterials. *Nanoscale.* **4**:1231.

2074. Powell M.C. and Kanarek M.S. (2006). Nanomaterial health effects-part 1: background and current knowledge. *Wisconsin Medical Journal.* **105**(2):16.

2075. Powell M.C. and Kanarek M.S. (2006). Nanomaterial health effects-part 2: uncertainties and recommendations for the future. *Wisconsin Medical Journal.* **105**(3):18–23.

2076. Powers K.W. et al. (2006). Research strategies for safety evaluation of nanomaterials. Part VI. characterization of nanoscale particles for toxicological evaluation. *Toxicological Sciences.* **90**(2):296–303.

2077. Ramesh K.T. (2009). *Nanomaterials: Mechanics and Mechanisms.* New York: Springer.

2078. Reynolds J.G. and Hart B.R. (2004). Nanomaterials and their application to defense and homeland security. *JOM.* **56**(1):36.

2079. Ripp S. (2009). Nanomaterials: risks and benefits. *Journal of Environmental Quality.* **38**(5):2160.

2080. Rittner M.N. (2004). Nanomaterials in nanoelectronics: who's who and what's next. *JOM.* **56**(6):22.

2081. Rushton E.K. et al. (2005). Toxicological profiles of nanomaterials. *MRS Proceedings.* **895**:G04.

2082. Santos J.S.S. (2015). Nanotoxicology: study of nanomaterials' genotoxic effects in cell lines. Available at ulfc115908_ tm_Joana_ Santos.pdf

2083. Savolainen K. and Vainio H. (2011). [Health risks of engineered nanomaterials and nanotechnologies]. *Duodecim.* **127**(11): 1097–1104.

2084. Savolainen K. et al. (2010). Risk assessment of engineered nanomaterials and nanotechnologies: a review. *Toxicology.* **269**(2–3):92–104.

2085. Scheringer M. (2008). Nanoecotoxicology: environmental risks of nanomaterials. *Nature Nanotechnology.* **3**(6):322–323.

2086. Schubauer-Berigan M. et al. (2011). Engineered carbonaceous nanomaterials manufacturers in the United States: workforce size, characteristics and feasibility of epidemiologic studies. *Journal of Occupational & Environmental Medicine.* **53**(Suppl 6):S62–S67.

2087. Schulte P.A. et al. (2010). Occupational exposure limits for nanomaterials: state of the art. *Journal of Nanoparticle Research.* **12**(6):1971–1987.

2088. Scott A. and Seewald N. (2004). Potential health effects cloud nanomaterials growth prospects. *Chemical Week.* **166**(41):25–26.

2089. Service R.F. (2008). Nanotechnology: can high-speed tests sort out which nanomaterials are safe? *Science.* **321**:1036–1037.

2090. Sharifi S. et al. (2012). Toxicity of nanomaterials. *Chemistry Society Reviews.* **41**(6):2323–2343.

2091. Smith M.J. et al. (2014). From immunotoxicity to nanotherapy: the effects of nanomaterials on the immune system. *Toxicological Sciences.* **138**:249–255.

2092. Sonawane S.K. et al. (2014). Use of nanomaterials in the detection of food contaminants. *European Journal of Food Research & Review.* **4**(4):301–317.

2093. Song Y. et al. (2011). Nanomaterials in humans: identification, characteristics, and potential damage. *Toxicological Pathology.* **39**:841–849.

2094. Stewart R. (2004). Promise of nanomaterials. *Plastics Engineering.* **60**(5):24.

2095. Stewart R. (2006). Nanomaterials: still climbing the steep curve of material development. *Plastics Engineering.* **62**(4): 12–18.

2096. Stone V. et al. (2010). Nanomaterials for environmental studies: classification, reference material issues, and strategies for physico-chemical characterisation. *Science of the Total Environment.* **408**:1745–1754.

2097. Swidwińska-Gajewska A.M. and Czerczak S. (2013). [Nanomaterials—proposals of occupational exposure limits in the world and hygiene standards in Poland]. *Medycyna Pracy.* **64**(6):829–845.

2098. Syded S. et al. (2013). Immune response to nanomaterials: implications for medicine and literature review. *Current Allergy Asthma Report.* **13**(1):50–57.

2099. Tervonen T. and Lahdelma R. (2007). Implementing stochastic multicriteria acceptability analysis. *European Journal of Operational Research.* **178**(2):500–513.

2100. Tervonen T. et al. (2009). Risk-based classification system of nanomaterials. *Journal of Nanoparticle Research.* **11**:757–766.

2101. Tetley T.D. (2007). Health effects of nanomaterials. *Biochemical Society Transactions* **35**(Pt. 3):527–531.

2102. Thomas K. and Sayre P. (2005). Research strategies for safety evaluation of nanomaterials, Part I: Evaluating the human health implications of exposure to nanoscale materials. *Toxicological Sciences.* **87**(2):316–321.

2103. Thomas T. et al. (2006). Research strategies for safety evaluation of nanomaterials, Part VII: Evaluating consumer exposures to nanoscale materials. *Toxicological Sciences.* **91**(1):14–19.

2104. Thomas T. et al. (2009). Moving toward exposure and risk evaluation of nanomaterials: challenges and future directions. *Wiley Interdisciplinary Reviews: Nanomedicine and Nanobiotechnology.* **1**(4):426–433.

2105. Tiwle R. (2012). Nanotoxicology an emerging tool used for the toxicity of nanomaterials. *Journal of Biomedical and Pharmaceutical Research.* **1**(3):66–83.

2106. Trout D.B. and Schulte P.A. (2010). Medical surveillance, exposure registries, and epidemiologic research for workers exposed to nanomaterials. *Toxicology.* **269**:128–135.

2107. Tsuda H. et al. (2009). Toxicology of engineered nanomaterials: a review of carcinogenic potential. *Asian Pacific Journal of Cancer Prevention.* **10**(6):975–980.

2108. Tsuji J.S. et al. (2006). Research strategies for safety evaluation of nanomaterials, part IV: risk assessment of nanoparticles. *Toxicological Sciences.* **89**(1):42–50.

2109. Tyshenko M.G. and Krewski D. (2008). A risk management framework for the regulation of nanomaterials. *International Journal of Nanotechnology.* **5**(1):143–160.

2110. Ugwekar R.P. and Lakhawat G.P. (2014). Nanomaterials and its application: a review. *Journal of Chemical, Biological & Physical Sciences.* **4**(1):607–616.

2111. U.S. Department of Energy (2005). Good practices for handling nanomaterials. *Safety & Health Bulletin.* Issue No. 2005-06.

2112. U.S. EPA (2009a). A conceptual framework for US EPA's National Exposure Research Laboratory (EPA/600/R-09/003). Washington, D.C.: Office of Research and Development.

2113. U.S. EPA (2009b). Nanomaterial research strategy (EPA/620/K-09/011). Washington, D.C.: Office of Research and Development.

2114. U.S. EPA (2010). State of the science report on predictive models and modeling approaches for characterizing and evaluating exposures to nanomaterials (EPA/600/R-10/129). Washington, D.C.: Office of Research & Development, National Exposure Research Laboratory.

2115. U.S. EPA (2013). Detection and characterization of engineered nanomaterials in the environment: current state-of-the-art and future directions report, annotated bibliography, and image library (EPA/600/R-14/244). Washington, D.C.: Office of Research & Development.

2116. Vajtai R. (ed.) (2013). *Handbook of Nanomaterials*. New York: Springer-Verlag.

2117. Velzeboer I. et al. (2008). Aquatic ecotoxicity tests of some nanomaterials. *Environmental Toxicology & Chemistry*. **27**(9):1942–1947.

2118. Wang Q. et al. (2014). A bibliometric analysis of research on the risk of engineering nanomaterials during 1999–2012. *Science of the Total Environment*. **473**:483–489.

2119. Wiesner M.R. et al. (2006). Assessing the risks of manufactured nanomaterials. *Environmental Science & Technology*. **40**(14): 4336–4345.

2120. Wood A. (2002). Nanomaterials: a big market potential. *Chemical Week*. **164**(41):17–21.

2121. Yokel R.A. and MacPhail R.C. (2011). Engineered nanomaterials: exposures, hazards, and risk prevention. *Journal of Occupational Medical Toxicology*. 6(7). doi:10.1186/1745-6673-6-7

2122. Zapata L.J. and Acevedo E.M. (2014). Environmental applications of nanomaterials: new technologies for sustainable development. *Nanotechnology Research Journal*. **7**(2):221–264.

2123. Zea H.R. (2012). Nanomaterials: health effects and legislation. *Ingeniería e Investigación*. **32**(1):36–41.

Nanomembranes

2124. Anonymous (2011). Nanomembranes promise new materials for advanced electronics. *Advanced Materials & Processes.* **169**(9):18–19.

2125. Cavallo F. and Lagally M.G. (2010). Semiconductors turn soft: inorganic nanomembranes. *Soft Matter.* **6**(3):439–455.

2126. Lagally M.G. (2007). Silicon nanomembranes. *MRS Bulletin.* **32**(01): 57–63.

2127. Kim D.H. and Rogers J.A. (2009). Bend, buckle, and fold: Mechanical engineering with nanomembranes. *ACS Nano.* **3**(3):498–501.

2128. Huang M. et al. (2011). Nanomechanical architecture of semiconductor nanomembranes. *Nanoscale.* **3**(1):96–120.

2129. Mchedlishvili B.V. et al. (2006). Problems and prospects of development of nanomembrane technology. *Crystallography Reports.* **51**(5):850–862.

2130. Ying M. et al. (2012). Silicon nanomembranes for fingertip electronics. *Nanotechnology.* **23**(34):344004.

Nanometals

General Overviews

2131. Akjouj A. et al. (2013). Nanometal plasmonpolaritons. *Surface Science Reports.* **68**(1):1–67.

2132. Faester S. et al. (2012). Nanometals: status and perspective. *Risoe International Symposium on Materials Science.* Roskilde, Denmark. September 3–7.

2133. Hinther A. et al. (2010). Nanometals induce stress and alter thyroid hormone action in amphibians at or below North American water quality guidelines. *Environmental Science & Technology.* **44**(21):8314.

2134. Jensen D.J. et al. (2012). Characterization of nanometals. Available online at http://www.emc2012.org.uk/documents /Abstracts/Abstracts/EMC2012_1134.pdf

2135. Liz-Marzán L.M. (2004). Nanometals: formation and color. *Materials Today.* **7**(2):26–31.

2136. Lowe T. (2002). The revolution in nanometals. *Advanced Materials & Processes.* **160**(1):63–65.

Aluminum

2137. Babuk V. et al. (2009). Nanoaluminum as a solid propellant fuel. *Journal of Propulsion and Power.* **25**(2):482–489.
2138. Tepper F. et al. (2001). Nanosized alumina fibers. *American Ceramic Society Bulletin.* **80**(6):57–60.

Brass

2139. Kassaee M.Z. et al. (2008). Media effects on nanobrass arc fabrications. *Journal of Alloys and Compounds.* **453**(1):229–232.

Copper

2140. Li H. et al. 2007. Preparation technology of metal nanocopper. *Chemical Engineer.* **12**:011.
2141. Wang G. et al. (2012). A simple way of preparing nanocopper powders and its catalytic application to synthesize carbon nanofibers. *Journal of Ovonic Research.* **8**(2):41–45.

Gold

2142. Alkilany A.M. and Murphy C.J. (2010). Toxicity and cellular uptake of gold nanoparticles: what we have learned so far? *Journal of Nanoparticle Research.* **12**(7):2313–2333.
2143. Brust M. et al. (1995). Synthesis and reactions of functionalised gold nanoparticles. *Journal of the Chemical Society, Chemical Communications.* (16):1655–1656.
2144. Daniel M.C. and Astruc D. (2004). Gold nanoparticles: assembly, supramolecular chemistry, quantum-size-related properties, and applications toward biology, catalysis, and nanotechnology. *Chemical Reviews.* **104**(1):293–346.
2145. Dreaden E.C. et al. (2012). The golden age: gold nanoparticles for biomedicine. *Chemical Society Reviews.* **41**(7):2740–2779.

2146. Eustis S. and El-Sayed M.A. (2006). Why gold nanoparticles are more precious than pretty gold: noble metal surface plasmon resonance and its enhancement of the radiative and nonradiative properties of nanocrystals of different shapes. *Chemical Society Reviews.* **35**(3):209–217.

2147. Giasuddin A.S.M. et al. (2012). Use of gold nanoparticles in diagnostics, surgery and medicine: a review. *Bangladesh Journal of Medical Biochemistry.* **5**(2):56–60.

2148. Giljohann D.A. et al. (2010). Gold nanoparticles for biology and medicine. *Angewandte Chemie International Edition.* **49**(19):3280–3294.

2149. Goodman C.M. et al. (2004). Toxicity of gold nanoparticles functionalized with cationic and anionic side chains. *Bioconjugate Chemistry.* **15**(4):897–900.

2150. Mingos D. (2014). Historical introduction to gold colloids, clusters and nanoparticles. In *Gold Clusters, Colloids and Nanoparticles I.* Switzerland: Springer International Publishing, pp. 1–47.

2151. Murphy C.J. et al. (2008). Gold nanoparticles in biology: beyond toxicity to cellular imaging. *Accounts of Chemical Research.* **41**(12):1721–1730.

2152. Newman L.T. (1938). Studies on gold colloids. PhD Dissertation: University of Toronto.

2153. Prati L. and Villa A. (2013). Gold colloids: from quasi-homogeneous to heterogeneous catalytic systems. *Accounts of Chemical Research.* **47**(3):855–863.

2154. Saha K. et al. (2012). Gold nanoparticles in chemical and biological sensing. *Chemical Reviews.* **112**(5):2739–2779.

2155. Sardar R. et al. (2009). Gold nanoparticles: past, present, and future. *Langmuir.* **25**(24):13840–13851.

2156. Schmid G. and Lehnert A. (1989). The complexation of gold colloids. *Angewandte Chemie International Edition in English.* **28**(6):780–781.

2157. Sperling R.A. et al. (2008). Biological applications of gold nanoparticles. *Chemical Society Reviews.* **37**(9):1896–1908.

2158. Tweney R.D. (2009). Faraday's gold colloids: nanoscience in 1856. In *Meeting Abstracts*. The Electrochemical Society, No. 3, pp. 202–202.

Iron

2159. Hilty-Vančura F. (2011). Novel nanoiron and nanozinc compounds: the next generation of food fortificants? Doctoral Dissertation: Eidgenössische Technische Hochschule ETH Zürich, Nr. 19537.
2160. Yang G.C. et al. (2007). Stability of nanoiron slurries and their transport in the subsurface environment. *Separation and Purification Technology*. **58**(1):166–172.

Silver

2161. AshaRani P.V. et al. (2009). Cytotoxicity and genotoxicity of silver nanoparticles in human cells. *ACS Nano*. **3**(2):279–290.
2162. Benn T. et al. (2010). The release of nanosilver from consumer products used in the home. *Journal of Environmental Quality*. **39**(6):1875.
2163. Chen D. et al. (2007). Biological effects induced by nanosilver particles: in vivo study. *Biomedical Materials*. **2**(3):S126–S128.
2164. Chen X. and Schluesener H.J. (2008). Nanosilver: a nanoproduct in medical application. *Toxicology Letters*. **176**(1): 1–12.
2165. Faunce T. and Watal A. (2010). Nanosilver and global public health: international regulatory issues. *Nanomedicine (Lond)*. **5**(4):617–632.
2166. Lansdown A. (2006). Silver in health care: antimicrobial effects and safety in use. *Current Problems in Dermatology*. **33**:17–34.
2167. Lansdown A. (2007). Critical observations on the neurotoxicity of silver. *Critical Reviews in Toxicology*. **37**(3):237–250.
2168. Liu J. et al. (2012). Chemical transformations of nanosilver in biological environments. *ACS Nano*. **6**(11):9887–9899.

2169. Lubick N. (2008). Nanosilver toxicity: ions, nanoparticles or both? *Environmental Science & Technology.* **42**(23):8617–8617.

2170. Nowack B. et al. (2011). 120 years of nanosilver history: implications for policy makers. *Environmental Science & Technology.* **45**:1177–1183.

2171. Quadros M.E. and Marr L.C. (2010). Environmental and human health risks of aerosolized silver nanoparticles. *Journal of the Air & Waste Management Association.* **60**(7):770–781.

2172. Rai M. et al. (2009). Silver nanoparticles as a new generation of antimicrobials. *Biotechnology Advances.* **27**(1):76–83.

2173. Schafer B. et al. (2011). Nanosilver in consumer products and human health: more information required! *Environmental Science & Technology.* **45**(17):7589–7590.

2174. Schluesener J. and Schluesener H. (2013). Nanosilver: application and novel aspects of toxicology. *Archives of Toxicology.* **87**(4):569–576.

2175. Sung J.H. et al. (2009). Subchronic inhalation toxicity of silver nanoparticles. *Toxicological Sciences.* **108**(2):452–461.

Zinc

2176. Clausen C.A. et al. (2009). Feasibility of nanozinc oxide as a wood preservative. *Proceedings of American Wood Protection Association.* **105**:255–260.

2177. Gazdowicz J. et al. (2003). Structure and susceptibility to rust of nonmetallic zinc. *Corrosion Engineering, Science and Technology.* **38**(2):139–146.

Nanoparticles

2178. Aitken R.J. et al. (2004). Nanoparticles: an occupational hygiene review (Research Report 274). Sudbury, England: Health and Safety Executive, Institute of Occupational Medicine.

2179. Albrecht M.A. et al. (2006). Green chemistry and the health implications of nanoparticles. *Green Chemistry.* **8**(5):417–432.

2180. Andujar P. et al. (2011). Respiratory effects of manufactured nanoparticles. *Revue des Maladies Respiratoires.* **28**(8):e66–e75.

2181. Arvizo R.R. et al. (2012). Intrinsic therapeutic applications of noble metal nanoparticles: past, present and future. *Chemical Society Reviews.* **41**(7):2943–2970.

2182. Astruc D. et al. (2005). Nanoparticles as recyclable catalysts: the frontier between homogeneous and heterogeneous catalysis. *Angewandte Chemie International Edition.* **44**(48):7852–7872.

2183. Auffan M. et al. (2009). Towards a definition of inorganic nanoparticles from an environmental, health and safety perspective. *Nature Nanotechnology.* **4**(10):634–641.

2184. Bakand S. et al. (2012). Nanoparticles: a review of particle toxicology following inhalation exposure. *Inhalation Toxicology.* **24**(2):125–135.

2185. Balbus J.M. et al. (2007). Meeting report: hazard assessment for nanoparticles: report from an interdisciplinary workshop. *Environmental Health Perspectives.* **115**(11):1654–1659.

2186. Bansal P. et al. (2014). Biogenesis of nanoparticles: a review. *African Journal of Biotechnology.* **13**(28):2778–2785.

2187. Bang J.J. and Murr L.E. (2002). Collecting and characterizing atmospheric nanoparticles. *JOM.* **54**(12):28.

2188. Barcikowski S. et al. (2009). Impact and structure of literature on nanoparticle generation by laser ablation in liquids. *Journal of Nanoparticle Research.* **11**(8):1883–1893.

2189. Baun A. et al. (2008). Ecotoxicity of engineered nanoparticles to aquatic invertebrates: a brief review and recommendations for future toxicity testing. *Ecotoxicology.* **17**:387–395.

2190. Belluci S. (ed.) (2009). *Nanoparticles and Nanodevices in Biological Applications* (The INFN Lectures, vol. 1). Berlin: Springer-Verlag.

2191. Berube D.M. et al. (2011). Comparing nanoparticle risk perceptions too other known EHS risks. *Journal of Nanoparticle Research.* **13**(8):3089–3099.

2192. Biskos G. and Schmidt-Ott A. (2012). Airborne engineered nanoparticles: potential risks and monitoring challenges for assessing their impacts on children. *Paediatric Respiratory Reviews.* **13**(2):79–83.

2193. Biswas P. and Wu CY. (2005). Critical review: nanoparticles and the environment. *Journal of the Air & Waste Management Association.* **55**(6):708–746.

2194. Bitounis D. et al. (2016). Detection and analysis of nanoparticles in patients: a critical review of the status quo of clinical nanotoxicology. *Biomaterials.* **76**:302–312.

2195. Borm P.J. and Kreyling W. (2004). Toxicological hazards of inhaled nanoparticles–potential implications for drug delivery. *Journal of Nanoscience & Nanotechnology.* **4**:521–531.

2196. Bouzigues C. et al. (2011). Biological applications of rare earth-based nanoparticles. *ACS Nano.* **5**(11):8488–8505.

2197. Brigger I. et al. (2002). Nanoparticles in cancer therapy and diagnosis. *Advanced Drug Delivery Reviews.* **54**(5):631–651.

2198. Brock J.R. (1998). Nanoparticle synthesis: a key process in the future of nanotechnology. In *Nanostructured Materials.* Netherlands: Springer, pp. 1–14.

2199. Bujak-Pietrek S. (2010). [Occupational exposure to nanoparticles. Assessment of workplace exposure]. *Medycyna Pracy.* **61**(2):183–189 (Polish).

2200. Burello E. and Worth A. (2011). Computational nanotoxicology: predicting toxicity of nanoparticles. *Nature Nanotechnology.* **6**(3):138–139.

2201. Burleson D.J. et al. (2004). On the characterization of environmental nanoparticles. *Journal of Environmental Science and Health. Part A, Toxic/Hazard Substances & Environmental Engineering.* **39**(10):2707–2753.

2202. Bystrzejewska-Piotrowska G. et al. (2009). Nanoparticles: their potential toxicity, waste and environmental management (literature review). *Waste Management.* **29**(9):2587–2595.

2203. Candeloro P. et al. (2011). Nanoparticle microinjection and Raman spectroscopy as tools for nanotoxicology studies. *Analyst.* **136**(21):4402–4408.

2204. Castranova V. (2011). Overview of current toxicological knowledge of engineered nanoparticles. *Journal of Occupational Medicine.* **53**(6 Suppl):S14–S17.

2205. Chan W.C.W. (ed.) (2007). *Bio-Applications of Nanoparticles.* New York: Springer.

2206. Chen Y. (2010). Lipid bilayer-nanoparticle interactions in nanotoxicology and nanomedicine. Doctoral Dissertation: University of Rhode Island.

2207. Chidambaram M. and Krishnasamy K. (2012). Nanotoxicology: toxicity of engineered nanoparticles and approaches to produce safer nanotherapeutics. *International Journal of Pharmaceutical Sciences.* **2**:117–122.

2208. Choksi A.N. et al. (2010). Nanoparticles: a closer look at their dermal effects. *Journal of Drugs in Dermatology.* **9**(5):475–481.

2209. Crosera M. et al. (2009). Nanoparticle dermal absorption and toxicity: a review of the literature. *International Archives of Occupational and Environmental Health.* **82**(9):1043–1055.

2210. Couvreur P. (2013). Nanoparticles in drug delivery: past, present and future. *Advanced Drug Delivery Review.* **65**(1):21–23.

2211. Chung D. (2003). Nanoparticles have health benefits too. *New Scientist.* **179**(2410):16.

2212. Crosera M. et al. (2009). Nanoparticle dermal absorption and toxicity: a review of the literature. *International Archives of Occupational and Environmental Health.* **82**(9):1043–1055.

2213. De M. et al. (2008). Applications of nanoparticles in biology. *Advanced Materials.* **20**(22):4225–4241.

2214. Devuyst O. and Schumann A. (2015). Peritoneal dialysis: nanoparticles have entered the game. *Peritoneal Dialysis International.* **35**:240.

2215. Diebold Y. and Calonge M. (2010). Applications of nanoparticles in ophthalmology. *Progress in Retinal & Eye Research.* **29**(6):596–609.

2216. Dieni C.A. et al. (2013). Spherical gold nanoparticles impede the function of bovine serum albumin in vitro: a new consideration for studies in nanotoxicology. *Journal of Nanomaterials & Molecular Nanotechnology.* **2**:30–90.

2217. Di Gioacchino M. et al. (2011). Immunotoxicity of nanoparticles. *International Journal of Immunopathology and Pharmacology.* **24**(1 Suppl):65S–71S.

2218. Dobrovolskaia M.A. et al. (2009). Evaluation of nanoparticle immunotoxicity. *Nature Nanotechnology.* **4**(7):411–414.

2219. Doktorovova S. et al. (2014). Nanotoxicology applied to solid lipid nanoparticles and nanostructured lipid carriers– a systematic review of in vitro data. *European Journal of Pharmaceutics and Biopharmaceutics.* **87**(1):1–18.

2220. Douglas S.J. et al. (1986). Nanoparticles in drug delivery. *Critical Reviews in Therapeutic Drug Carrier Systems.* **3**(3):233–261.

2221. Dreher K.L. (2004). Health and environmental impact of nanotechnology: toxicological assessment of manufactured nanoparticles. *Toxicological Sciences.* **77**(1):3–5.

2222. Duffin R. et al. (2007). Nanoparticles-a thoracic toxicology perspective. *Yonsei Medical Journal.* **48**:561–572.

2223. Dwivedi A.D. and Ma L.Q. (2014). Biocatalytic synthesis pathways, transformation, and toxicity of nanoparticles in the environment. *Critical Reviews in Environmental Science & Technology.* **44**(15):1679–1739.

2224. Dwivedi P.D. et al. (2011). Impact of nanoparticles on the immune system. *Journal of Biomedical Nanotechnology.* **7**(1):193–194.

2225. Ekambaram P. et al. (2012). Solid lipid nanoparticles: a review. *Scientific Reviews & Chemical Communications.* **2**(1):80–102.

2226. El-Ansary A. and Faddah L.M. (2010). Nanoparticles as biochemical sensors. *Nanotechnology, Science & Applications.* **3**:65–76.

2227. Elsaesser A. and Howard C.V. (2012). Toxicology of nanoparticles. *Advanced Drug Delivery Reviews.* **64**(2):129–137.

2228. Fadeel B. (2012). Clear and present danger? Engineered nanoparticles and the immune system. *Swiss Medical Weekly.* **142**:13609.

2229. Farmer L. and Graff A. (2007). Toxicological effects of nanoparticles on human cells. *Specialty Chemicals.* **27**(7):36.

2230. Feng X. et al. (2015). Central nervous system toxicity of metallic nanoparticles. *International Journal of Nanomedicine.* **10**:4321–4340.

2231. Ferreira L. (2009). Nanoparticles as tools to study and control stem cells. *Journal of Cellular Biochemistry.* **108**(4):746–752.

2232. Ferreira A.J. et al. (2013). Nanoparticles, nanotechnology and pulmonary nanotoxicology. *Revista Portuguesa de Pneumologia (English Edition).* **19**(1):28–37.

2233. Frimmel F.H. and Niessner R. (eds.) (2010). *Nanoparticles in the Water Cycle: Properties, Analysis and Environmental Relevance.* New York: Springer.

2234. Fröhlich E. et al. (2014). Use of whole genome expression analysis in the toxicity screening of nanoparticles. *Toxicology and Applied Pharmacology.* **280**(2):272–284.

2235. Gade A. et al. (2010). Mycogenic nanoparticles: progress and applications. *Biotechnology Letters.* **32**(5):593–600.

2236. Grillo R. et al. (2015). Engineered nanoparticles and organic matter: a review of the state-of-the-art. *Chemosphere.* **119**:608–619.

2237. Guarnieri D. et al. (2014). Transport across the cell-membrane dictates nanoparticle fate and toxicity: a new paradigm in nanotoxicology. *Nanoscale.* **6**(17):10264–10273.

2238. Gwinn M.R. and Vallyathan V. (2006). Nanoparticles: health effects—pros and cons. *Environmental Health Perspectives.* **114**:1818–1825.

2239. Hadrup N. et al. (2015). Toxicological risk assessment of elemental gold following oral exposure to sheets and nanoparticles: a review. *Regulatory Toxicology & Pharmacology.* **72**(2):216–221.

2240. Hagens W.I. et al. (2007). What do we (need to) know about the kinetic properties of nanoparticles in the body? *Regulatory Toxicology and Pharmacology.* **49**(3):217–229.

2241. Halford B. (2005). Questioning common perceptions about nanoparticle toxicity. *Chemical & Engineering News.* **83**(51): 52.

2242. Handy R.D. and Shaw B.J. (2007). Toxic effects of nanoparticles and nanomaterials: implications for public health, risk assessment and the public perception of nanotechnology. *Health, Risk & Society.* **9**(2):125–144.

2243. Handy R.D. et al. (2008). The ecotoxicology and chemistry of manufactured nanoparticles. *Ecotoxicology.* **17**(4):287–314.

2244. Hansen S.F. et al. (2007). Categorization framework to aid Hazard Identification of Nanomaterials. *Nanotoxicology.* **1**:243–250.

2245. Hansen S.F. et al. (2008). Categorization framework to aid exposure assessment of nanomaterials in consumer products. *Ecotoxicology.* **17**:438–447.

2246. Hardman R. (2006). A toxicologic review of quantum dots: toxicity depends on physicochemical and environmental factors. *Environmental Health Perspectives.* **114**:165–172.

2247. Hartman J. (2002). Nanoparticles help cut groundwater cleanup costs. *Civil Engineering.* **72**(5):33.

2248. Hoet P.H. et al. (2004). Nanoparticles–known and unknown health risks. *Journal of Nanobiotechnology.* **2**(1):12.

2249. Hristozov D. and Malsch I. (2009). Hazards and risks of engineered nanoparticles for the environment and human health. *Sustainability.* **1**:1161–1194.

2250. Hu Y.L. and Gao J.Q. (2010). Potential neurotoxicity of nanoparticles. *International Journal of Pharmaceutics.* **394**(1–2): 115–121.

2251. Indira T.K. and Lakshmi P.K. (2010). Magnetic nanoparticles: a review. *International Journal of Pharmaceutical Science and Nanotechnology.* **3**(3):1035–1042.

2252. Ingle A. P. et al. (2014). Bioactivity, mechanism of action, and cytotoxicity of copper-based nanoparticles: a review. *Applied Microbiology & Biotechnology.* **98**(3):1001–1009.

2253. Jadhav N. et al. (2014). Herbal nanoparticles: a patent review. *Asian Journal of Pharmaceutics.* **8**(1):58–69.

2254. Jennifer M. and Maciej W. (2013). Nanoparticle technology as a double-edged sword: cytotoxic, genotoxic and epigenetic effects on living cells. *Journal of Biomaterials & Nanobiotechnology.* **4**(1): Article ID 26984.

2255. Jennings T. and Strouse G. (2007). Past, present and future of gold nanoparticles. *Advances in Experimental Medicine & Biology.* **620**:34–47.

2256. Journeay W.S. et al. (2008). High-aspect ratio nanoparticles in nanotoxicology. *Integrated Environmental Assessment and Management.* **4**(1):128–129.

2257. Ju-Nam Y. and Lead J. (2008). Manufactured nanoparticles: an overview of their chemistry, interactions and potential environmental implications. *Science of the Total Environment.* **400**:396–414.

2258. Kanagesan S. et al. (2013). Synthesis, characterization and cytotoxicity of iron oxide nanoparticles. *Advances in Materials Science and Engineering.* **2013**: Article ID 710432.

2259. Kandlikar M. et al. (2007). Health risk assessment for nanoparticles: a case for using expert judgement. *Journal of Nanoparticle Research.* **9**:137–156.

2260. Kelly R.J. (2009). Occupational medicine implications of engineered nanoscale particulate matter. *Journal of Chemical Health and Safety.* **16**:24–39.

2261. Kessler R. (2011). Engineered nanoparticles in consumer products. *Environmental Health Perspectives.* **119**(3):A120–A125.

2262. Khan F.H. (2013). Chemical hazards of nanoparticles to human and environment. *Oriental Journal of Chemistry.* **29**(4):1399–1408.

2263. Khanna V. and Bakshi B.R. (2006). Towards a systems view in nanotechnology-life cycle assessment of nanoparticles synthesis. In *American Institute of Chemical Engineers Annual Meeting*, vol. 15.

2264. Kim H.A. et al. (2012). Nanoparticles in the environment: stability and toxicity. *Reviews on Environmental Health.* **27**(4):175–179.

2265. Kittlelson D.B. (1998). Engines and nanoparticles: a review. *Journal of Aerosol Science.* **29**(5–6):575–588.

2266. Klaine S. et al. (2008). Critical review: nanomaterials in the environment: behavior, fate, bioavailability, and effects. *Environmental Toxicology and Chemistry.* **27**(9):1825–1851.

2267. Kreuter J. (1988). Possibilities of using nanoparticles as carriers for drugs and vaccines. *Journal of Microencapsulation.* **5**(2):115–127.

2268. Kreuter J. (2007). Nanoparticles: a historical perspective. *International Journal of Pharmaceutics.* **331**(1):1–10.

2269. Krüger G. (2005). Nanoparticles in contact adhesives. *International Polymer Science & Technology.* **32**(11):T14–T17.

2270. Lademann J. et al. (2009). [How safe are nanoparticles?]. *Der Hautarzt.* **60**(4):305–309 (German).

2271. Lau B.L.T. (2011). Understanding how nanoparticles behave in natural and engineered waters. *AWWA Journal.* **103**(11):20–22.

2272. Le Corre D. et al. (2010). Starch nanoparticles: a review. *Biomacromolecules.* **11**(5):1139–1153.

2273. Lee S.B. (2011). Nanotoxicology: toxicity and biological effects of nanoparticles for new evaluation standards. *Nanomedicine.* **6**(5):759.

2274. Liu W.T. (2006). Nanoparticles and their biological and environmental applications. *Journal of Bioscience and Bioengineering.* **102**(1):1–7.

2275. Ma H. et al. (2013). Ecotoxicity of manufactured ZnO nanoparticles: a review. Environmental Pollution. **172**:76–85.

2276. MacPhail R.C. et al. (2013). Assessing nanoparticle risk poses prodigious challenges. *Wiley Interdisciplinary Reviews: Nanomedicine and Nanobiotechnology.* **5**(4):374–387.

2277. Madl A.K. and Pinkerton K.E. (2009). Health effects of inhaled engineered and incidental nanoparticles. *Critical Reviews in Toxicology.* **39**(8):629–658.

2278. Maeng S.H. and Yu I.J. (2005). The concepts of nanotoxicology and risk assessment of the nanoparticles. *Toxicological Research.* **21**(2):87–98.

2279. Malam Y. et al. (2011). Current trends in the application of nanoparticles in drug delivery. *Current Medicinal Chemistry.* **18**(7):1067–1078.

2280. Marano F. et al. (2011). Nanoparticles: molecular targets and cell signaling. *Archives of Toxicology.* **85**(7):733–741.

2281. Marijnissen J.C. and Gradon J. (eds.) (2009). *Nanoparticles in Medicine and Environment: Inhalation and Health Effects.* New York: Springer.

2282. Martin J. et al. (2015). Occupational exposure to nanoparticles at commercial photocopy centers. *Journal of Hazardous Materials.* **298**:351–360.

2283. Matsui I. (2005). Nanoparticles for electronic device applications: a brief review. *Journal of Chemical Engineering of Japan.* **38**(8):535–546.

2284. Maurer-Jones M.A. et al. (2013). Toxicity of engineered nanoparticles in the environment. *Analytical Chemistry.* **85**(6):3036–3049.

2285. Maynard A.D. and Pui D.Y.H. (eds.) (2012). *Nanoparticles and Occupational Health.* Dordrecht, The Netherlands: Springer.

2286. Medina C. et al. (2007). Nanoparticles: pharmacological and toxicological significance. *British Journal of Pharmacology.* **150**(5):552–558.

2287. Messing M. et al. (2011). Spark generated particles for nano-toxicology studies. In *European Aerosol Conference (EAC)*, pp. 1–1.

2288. Misra R. et al. (2013). Nanoparticles as carriers for chemotherapeutic drugs: a review. *Journal of Nanopharmaceutics and Drug Delivery.* **1**(2):103–137.

2289. Mody V. et al. (2010). Introduction to metallic nanoparticles. *Journal of Pharmacy & Bioallied Sciences.* **2**(4):282–289.

2290. Mohan D. et al. (2013). Toxicity of exhaust nanoparticles. *African Journal of Pharmacy and Pharmacology.* **7**(7):318–331.

2291. Mohanraj V.J. and Chen Y. (2006). Nanoparticles: a review. *Tropical Journal of Pharmaceutical Research.* **5**(1):561–573.

2292. Mohnen V. and Hidy G.M. (2010). Measurements of atmospheric nanoparticles (1875–1980). *Bulletin of the American Meteorological Society.* **91**(11):1525–1539.

2293. Moore M.N. (2006). Do nanoparticles present ecotoxicological risks for the health of the aquatic environment? *Environment International.* **32**(8):967–976.

2294. Morawska L. et al. (2009). JEM spotlight: environmental monitoring of airborne nanoparticles. *Journal of Environmental Monitoring.* **11**(10):1758–1773.

2295. Morgan K. (2005). Development of a preliminary framework for informing the risk analysis and risk management of nanoparticles. *Risk Analysis.* **25**(6):1621–1635.

2296. Mueller N. and Nowack B. (2008). Exposure modeling of engineered nanoparticles in the environment. *Environmental Science & Technology.* **42**:4447–4453.

2297. Murthy S.K. (2007). Nanoparticles in modern medicine: state of the art and future challenges. *International Journal of Nanomedicine.* **2**(2):129–141.

2298. Musumeci A.W. et al. (2009). Synthesis and characterization of dual radiolabeled layered double hydroxide nanoparticles for use in in vitro and in vivo nanotoxicology studies. *The Journal of Physical Chemistry C.* **114**(2):734–740.

2299. Naha P.C. et al. (2015). Systematic in vitro toxicological screening of gold nanoparticles designed for nanomedicine applications. *Toxicol In Vitro.* **29**(7):1445–1453.

2300. Navarro E. et al. (2008). Environmental behavior and ecotoxicity of engineered nanoparticles to algae, plants and fungi. *Ecotoxicology.* **17**(5):372–386.

2301. NIOSH (2009). Current intelligence bulletin 60: interim guidance for medical screening and hazard surveillance for workers potentially exposed to engineered nanoparticles. Cincinnati, OH: U.S. Department of Health and Human Services, Centers for Disease Control and Prevention, National Institute for Occupational Safety and Health.

2302. Norwegian Pollution Control Authority (2008). Environmental fate and ecotoxicity of engineered nanoparticles (Report No. TA 2304/2007).

2303. Nowack B. and Bucheli T.D. (2007). Occurrence, behavior and effects of nanoparticles in the environment. *Environmental Pollution.* **150**(1):5–22.

2304. Nowack B. and Mueller N. (2008). Is anything out there? What life cycle perspectives of nanoproducts can tell us about nanoparticles in the environment. *Environmental Science & Technology.* **42**:4447–4453.

2305. Oberdörster G. et al. (2007a). Concepts of nanoparticle dose metric and response metric. *Environmental Health Perspectives.* **115**(6):A290.

2306. Oberdörster G. et al. (2007b). Toxicology of nanoparticles: a historical perspective. *Nanotoxicology.* **1**(1):2–25.

2307. Ogden L.E. (2013). Nanoparticles in the environment: tiny size, large consequences? *BioScience.* **63**(3):236.

2308. Pankhurst Q.A. et al. (2003). Applications of magnetic nanoparticles in biomedicine. *Journal of Physics D: Applied Physics.* **36**(13):R167–R181.

2309. Papakostas D. et al. (2011). Nanoparticles in dermatology. *Archives of Dermatological Research.* **303**(8):533–550.

2310. Park S. et al. (2014). Regulatory ecotoxicity testing of engineered nanoparticles: are the results relevant to the natural environment. *Nanotoxicology.* **8**(5):583–592.

2311. Parveen S. et al. (2012). Nanoparticles: a boon to drug delivery, therapeutics, diagnostics and imaging. *Nanomedicine.* **8**(2):147–166.

2312. Pauwels E.K. and Erba P. (2007). Towards the use of nanoparticles in cancer therapy and imaging. *Drug News & Perspectives.* **20**(4):213–220.

2313. Pelaz B. et al. (2012). The state of nanoparticle-based nanoscience and biotechnology: progress, promises, and challenges. *ACS Nano.* **6**(10):8468–8483.

2314. Penn S.G. et al. (2003). Nanoparticles for bioanalysis. *Current Opinion in Chemical Biology.* **7**(5):609–615.

2315. Pietroiusti A. and Magrini A. (2015). Engineered nanoparticles at the workplace: current knowledge about worker's risk. *Occupational Medicine.* **65**:171–173.

2316. Poater A. et al. (2010). Computational methods to predict the reactivity of nanoparticles through structure property relationships. *Expert Opinion on Drug Delivery.* 7:295–305.

2317. Powers K.W. et al. (2006). Research strategies for safety evaluation of nanomaterials. part vi. characterization of nanoscale particles for toxicological evaluation. *Toxicological Sciences.* **90**(2):296–303.

2318. Prosie F. et al. (2008). [Nanoparticles: structures, utilizations and health impacts]. *Presse Medicale.* **37**(10):1431–1437.

2319. Quadros M.E. and Marr L.C. (2010). Environmental and human health risks of aerosolized silver nanoparticles. *Journal of the Air & Waste Management Association.* **60**(7):770–781.

2320. Quan M. (2000). Research into silicon nanoparticles heats up. *Electronic Engineering Times.* **1105**:73.

2321. Quebec Institue de Recherce (2008). Best practices guide to synthetic nanoparticle risk management (Guide de bonnes pratiques favorisant la gestion des risques reliés aux nanoparticules de synthèse). Available online at

http://www.irsst.qc.ca/en/publications-tools/publication/
i/100432/n/best-practices-guide-to-synthetic-nanoparticle-
risk-management-r-599

2322. Richter V. et al. (2008). Evaluation of health risks of nano-and
microparticles. *Powder Metallurgy.* **51**(1):8–9.

2323. Rössler A. et al. (2001). Nanopartikel - Materialien der
Zukunft. *Chemie in unserer Zeit.* **35**(1):32–41.

2324. Rushton E.K. et al. (2010). Concept of assessing nanoparticle
hazards considering nanoparticle dosemetric and chemi-
cal/biological response metrics. *Journal of Toxicology and
Environmental Health A.* **73**:445–461.

2325. Robalo-Cordeiro C. et al. (2012). Nanoparticles, nanotechnology and pulmonary nanotoxicology. *Revista Portugesa de
Pneumologia.* **120**:1–10.

2326. Sa L.T.M. et al. (2012). Biodistribution of nanoparticles:
Initial considerations. *Journal of Pharmaceutical & Biomedical
Analysis.* **70**:602–604.

2327. Salata O.V. (2004). Applications of nanoparticles in biology
and medicine. *Journal of Nanobiotechnology.* **2**:3.

2328. Saliner A.G. et al. (2008). Review of computational
approaches for predicting the physicochemical and biological
properties of nanoparticles. *Scientific and Technical Research
Reports.* Available at http://publications.jrc.ec.europa.eu/
repository/handle/JRC52120

2329. Sanvicens N. and Marco M.P. (2008). Multifunctional
nanoparticles—properties and prospects for their use in
human medicine. *Trends in Biotechnology.* **26**(8):425–433.

2330. Sardar R. et al. (2009). Gold nanoparticles: past, present, and
future. *Langmuir.* **25**(24):13840–13851.

2331. Sattler K.D. (ed.) (2010). *Handbook of Nanophysics: Nanoparticles and Quantum Dots.* Boca Raton, FL: CRC Press.

2332. Schmid G. (ed.) (2011). *Nanoparticles: From Theory to
Application*, 2nd edition. New York: Wiley.

2333. Schmid K. et al. (2010). Nanoparticle usage and protection
measures in the manufacturing industry: a representative

survey. *Journal of Occupational and Environmental Hygiene.* **7**:224–232.

2334. Schneider T. (2008). Editorial: Role of occupational hygiene research in the control of occupational health risks from engineered nanoparticles. *Scandinavian Journal of Work, Environment & Health.* **34**(6):407–409.

2335. Schrand A.M. et al. (2010). Metal-based nanoparticles and their toxicity assessment. *Wiley Interdisciplinary Reviews: Nanomedicine and Nanobiotechnology.* **2**(5):544–568.

2336. Schulte P. et al. (2008). Occupational risk management of engineered nanoparticles. *Journal of Occupational & Environmental Hygiene.* **5**(4):239–249.

2337. Scown T.M. et al. (2010). Review: do engineered nanoparticles pose a significant threat to the aquatic environment? *Critical Reviews in Toxicology.* **40**(7):653–670.

2338. Seipenbusch M. et al. (2008). Temporal evolution of nanoparticle aerosols in workplace exposure. *Annals of Occupational Hygiene.* **52**:707–716.

2339. Simonet B.M. and Valcarcel M. (2009). Monitoring nanoparticles in the environment. *Analytical & Bioanalytical Chemistry.* **393**(1):17–21.

2340. Soloviev M. (2007). Nanobiotechnology today: focus on nanoparticles. *Journal of Nanobiotechnology.* **5**:11.

2341. Soni D. et al. (2015). Release, transport and toxicity of engineered nanoparticles. *Reviews of Environmental Contamination & Toxicology.* **234**:1–47.

2342. Strand R. and Kjølberg K.L. (2011). Regulating nanoparticles: the problem of uncertainty. *European Journal of Law & Technology.* **2**(3).

2343. Stratmeyer M.E. et al. (2010). What we know and don't know about the bioeffects of nanoparticles: developing experimental approaches for safety assessment. *Biomedical Microdevices.* **12**(4):569–573.

2344. Subbiah R. et al. (2010). Nanoparticles: functionalization and multifunctional applications in biomedical sciences. *Current Medicinal Chemistry.* **17**(36):4559–4577.

2345. Swidwińska-Gajewska A.M. (2007a). [Nanoparticles (part 1)—the product of modern technology and new hazards in the work environment]. *Medycyna Pracy.* **58**(3):243–251 (Polish).

2346. Swidwińska-Gajewska A.M. (2007b). [Nanoparticles (part 2)—advantages and health risk]. *Medycyna Pracy.* **58**(3): 253–263 (Polish).

2347. Tervonen T. et al. (2009). Risk-based classification system of nanomaterials. *Journal of Nanoparticle Research.* **11**:757–766.

2348. Tiede K. et al. (2008). Detection and characterization of engineered nanoparticles in food and the environment. *Food Additives & Contaminants. Part A, Chemistry, Analysis Control, Exposure & Risk Assessment.* **7**:795–821.

2349. Tonga G.Y. et al. (2014). 25th anniversary article: interfacing nanoparticles and biology: new strategies for biomedicine. *Advanced Materials.* **26**(3):359–370.

2350. Toth-Fejel T. (2007). From nanoparticles to productive nanosystems: bridging the gap. *Rapid Prototyping.* **13**(3):1–24.

2351. Toxics Use Reduction Institute (2012). Best practices for working safely with nanoparticles in university research laboratories. Lowell, MA.

2352. Tsao T.M. et al. (2011). Origin, separation and identification of environmental nanoparticles: a review. *Journal of Environmental Monitoring.* **13**(5):1156–1163.

2353. Ugarte D. (1994). Graphitic nanoparticles. *MRS Bulletin.* **19**(11):39–42.

2354. van Broekhuizen P. et al. (2012). Exposure limits for nanoparticles: report of an international workshop on nano reference values. *Annals of Occupational Hygiene.* **56**(5):515–524.

2355. Vishwakarma V. et al. (2010). Safety and risk associated with nanoparticles-a review. *Journal of Minerals and Materials Characterization and Engineering.* **9**(5):455.

2356. Voight N. et al. (2014). Toxicity of polymeric nanoparticles in vivo and in vitro. *Journal of Nanoparticle Research.* **16**(6):1–13.

2357. Warheit D.B. et al. (2008). Health effects related to nanoparticle exposures: environmental, health and safety considerations for assessing hazards and risks. *Pharmacology & Therapeutics.* 120(1):35–42.

2358. Webster T.J. (ed.) (2009). *Safety of Nanoparticles: From Manufacturing to Medical Applications.* New York: Springer Science.

2359. Wilkinson C. et al. (2007). From uncertainty to risk? Scientific and news media portrayals of nanoparticle safety. *Health, Risk & Society.* **9**(2):145–157.

2360. Win-Shwe T.T. and Fujimaki H. (2011). Nanoparticles and neurotoxicity. *International Journal of Molecular Sciences.* **12**(9):6267–6280.

2361. Worth A. (2007). Computational nanotoxicology–towards a structure-activity based paradigm for investigation the activity of nanoparticles. In *Icon Workshop. Towards Predicting Nano-Bio Interactions,* Zurich, Switzerland.

2362. Xia T. et al. (2009). Potential health impact of nanoparticles. *Annual Review of Public Health.* **30**:137–150.

2363. Xie H. et al. (2011). Genotoxicity of metal nanoparticles. *Reviews on Environmental Health.* **26**(4):251–268.

2364. Xu G. and Zhang N. (2009). Nanoparticles for gene delivery: a brief patent review. *Recent Patents on Drug Delivery & Formulation.* **3**(2):125–136.

2365. Xu M. et al. (2013). Challenge to assess the toxic contribution of metal cation released from nanomaterials for nanotoxicology–the case of ZnO nanoparticles. *Nanoscale.* **5**(11):4763–4769.

2366. Yah C.S. et al. (2012). Nanoparticles toxicity and their routes of exposures. *Pakistan Journal of Pharmaceutical Sciences.* **25**(2):477–491.

2367. Yang B. et al. (2010). Systems toxicology used in nanotoxicology: mechanistic insights into the hepatotoxicity of nano-copper particles from toxicogenomics. *Journal of Nanoscience and Nanotechnology.* **10**(12):8527–8537.

2368. Yang R.S. et al. (2010). Pharmacokinetics and physiologically-based pharmacokinetic modeling of nanoparticles. *Journal of Nanoscience and Nanotechnology.* **10**(12):8482–8490.

2369. Yang W. et al. (2008). Inhaled nanoparticles: a current review. *International Journal of Pharmaceutics.* **356**(1/2):239–247.

2370. Yarnell A. (2002). Nanoparticle guided missiles. *Chemical & Engineering News.* **80**(26):6.

2371. Youns M. et al. (2011). Therapeutic and diagnostic applications of nanoparticles. *Current Drug Targets.* **12**(3):357–365.

2372. Zalk D.M. et al. (2009). Evaluating the control banding nanotool: a qualitative risk assessment method for controlling nanoparticle exposure. *Journal of Nanoparticle Research.* **11**:1685–1704.

2373. Zhang H. et al. (2013). [Eco-toxicological effect of metal-based nanoparticles on plants: research progress]. *Ying Yong Sheng Tai Xue Bao.* **24**(3):885–892 (Chinese).

2374. Zolnik B.S. et al. (2010). Minireview: nanoparticles and the immune system. *Endocrinology.* **151**(2):458–465.

Nanoplates

2375. Assadi A., Farshi B. and Alinia-Ziazi A. (2010). Size dependent dynamic analysis of nanoplates. *Journal of Applied Physics.* **107**(12):124310.

2376. Chen S. and Carroll D.L. (2002). Synthesis and characterization of truncated triangular silver nanoplates. *Nano Letters.* **2**(9):1003–1007.

2377. Pastoriza-Santos I. and Liz-Marzán L.M. (2008). Colloidal silver nanoplates. State of the art and future challenges. *Journal of Materials Chemistry.* **18**(15):1724–1737.

2378. Xie J., Lee J.Y., Wang D.I. and Ting Y.P. (2007). Silver nanoplates: from biological to biomimetic synthesis. *ACS Nano.* **1**(5):429–439.

Nanopolymers

2379. Dangtungee R. et al. (2015). Silver nanopolymer composites: production and efficiency. *Mechanics of Composite Materials.* **51**(2):239–244.

2380. Kambe N. et al. (2001). Refractive index engineering of nano-polymer composites. *MRS Proceedings.* **676**:Y8–Y22.

2381. Larena A. et al. (2008). Classification of nanopolymers. *Journal of Physics: Conference Series.* **100** (1):012023.

2382. Tong R. et al. (2009). Nanopolymeric therapeutics. *MRS Bulletin.* **34**(06):422–431.

2383. Xavier R.H. and Xavier D.G. (2015). Nano polymer composites for engineering application. *Journal of Chemical and Pharmaceutical Sciences.* **11**:50–51.

Nanopores

2384. Baker L.A. and Bird S.P. (2008). Nanopores: a makeover for membranes. *Nature Nanotechnology.* **3**(2):73–74.

2385. Beckstein O. et al. (2001). A hydrophobic gating mechanism for nanopores. *The Journal of Physical Chemistry B.* **105**(51): 12902–12905.

2386. Bhatia S.K. et al. (2011). Molecular transport in nanopores: a theoretical perspective. *Physical Chemistry Chemical Physics.* **13**(34):15350–15383.

2387. Branton D. (2008). The potential and challenges of nanopore sequencing. *Nature Biotechnology.* **26**:1146–1153.

2388. Dekker C. (2007). Solid-state nanopores. *Nature Nanotechnology.* **2**(4):209–215.

2389. Demming A. (2012). Nanopores—the 'Holy Grail' in nanotechnology research. *Nanotechnology.* **23**(25):250201.

2390. Gyurcsányi R.E. (2008). Chemically-modified nanopores for sensing. *Trends in Analytical Chemistry.* **27**(7):627–639.

2391. Howorka S. and Siwy Z.S. (2012). Nanopores as protein sensors. *Nature Biotechnology.* **30**(6):506–507.

2392. Iqbal S.M. and Bashir R. (eds.) (2011). *Nanopores: Sensing and Fundamental Biological Interactions.* New York: Springer Science & Business Media.

2393. Keyser U.F. (2011). Controlling molecular transport through nanopores. *Journal of the Royal Society Interface.* **8**:369–1378.

2394. Kowalczyk S.W. et al. (2011). Biomimetic nanopores: learning from and about nature. *Trends in Biotechnology.* **29**(12):607–614.

2395. Kumar H. et al. (2011). Biopolymers in nanopores: challenges and opportunities. *Soft Matter.* **7**(13):5898–5907.

2396. Mafe S. et al. (2010). Gating of nanopores: modeling and implementation of logic gates. *The Journal of Physical Chemistry C.* **114**(49):21287–21290.

2397. Miles B.N. et al. (2013). Single molecule sensing with solid-state nanopores: novel materials, methods, and applications. *Chemical Society Reviews.* **42**(1):15–28.

2398. Schneider G.F. and Dekker C. (2012). DNA sequencing with nanopores. *Nature Biotechnology.* **30**(4):326–328.

2399. Siwy Z.S. and Howorka S. (2010). Engineered voltage-responsive nanopores. *Chemical Society Reviews.* **39**(3):1115–1132.

2400. Stein D. (2007). Nanopores: molecular ping-pong. *Nature Nanotechnology.* **2**(12):741–742.

2401. Stoloff D.H. and Wanunu M. (2013). Recent trends in nanopores for biotechnology. *Current Opinion in Biotechnology.* **24**(4):699–704.

2402. Wanunu M. (2012). Nanopores: a journey towards DNA sequencing. *Physics of Life Reviews.* **9**(2):125–158.

2403. Wanunu M. and Meller A. (2007). Chemically modified solid-state nanopores. *Nano Letters.* **7**(6):1580–1585.

Nanopowders

2404. Bouillard J. et al. (2010). Ignition and explosion risks of nanopowders. *Journal of Hazardous Materials.* **181**(1):873–880.

2405. Brunner T.J. et al. (2006). Glass and bioglass nanopowders by flame synthesis. *Chemical Communications.* (13):1384–1386.

2406. Delplancke J.L. et al. (1996). Production of magnetic nanopowders by pulsed sonoelectrochemistry. *MRS Proceedings.* **451**:383.

2407. Dobashi R. (2009). Risk of dust explosions of combustible nanomaterials. *Journal of Physics: Conference Series.* **170**(1).

2408. Dufaud O. et al. (2011). Ignition and explosion of nanopowders: something new under the dust. *Journal of Physics: Conference Series.* **304**(1). 012076.

2409. Holbrow P. et al. (2010). Fire and explosion properties of nanopowders (RR–782). London: UK Health and Safety Executive.

2410. Jones D.E. et al. (2003). Hazard characterization of aluminum nanopowder compositions. *Propellants, Explosives, Pyrotechnics.* **28**(3):120–131.

2411. Kalyanaraman R. et al. (1998). Synthesis and consolidation of iron nanopowders. *Nanostructured Materials.* **10**(8):1379–1392.

2412. Kim K.C. et al. (2008). Synthesis and characterization of magnetite nanopowders. *Current Applied Physics.* **8**(6):758–760.

2413. Kwok Q.S. et al. (2002). Characterization of aluminum nanopowder compositions. *Propellants Explosives Pyrotechnics.* **27**(4):229–240.

2414. Ma H. et al. (2013). Ecotoxicity of manufactured ZnO nanoparticles: a review. *Environmental Pollution.* **172**:76–85.

2415. McCormick P.G. et al. (2001). Nanopowders synthesized by mechanochemical processing. *Advanced Materials.* **13**(12–13):1008–1010.

2416. Osipov V.V. et al. (2006). Laser synthesis of nanopowders. *Laser Physics.* **16**(1):116–125.

2417. Pritchard D.K. (2004). Literature review: explosion hazards associated with nanopowders. Buxton, England: Health and Safety Laboratory.

2418. Sin A. et al. (2003). Nanopowders by organic polymerisation. *Journal of Sol-Gel Science & Technology.* **26**(1–3):541–545.

2419. Sohn H.Y. et al. (2007). The chemical vapor synthesis of inorganic nanopowders. *JOM.* **59**(12):44–49.

2420. Terekhov V.A. et al. (2008). Structure and optical properties of silicon nanopowders. *Materials Science and Engineering: B.* **147**(2):222–225.

2421. Tong L. and Reddy R.G. (2006). The processing of nanopowders by thermal plasma technology. *JOM.* **58**(4):62–66.

2422. van Ommen J.R. et al. (2012). Fluidization of nanopowders: a review. *Journal of Nanoparticle Research.* **14**(3):1–29.

2423. Vignes A. et al. (2012). Risk assessment of the ignitability and explosivity of aluminum nanopowders. *Process Safety and Environmental Protection.* **90**(4):304–310.

2424. Vissokov G. et al. (2003). On the plasma-chemical synthesis of nanopowders. *Plasma Science and Technology.* **5**(6):2039.

2425. Vollath D. (2008). Plasma synthesis of nanopowders. *Journal of Nanoparticle Research.* **10**(1):39–57.

2426. Wejrzanowski T. et al. (2006). Quantitative methods for nanopowders characterization. *Applied Surface Science.* **253**(1):204–208.

2427. Won C.W. et al. (2010). Refractory metal nanopowders: synthesis and characterization. *Current Opinion in Solid State and Materials Science.* **14**(3):53–68.

2428. Worsfold S.M. et al. (2012). Review of the explosibility of nontraditional dusts. *Industrial & Engineering Chemistry Research.* **51**(22):7651–7655.

2429. Wu H.C. et al. (2010). Explosion characteristics of aluminum nanopowders. *Aerosol and Air Quality Research.* **10**(1):38–42.

2430. Zheng C. et al. (2001). Preparation and characterization of VO2 nanopowders. *Journal of Solid State Chemistry.* **156**(2):274–280.

2431. Jillavenkatesa A. and Kelly J.F. (2002). Nanopowder characterization: challenges and future directions. *Journal of Nanoparticle Research.* **4**(5):463–468.

2432. Jones D.E. et al. (2003). Hazard characterization of aluminum nanopowder compositions. *Propellants, Explosives, Pyrotechnics.* **28**(3):120–131.

2433. Kiminami R.H. (2001). Combustion synthesis of nanopowder ceramic powders. *KONA Powder and Particle Journal.* **19**(0):156–165.

2434. Kwok Q.S. et al. (2002). Characterization of aluminum nanopowder compositions. *Propellants Explosives Pyrotechnics.* **27**(4):229–240.

Nanorings

2435. Aizpurua J. et al. (2003). Optical properties of gold nanorings. *Physical Review Letters.* **90**(5):057401.

2436. Cuesta I.G. et al. (2006). Carbon nanorings: a challenge to theoretical chemistry. *ChemPhysChem.* **7**(12):2503–2507.

2437. Feng C. and Liew K.M. (2009). Energetics and structures of carbon nanorings. *Carbon.* **47**(7):1664–1669.

2438. Hughes W.L. and Wang Z.L. (2004). Formation of piezoelectric single-crystal nanorings and nanobows. *Journal of the American Chemical Society.* 126(21):6703–6709.

2439. Kim M.W. and Ku P.C. (2011). Semiconductor nanoring lasers. *Applied Physics Letters.* **98**(20):201105.

2440. Somaschini C. et al. (2009). Fabrication of multiple concentric nanoring structures. *Nano Letters.* **9**(10):3419–3424.

2441. Vaz C.A. et al. (2007). Ferromagnetic nanorings. *Journal of Physics: Condensed Matter.* **19**(25):255207.

2442. Zhou L. et al. (2009). Crystal structure and optical properties of silver nanorings. *Applied Physics Letters.* **94**(15):153102.

Nanorods

2443. Cordente N. et al. (2001). Synthesis and magnetic properties of nickel nanorods. *Nano Letters.* **1**(10):565–568.

2444. Dai H. et al. (1995). Synthesis and characterization of carbide nanorods. *Nature.* **375**:769–772.

2445. Gole A. and Murphy C.J. (2005). Polyelectrolyte-coated gold nanorods: synthesis, characterization and immobilization. *Chemistry of Materials.* **17**(6):1325–1330.

2446. Huang X. et al. (2009). Gold nanorods: from synthesis and properties to biological and biomedical applications. *Advanced Materials.* **21**(48):4880.

2447. Huang Y.F. et al. (2008). Cancer cell targeting using multiple aptamers conjugated on nanorods. *Analytical Chemistry.* **80**(3):567–572.

2448. Jiang X.C. et al. (2006). Gold nanorods: limitations on their synthesis and optical properties. *Colloids and Surfaces A: Physicochemical and Engineering Aspects.* **277**(1):201–206.

2449. Kim F. et al. (2002). Photochemical synthesis of gold nanorods. *Journal of the American Chemical Society.* **124**(48): 14316–14317.

2450. Kislyuk V.V. and Dimitriev O.P. (2008). Nanorods and nanotubes for solar cells. *Journal of Nanoscience and Nanotechnology.* **8**(1):131–148.

2451. Li J.Y. et al. (2001). Fabrication of zinc oxide nanorods. *Journal of Crystal Growth.* **233**(1):5–7.

2452. Liang J. and Li Y. (2004). Synthesis and characterization of lead chromate uniform nanorods. *Journal of Crystal Growth.* **261**(4):577–580.

2453. Liu Z. and Bando Y. (2003). A novel method for preparing copper nanorods and nanowires. *Advanced Materials.* **15**(4):303–305.

2454. Sharma V. et al. (2009). Colloidal dispersion of gold nanorods: historical background, optical properties, seed-mediated synthesis, shape separation and self-assembly. *Materials Science and Engineering: R: Reports.* **65**(1):1–38.

2455. Tüzün B. and Erkoç Ş. (2012). Structural and electronic properties of unusual carbon nanorods. *Quantum Matter.* **1**(2):136–148.

2456. Vigderman L. et al. (2012). Functional gold nanorods: synthesis, self-assembly, and sensing applications. *Advanced Materials.* **24**(36):4811–4841.

2457. Yi G.C. et al. (2005). ZnO nanorods: synthesis, characterization and applications. *Semiconductor Science and Technology.* **20**(4):S22.

2458. Yu Y.Y. et al. (1997). Gold nanorods: electrochemical synthesis and optical properties. *The Journal of Physical Chemistry.* **101**(34):6661–6664.

Nanosheets

2459. Alwarappan S. et al. (2009). Probing the electrochemical properties of graphene nanosheets for biosensing applications. *The Journal of Physical Chemistry C.* **113**(20):8853–8857.

2460. Garcia J.C. et al. (2011). Group IV graphene-and graphene-like nanosheets. *The Journal of Physical Chemistry C.* **115**(27): 13242–13246.

2461. Guo S. and Dong S. (2011). Graphene nanosheet: synthesis, molecular engineering, thin film, hybrids, and energy and analytical applications. *Chemical Society Reviews.* **40**(5):2644–2672.

2462. Huang X. et al. (2013). Metal dichalcogenide nanosheets: preparation, properties and applications. *Chemical Society Reviews.* **42**(5):1934–1946.

2463. Seo J.W. et al. (2007). Two-dimensional nanosheet crystals. *Angewandte Chemie International Edition.* **46**(46):8828–8831.

2464. Shen J.M. and Feng Y.T. (2008). Formation of flower-like carbon nanosheet aggregations and their electrochemical application. *The Journal of Physical Chemistry C.* **112**(34):13114–13120.

2465. Wang G. et al. (2009). Synthesis and characterisation of hydrophilic and organophilic graphene nanosheets. *Carbon.* **47**(5):1359–1364.

2466. Yao Y. et al. (2012). Large-scale production of two-dimensional nanosheets. *Journal of Materials Chemistry.* **22**(27):13494–13499.

Nanoshells

2467. Averitt R.D. et al. (1999). Linear optical properties of gold nanoshells. *Journal of the Optical Society of America B.* **16**(10): 1824–1832.

2468. Brongersma M.L. (2003). Nanoscale photonics: nanoshells: gifts in a gold wrapper. *Nature Materials.* **2**(5):296–297.

2469. Gheorghe D.E. et al. (2011). Gold-silver alloy nanoshells: a new candidate for nanotherapeutics and diagnostics. *Nanoscale Research Letters.* **6**(1):1–12.

2470. Halas N. (2002). The optical properties of nanoshells. *Optics and Photonics News.* **13**(8):26–30.

2471. Hirsch L.R. et al. (2006). Metal nanoshells. *Annals of Biomedical Engineering.* **34**(1):15–22.

2472. Jackson J.B. and Halas N.J. (2001). Silver nanoshells: variations in morphologies and optical properties. *The Journal of Physical Chemistry B.* **105**(14):2743–2746.

2473. Knight M.W. and Halas N.J. (2008). Nanoshells to nanoeggs to nanocups: optical properties of reduced symmetry core-shell nanoparticles beyond the quasistatic limit. *New Journal of Physics.* **10**(10):105006.

Nanosprings

2474. Ben S. et al. (2015). Does Hooke's law work in helical nanosprings? *Physical Chemistry Chemical Physics.* **17**(32): 20990–20997.

2475. Chang I.L. and Yeh M.S. (2008). An atomistic study of nanosprings. *Journal of Applied Physics.* **104**(2):024305.

2476. da Fonseca A.F. and Galvão D.S. (2004). Mechanical properties of nanosprings. *Physical Review Letters.* **92**(17):175502.

2477. da Fonseca A.F. and Malta C.P. (2007). Is it possible to grow amorphous normal nanosprings? *Nanotechnology.* **18**(43): 435606.

2478. da Fonseca A.F. et al. (2006). Mechanical properties of amorphous nanosprings. *Nanotechnology.* **17**(22):5620.

2479. Khudiyev T. and Bayindir M. (2015). Nanosprings harvest light more efficiently. *Applied Optics.* **54**(26):8018–8023.

2480. Korgel B.A. (2005). Nanosprings take shape. *Science.* **309**(5741): 1683–1684.

2481. McIlroy D.N. et al. (2001). Nanosprings. *Applied Physics Letters.* **79**(10):1540–1542.

2482. Wang L. et al. (2006). High yield synthesis and lithography of silica-based nanospring mats. *Nanotechnology.* **17**(11):S298.

2483. Zhang D. et al. (2003). Silicon carbide nanosprings. *Nano Letters.* **3**(7):983–987.

2484. Zheng Y. et al. (2009). Torsional properties of metallic nanosprings. *Acta Mechanica Solida Sinica.* 22(6):657–664.

Nanostructures

2485. Anonymous (1992). Nanostructures come of age. *Nature.* **359**(6396):591.

2486. Anonymous (2013). A 3-D printer creates nanostructures. *Mechanical Engineering.* **135**(5):19.

2487. Bennemann K. (2010). Magnetic nanostructures. *Journal of Physics: Condensed Matter.* **22**(24):243201.

2488. Cahn R. (1992). Nanostructures come of age. *Nature.* **359**(6396): 591.

2489. Chen A. and Holt-Hindle P. (2010). Platinum-based nanostructured materials: synthesis, properties, and applications. *Chemical Reviews.* **110**(6):3767–3804.

2490. Chow G.-M. and Noskova N.I. (eds). (2001). *Nanostructured Materials Science & Tehcnology.* Berlin: Springer Science+Business Media, B.V.

2491. Dongsheng Y. (1995). Synthesis and fabrication of nanostructured materials. *Journal of Inorganic Materials.* **1.**

2492. Gleiter H. (1992). Nanostructured materials. *Advanced Materials.* **4**(7–8):474–481.

2493. Gleiter H. (1997). Nanostructured materials. *Acta Metallurgica Sinica.* **33**(2):165–174.

2494. Gleiter H. (2000). Nanostructured materials: basic concepts and microstructure. *Acta Materialia.* **48**(1):1–29.

2495. Kharisov B.I. et al. (2012). *Handbook of Less-Common Nanostructures.* Boca Raton, FL: CRC Press.

2496. Knauth P. and Schoonman J. (eds.) (2006). *Nanostructured Materials: Selected Synthesis Methods, Properties and Applications.* New York: Springer Science & Business Media.

2497. Koch C.C. (1997). Synthesis of nanostructured materials by mechanical milling: problems and opportunities. *Nanostructured Materials.* **9**(1):13–22.

2498. Koch C.C. et al. (1999). Ductility of nanostructured materials. *MRS Bulletin.* **24**(02):54–58.

2499. Kuchibhatla S.V. et al. (2007). One dimensional nanostructured materials. *Progress in Materials Science.* **52**(5):699–913.

2500. Kung H. and Foecke T. (1999). Mechanical behavior of nanostructured materials. *MRS Bulletin.* **24**(2):14–15.

2501. Leslie-Pelecky D.L. and Rieke R.D. (1996). Magnetic properties of nanostructured materials. *Chemistry of Materials.* **8**(8):1770–1783.

2502. Levine L.A. and Williams M.E. (2009). Inorganic biometric nanostructures. *Current Opinion in Chemical Biology.* **13**(5–6):669–677.

2503. Liu Y. et al. (2013). Templated synthesis of nanostructured materials. *Chemical Society Reviews.* **42**(7):2610–2653.

2504. Lu Y. and Liaw P.K. (2001). The mechanical properties of nanostructured materials. *JOM.* **53**(3):31–35.

2505. Lyons M.E. (2009). Paving the way to the integration of smart nanostructures. Part 1: nanotethering and nanowiring via material nanoengineering and electrochemical identification. *International Journal of Electrochemical Science.* **4**:481–515.

2506. Moriarty P. (2001). Nanostructured materials. *Reports on Progress in Physics.* **64**(3):297.

2507. Myhra S. and Rivière J.C. (2013). *Characterization of Nanostructures.* Boca Raton, FL: CRC Press.

2508. Noyce S. (2010). Building from the ground up: nanostructures to microstructures. *Young Scientists Journal.* **3**(8):42–45.

2509. Padmanabhan K.A. (2001). Mechanical properties of nanostructured materials. *Materials Science and Engineering.* **304**:200–205.

2510. Quintero C.M. et al. (2014). Hybrid spin-crossover nanostructures. *Beilstein Journal of Nanotechnology.* **5**:2230–2239.

2511. Rehr J.J. (2006). Materials science: nanostructures in a new league. *Nature.* **440**(7084):618–619.

2512. Shaw L.L. (2000). Processing nanostructured materials: an overview. *JOM.* **52**(12):41–45.

2513. Siegel R.W. (1993). Exploring mesocopia: the bold new world of nanostructures. *Physics Today.* **46**(10):64.

2514. Singh S. and Nalwa H.S. (2007). Nanotechnology and health safety–toxicity and risk assessments of nanostructured materials on human health. *Journal of Nanoscience and Nanotechnology.* **7**(9):3048–3070.

2515. Suryanarayana C. (2005). Recent developments in nanostructured materials. *Advanced Engineering Materials.* **7**(11):983–992.

2516. Vasilevska N. et al. (2015). Risks and health effects from exposure to engineered nanostructures: a critical review. *Journal of Chemical Technology and Metallurgy.* **50**(2):117–134.

2517. Wang Z.L. (2002). *Handbook of Nanophase and Nanostructured Materials: Synthesis/Characterization/Materials Systems and Applications I/Materials Systems and Applications II.* New York: Springer Science & Business Media.

2518. Willander M. et al. (2005). Solid and soft nanostructured materials: fundamentals and applications. *Microelectronics Journal.* **36**(11):940–949.

2519. Wilde G. (ed.) (2009). *Nanostructured Materials.* Oxford: Elsevier.

2520. Xu Q. et al. (2008). Nanoskiving: a new method to produce arrays of nanostructures. *Accounts of Chemical Research.* **41**(12):1566–1577.

2521. Yang P. (2003). *The Chemistry of Nanostructured Materials.* Singapore: World Scientific.
2522. Yeom S.H. et al. (2011). Nanostructures in biosensors: a review. *Frontiers in Bioscience (Landmark Edition).* **16**:997–1023.

Nanotubes (Carbon and Inorganic, C60-buckminsterfullerene, Buckyballs, Buckycones, Buckycubes, Buckytubes, C60 fullerene, Fullerenes)

2523. Ajayan P.M. (1997). Carbon nanotubes: novel architecture in nanometer space. *Progress in Crystal Growth and Characterization of Materials.* **34**(1):37–51.
2524. Ajayan P.M. (1999). Nanotubes from carbon. *Chemical Reviews.* **99**(7):1787–1800.
2525. Ali-Boucetta H. and Kostaleros K. (2013). Pharmacology of nanotubes: toxicokinetics, excretion and tissue accumulation. *Advanced Drug Delivery Reviews.* **65**(15):2111–2119.
2526. Andrievsky G. et al. (2005). Is the C60 fullerene molecule toxic? *Fullerenes, Nanotubes, and Carbon Nanostructures.* **13**(4):363–376.
2527. Applewhite E.J. (2015). The naming of buckminsterfullerene. In *Culture of Chemistry.* New York: Springer, pp. 21–23.
2528. Arrais A. and Diana E. (2003). Highly water soluble C60 derivatives: a new synthesis. *Fullerenes, Nanotubes and Carbon Nanostructures.* **11**(1):35–46.
2529. Avanasi R. et al. (2014). C60 fullerene soil sorption, biodegradation, and plant uptake. *Environmental Science & Technology.* **48**(5):2792–2797.
2530. Avent A.G. et al. (1994). The structure of buckminsterfullerene compounds. *Journal of Molecular Structure.* **325**:1–11.
2531. Avouris P. et al. (1999). Carbon nanotubes: nanomechanics, manipulation, and electronic devices. *Applied Surface Science.* **141**(3):201–209.

2532. Baker G.L. et al. (2008). Inhalation toxicity and lung toxicokinetics of C60 fullerene nanoparticles and microparticles. *Toxicological Sciences.* **101**(1):122–131.

2533. Bakry R. et al. (2007). Medicinal applications of fullerenes. *International Journal of Nanomedicine.* **2**(4):639.

2534. Baughman R.H. et al. (2002). Carbon nanotubes: the route toward applications. *Science.* **297**(5582):787–792.

2535. Belin T. and Epron F. (2005). Characterization methods of carbon nanotubes: a review. *Materials Science and Engineering: B.* **119**(2):105–118.

2536. Benjamin S.C. et al. (2006). Towards a fullerene-based quantum computer. *Journal of Physics: Condensed Matter.* **18**(21):S867.

2537. Berhanu D. et al. (2009). Characterisation of carbon nanotubes in the context of toxicity studies. *Environment Health: A Global Access Science Source.* **8**(Suppl 1):1–4.

2538. Bernholc J. et al. (1994). Structural transformations, reactions, and electronic properties of fullerenes, onions, and buckytubes. *Computational Materials Science.* **2**(3):547–556.

2539. Bianco A. and Prato M. (2003). Can carbon nanotubes be considered useful tools for biological applications? *Advanced Materials.* **15**(20):1765–1768.

2540. Bianco A. et al. (2005). Applications of carbon nanotubes in drug delivery. *Current Opinion in Chemical Biology.* **9**(6):674–679.

2541. Bianco A. et al. (2005). Biomedical applications of functionalised carbon nanotubes. *Chemical Communications.* (5):571–577.

2542. Bichoutskaia E. et al. (2006). Multi-walled nanotubes: commensurate-incommensurate phase transition and NEMS applications. *Fullerenes, Nanotubes and Carbon Nanostructures.* **14**:131–140.

2543. Birkett P.R. et al. (1993). The structural characterization of buckminsterfullerene compounds. *Journal of Molecular Structure.* **292**:1–8.

2544. Bonard J.M. et al. (2001). Field emission from carbon nanotubes: the first five years. *Solid-State Electronics*. **45**(6):893–914.

2545. Bong D.T. et al. (2001). Self-assembling organic nanotubes. *Angewandte Chemie International Edition*. **40**(6):988–1011.

2546. Boo W.O.J. (1992). An introduction to fullerene structures. *Journal of Chemical Education*. **92**(69):605.

2547. Buseck P.R. (2002). Geological fullerenes: review and analysis. *Earth and Planetary Science Letters*. **203**(3):781–792.

2548. Buseck P. et al. (1992). Fullerenes from the geological environment. *Science*. **257**(5067):215–217.

2549. Bussy C. et al. (2013). Hemotoxicity of carbon nanotubes. *Advanced Drug Delivery Reviews*. **65**(15):2127–2134.

2550. Cadenas C. et al. (2012). Buckyballs (fullerenes): free radical sponges or inflammatory agents? *Archives of Toxicology*. **86**(12):1807–1808.

2551. Cahill P.A. and Rohlfing C.M. (1996). Theoretical studies of derivatized buckyballs and buckytubes. *Tetrahedron*. **52**(14):5247–5256.

2552. Chamberlain G. (1993). Buckyballs could shape future of copier images. *Design News*.

2553. Charlier J.C. et al. (2007). Electronic and transport properties of nanotubes. *Reviews of Modern Physics*. **79**(2):677.

2554. Chibante L.P.F. and Smalley R.E. (1992). Complete buckminsterfullerene bibliography.

2555. Che J. et al. (2000). Thermal conductivity of carbon nanotubes. *Nanotechnology*. **11**(2):65.

2556. Chen Z. et al. (2008). Quantification of C60 fullerene concentrations in water. *Environmental Toxicology and Chemistry*. **27**(9):1852–1859.

2557. Chopra N.G. et al. (1995). Boron nitride nanotubes. *Science*. **269**(5226):966–967.

2558. Christou A. et al. (2016). A review of exposure and toxicological aspects of carbon nanotubes, and as additives to fire retardants in polymers. *Critical Reviews in Toxicology*. **46**(1):74–95.

2559. Colbert D.T. and Smalley R.E. (2002). Past, present and future of fullerene nanotubes: buckytubes. In *Perspectives of Fullerene Nanotechnology*. Amsterdam, Netherlands: Springer, pp. 3–10.

2560. Collins P.G. and Avouris P. (2000). Nanotubes for electronics. *Scientific American*. **283**(6):62–69.

2561. Copley J.R.D. et al. (1993). Structure and dynamics of buckyballs. *Neutron News*. **4**(4):20–28.

2562. Creasy W.R. (1993). A survey of the research areas related to buckminsterfullerene. *Fullerenes, Nanotubes, and Carbon Nanostructures*. **1**(1):23–44.

2563. Culotta E. and Koshland Jr, D.E. (1991). Buckyballs: wide open playing field for chemists. *Science*. **254**(5039):1706.

2564. Curl R.F. (1997). Dawn of the fullerenes: experiment and conjecture. *Reviews of Modern Physics*. **69**(3):691.

2565. Dai H. (2002). Carbon nanotubes: opportunities and challenges. *Surface Science*. **500**(1):218–241.

2566. Dai H. (2002). Carbon nanotubes: synthesis, integration, and properties. *Accounts of Chemical Research*. **35**(12):1035–1044.

2567. Da Ros T. et al. (2001). Biological applications of fullerene derivatives: a brief overview. *Croatica Chemica Acta*. **74**:743–755.

2568. Deepak F.L. et al. (2002). Boron nitride nanotubes and nanowires. *Chemical Physics Letters*. **353**(5):345–352.

2569. De Voider M.F. et al. (2013). Carbon nanotubes: present and future commercial applications. *Science*. **339**(6119):535–539.

2570. Dimitrakopulos G.P. et al. (1997). The defect character of carbon nanotubes and nanoparticles. *Acta Crystallographica Section A: Foundations of Crystallography*. **53**(3):341–351.

2571. Donaldson K. et al. (2006a). Carbon nanotubes: a review of their properties in relation to pulmonary toxicology and workplace safety. *Toxicological Sciences*. **92**(1):5–22.

2572. Djordjević A. et al. (2006b). Fullerenes in biomedicine. *Journal of Balkan Union of Oncology*. **11**(4):391–404.

2573. Donaldson K. et al. (2006). Carbon nanotubes: a review of their properties in relation to pulmonary toxicology and workplace safety. *Toxicological Sciences.* **92**(1):5–22.

2574. Dresselhaus M.S. et al. (1993). Fullerenes. *Journal of Materials Research.* **8**(08):2054–2097.

2575. Dresselhaus M.S. et al. (1995). Physics of carbon nanotubes. *Carbon.* **33**(7):883–891.

2576. Dresselhaus M.S. et al. (1996). *Science of Fullerenes and Carbon Nanotubes: Their Properties and Applications.* New York: Academic Press.

2577. Dugan L.L. et al. (1997). Carboxyfullerenes as neuroprotective agents. *Proceedings of the National Academy of Sciences of the United States of America.* **94**:9434–9439.

2578. Ebbesen T.W. (1996a). Carbon nanotubes. *Physics Today.* **49**(6):26–35.

2579. Ebbesen T.W. (1996b). *Carbon Nanotubes: Preparation and Properties.* Boca Raton, FL: CRC Press.

2580. Ebbesen T.W. and Ajayan P.M. (1992). Large-scale synthesis of carbon nanotubes. *Nature.* **358**(6383):220–222.

2581. Ehrenfreund P. and Foing B.H. (1997). Fullerenes in space. *Advances in Space Research.* **19**(7):1033–1042.

2582. Elder A. (2009). Nanotoxicology: how do nanotubes suppress T cells? *Nature Nanotechnology.* **4**(7):409–410.

2583. Emsley J. (1993). The weird and wonderful world of buckyballs. *New Scientist.* **138**(1872):13.

2584. Fann Y.C. et al. (1992). From buckyballs to bunnyballs: a theoretical analysis of adduct-induced electronic effects. *The Journal of Physical Chemistry.* 96(14):5817–5818.

2585. Firme C.P. and Bandaru P.R. (2010). Toxicity issues in the application of carbon nanotubes to biological systems. *Nanomedicine: Nanotechnology, Biology and Medicine.* **6**(2):245–256.

2586. Foldvari M. and Bagonluri M. (2008a). Carbon nanotubes as functional excipients for nanomedicines: I. Pharmaceutical properties. *Nanomedicine: Nanotechnology, Biology and Medicine.* **4**(3):173–182.

2587. Foldvari M. and Bagonluri M. (2008b). Carbon nanotubes as functional excipients for nanomedicines: II. Drug delivery and biocompatibility issues. *Nanomedicine: Nanotechnology, Biology and Medicine.* **4**(3):183–200.

2588. Geckeler K.E. and Samal S. (1999). Syntheses and properties of macromolecular fullerenes, a review. *Polymer International.* **48**(9):743–757.

2589. Georgakilas V. et al. (2002). Organic functionalization of carbon nanotubes. *Journal of the American Chemical Society.* **124**(5): 760–761.

2590. Gharbi N. et al. (2005). Fullerene is a powerful antioxidant in vivo with no acute or subacute toxicity. *Nano Letters.* **5**(12):2578–2585.

2591. Giacalone F. and Martín N. (2010). New concepts and applications in the macromolecular chemistry of fullerenes. *Advanced Materials.* **22**(38):4220–4248.

2592. Girifalco L.A. et al. (2000). Carbon nanotubes, buckyballs, ropes, and a universal graphitic potential. *Physical Review B.* **62**(19):13104.

2593. Godly E.W. and Taylor R. (1997). Nomenclature and terminology of fullerenes: a preliminary survey. *Fullerenes, Nanotubes, and Carbon Nanostructures.* **5**(7):1667–1708.

2594. Goho A. (2004). Buckyballs at bat: toxic nanomaterials get a tune-up. *Science News.* **166**(14):211–211.

2595. Goho A. (2005). Nanowaste: predicting the environmental fate of buckyballs. *Science News.* **167**(19):292–293.

2596. Good R.H. (1992). Buckyballs, anyone? *The Physics Teacher.* **30**(1):20.

2597. Govindjee S. and Sackman J.L. (1999). On the use of continuum mechanics to estimate the properties of nanotubes. *Solid State Communications.* **110**(4):227–230.

2598. Gupta V.K. and Saleh T.A. (2013). Sorption of pollutants by porous carbon, carbon nanotubes and fullerene: an overview. *Environmental Science and Pollution Research.* **20**(5):2828–2843.

2599. Hardman R. (2006). A toxicologic review of quantum dots: toxicity depends on physicochemical and environmental factors. *Environmental Health Perspectives.* **114**(2):165–172.

2600. Harris D.L. and Bawa R. (2007). The carbon nanotube patent landscape in nanomedicine: an expert opinion. *Expert Opinion on Therapeutic Patents.* **17**(9):1–11

2601. Heath J.R. (1998). Fullerenes: C60's smallest cousin. *Nature.* **393**(6687):730–731.

2602. Hebard A.F. (1993). Buckminsterfullerene. *Annual Review of Materials Science.* **23**(1):159–191.

2603. Helland A. et al. (2007). Reviewing the environmental and human health knowledge base of carbon nanotubes. *Environmental Health Perspectives.* **115**(8):1125–1131.

2604. Herbst M.H. et al. (2004). Tecnologia dos nanotubos de carbono: tendências e perspectivas de uma área multidisciplinar. *Química Nova.* **27**(6):986–992.

2605. Heymann D. et al. (2003). Terrestrial and extraterrestrial fullerenes. *Fullerenes, Nanotubes and Carbon Nanostructures.* **11**(4):333–370.

2606. Hirsch A. (1994). *The Chemistry of the Fullerenes.* Stuttgart: G. Thieme Verlag.

2607. Holt William W. (1993). Buckyballs of yore? *Physics Today.* **46**:97.

2608. Hoshino A. et al. (2011). Toxicity of nanocrystal quantum dots: the relevance of surface modifications. *Archives of Toxicology.* **85**(7):707–720.

2609. Huffman D.R. (1994). Materials update: applications of fullerenes. *Materials Letters.* **21**(2):127–129.

2610. Hurt R.H. et al. (2006). Toxicology of carbon nanomaterials: status, trends, and perspectives on the special issue. *Carbon.* **44**(6):1028–1033.

2611. Iijima S. (2002). Carbon nanotubes: past, present, and future. *Physica B: Condensed Matter.* **323**(1):1–5.

2612. Iijima S. et al. (1996). Structural flexibility of carbon nanotubes. *Journal of Chemical Physics.* **104**(5):2089–2092.

2613. Jackson P. et al. (2013). Bioaccumulation and ecotoxicity of carbon nanotubes. *Chemistry Central Journal.* **7**.

2614. James G. (1994). The representation-theory for buckminster-fullerene. *Journal of Algebra.* **167**(3):803–820.

2615. Jensen A.W. et al. (1996). Biological applications of fullerenes. *Bioorganic & Medicinal Chemistry.* **4**(6):767–779.

2616. Jonsson D. et al. (1998). Electric and magnetic properties of fullerenes. *Journal of Chemical Physics.* **109**(2):572–577.

2617. Journet C. and Bernier P. (1998). Production of carbon nanotubes. *Applied physics A: Materials Science & Processing.* **67**(1):1–9.

2618. Kadish K.M. and Ruoff R.S. (2000). *Fullerenes: Chemistry, Physics, and Technology.* New York: John Wiley & Sons.

2619. Kar K.K. (2011). *Carbon Nanotubes: Synthesis, Characterization and Applications.* Singapore: Research Publishing Service.

2620. Karaulova E.N. and Bagrii E.I. (1999). Fullerenes: functionalisation and prospects for the use of derivatives. *Russian Chemical Reviews.* **68**(11):889.

2621. Katz E.A. (2002). Potential of fullerene-based materials for the utilization of solar energy. *Physics of the Solid State.* **44**(4):647–651.

2622. Ketterson J.B. and Chang R.P.H. (1990). Buckytubes and derivatives: their growth and implications for buckyball formation. *Chemical Physics Letters.* **170**:167.

2623. Kimoto Y. et al. (2005). Molecular dynamics study of double-walled carbon nanotubes for nano-mechanical manipulation. *Japanese Journal of Applied Physics.* **44**(4A):1641–1647.

2624. Kniaz K. et al. (1993). Fluorinated fullerenes: synthesis, structure, and properties. *Journal of the American Chemical Society.* **115**(14):6060–6064.

2625. Kolosnjaj J. et al. (2007). Toxicity studies of fullerenes and derivatives. In *Bio-Applications of Nanoparticles.* New York: Springer, pp. 168–180.

2626. Kostarelos K. (2008). The long and short of carbon nanotube toxicity. *Nature Biotechnology.* **26**(7):774–776.

2627. Kostarelos K. et al. (2009). Promises, facts and challenges for carbon nanotubes in imaging and therapeutics. *Nature Nanotechnology.* **4**(10):627–633.

2628. Kotsalis E.M. et al. (2001). Buckyballs in water: structural characteristics and energetics. NASA Center for Turbulence Research Annual Research Briefs.

2629. Kroto H. (1993). The birth of C60: buckminsterfullerene. In *Electronic Properties of Fullerenes.* Berlin Heidelberg: Springer, pp. 1–7.

2630. Kroto H.W. et al. (1985). C60: buckminsterfullerene. *Nature.* **318**(6042):162–163.

2631. Kroto H.W. et al. (1991). C60: buckminsterfullerene. *Chemical Reviews.* **91**(6):1213–1235.

2632. Lacerda L. et al. (2006). Carbon nanotubes as nanomedicines: from toxicology to pharmacology. *Advanced Drug Delivery Reviews.* **58**(14):1460–1470.

2633. Lam C.W. et al. (2006). A review of carbon nanotube toxicity and assessment of potential occupational and environmental health risks. *Critical Reviews in Toxicology.* **36**(3):189–217.

2634. Lan Y. et al. (2011). Physics and applications of aligned carbon nanotubes. *Advances in Physics.* **60**(4):553–678.

2635. Langa F. and Nierengarten J.F. (2007). *Fullerenes: Principles and Applications.* Cambridge: Royal Society of Chemistry.

2636. Laplaze D. et al. (1996). Solar energy: application to the production of fullerenes. *Journal of Physics B: Atomic, Molecular and Optical Physics.* **29**(21):4943.

2637. Lindberg J.D. et al. (1993). Imaginary refractive index of buckyballs. *Applied Optics.* **32**(21):3921–3922.

2638. Liu X. et al. (2013). Application of potential carbon nanotubes in water treatment: a review. *Journal of Environmental Sciences.* **25**(7):1263–1280.

2639. Loiseau A. et al. (2006). Understanding carbon nanotubes. *Lecture Notes in Physics.* **677**:495–543.

2640. Lu F. et al. (2009). Advances in bioapplications of carbon nanotubes. *Advanced Materials.* **21**(2):139–152.

2641. Lu J.P. (1995). Novel magnetic properties of carbon nanotubes. *Physical Review Letters.* **74**(7):1123.

2642. Mamalis A.G. et al. (2004). Nanotechnology and nanostructured materials: trends in carbon nanotubes. *Precision Engineering.* **28**(1):16–30.

2643. Marbach W.D. (1991). The buckyballs throws a new curve at scientists. *Business Week.* **3211**:86.

2644. Meyyappan M. (ed). (2004). *Carbon Nanotubes Science an Applications.* Boca Raton, FL: CRC Press.

2645. Mintmire J.W. and White C.T. (1995). Electronic and structural properties of carbon nanotubes. *Carbon.* **33**(7):893–902.

2646. Mondini P. and Cataldo F. (2001). Chlorinated fullerenes: a theoretical study. *Fullerene Science & Technology.* **9**(1):25.

2647. Moussa F. et al. (1997). C60 fullerene toxicity: preliminary account of an in vivo study. *Fullerenes.* **97**(42):332–336.

2648. Mukhopadhyay K. et al. (1994). Fullerenes from camphor: a natural source. *Physical Review Letters.* **72**(20):3182.

2649. Muller J. et al. (2005). Respiratory toxicity of multi-wall carbon nanotubes. *Toxicology and Applied Pharmacology.* **207**(3):221–231.

2650. Muller J. et al. (2006). Toxicology of carbon nanomaterials. *Carbon.* **44**(6):1048–1056.

2651. Nakamura E. and Isobe H. (2003). Functionalized fullerenes in water. The first 10 years of their chemistry, biology, and nanoscience. *Accounts of Chemical Research.* **36**(11):807–815.

2652. National Institute for Occupational Safety and Health (2013). Occupational exposure to carbon nanotubes and nanofibers. (Current Intelligence Bulletin 65 [NIOSH] Publication No. 2013-145). Cincinnati, Ohio.

2653. Neretin I.S. and Slovokhotov Y.L. (2004). Chemical crystallography of fullerenes. *Russian Chemical Reviews.* **73**(5):455–486.

2654. Niyogi S. et al. (2002). Chemistry of single-walled carbon nanotubes. *Accounts of Chemical Research.* **35**(12):1105–1113.

2655. O'Connell M.J. (2006). *Carbon Nanotubes: Properties and Applications.* Boca Raton, FL: CRC Press.

2656. Osawa E. (2002). *Perspectives of Fullerene Nanotechnology.* New York: Springer Science & Business Media.

2657. Osawa E. et al. (1994). Shape and fantasy of fullerenes. *MRS Bulletin.* **19**(11):33–36.

2658. Palmer W.P. (1994). Buckminster fuller, buckyballs and the teaching of chemistry. *Online Submission.* **61**:4.

2659. Parilla P.A. et al. (1999). The first true inorganic fullerenes? *Nature.* **397**(6715):114.

2660. Patole S.P. et al. (2008). Alignment and wall control of ultra long carbon nanotubes in water assisted chemical vapour deposition. *Journal of Physics D: Applied Physics.* **41**:155311.

2661. Pool R. (1990). All worked up about buckyballs. *Science.* 250(4978):209.

2662. Popov V.N. (2004). Carbon nanotubes: properties and application. *Materials Science and Engineering: R: Reports.* **43**(3):61–102.

2663. Porter D.W. et al. (2012). Acute pulmonary dose-responses to inhaled multi-walled carbon nanotubes. *Nanotoxicology.* **7**(7):1179–1194.

2664. Pradeep T. and Rao C.N.R. (1991). Preparation of buckminsterfullerene, C60. *Materials Research Bulletin.* **26**(10):1101–1105.

2665. Prato M. and Maggini M. (1998). Fulleropyrrolidines: a family of full-fledged fullerene derivatives. *Accounts of Chemical Research.* **31**(9):519–526.

2666. Qian D. et al. (2002). Mechanics of carbon nanotubes. *Applied Mechanics Reviews.* **55**(6):495–533.

2667. Ramierz A.P. (2005). Carbon nanotubes for science and technology. *Bell Labs Technical Journal.* **10**(3):171–185.

2668. Ray A. (2012). Fullerene (C60) molecule: a review. *Asian Journal of Pharmaceutical Research.* **2**(2):47–50.

2669. Reich S. et al. (2008). *Carbon Nanotubes: Basic Concepts and Physical Properties.* New York: John Wiley & Sons.

2670. Rivas G.A. et al. (2007). Carbon nanotubes for electrochemical biosensing. *Talanta*. **74**(3):291–307.

2671. Ruoff R.S. and Lorents D.C. (1995). Mechanical and thermal properties of carbon nanotubes. *Carbon*. **33**(7):925–930.

2672. Salvetat J.P. et al. (1999). Mechanical properties of carbon nanotubes. *Applied Physics A*. **69**(3):255–260.

2673. Salvetat-Delmotte J.P. and Rubio A. (2002). Mechanical properties of carbon nanotubes: a fiber digest for beginners. *Carbon*. **40**(10):1729–1734.

2674. Sapmaz S. et al. (2003). Carbon nanotubes as nanoelectromechanical systems. *Physical Review B*. **67**(23):235414.

2675. Sattler K. (1996). Growth and scanning probe microscopy of buckytubes and buckycones. *Surface Review and Letters*. **3**(01):813–818.

2676. Sayes C. et al. (2004). The differential cytotoxicity of water-soluble fullerenes. *Journal of the American Chemical Society*. **4**:1881–1887.

2677. Schipper M.L. et al. (2008). A pilot toxicology study of single-walled carbon nanotubes in a small sample of mice. *Nature Nanotechnology*. **3**(4):216–221.

2678. Sergio M. et al. (2013). Fullerenes toxicity and electronic properties. *Environmental Chemistry Letters*. **11**(2):105–118.

2679. Sharma M. (2010). Understanding the mechanism of toxicity of carbon nanoparticles in humans in the new millennium: a systematic review. *Indian Journal of Occupational & Environmental Medicine*. **14**(1):3–5.

2680. Shiraishi M. and Ata M. (2001). Work function of carbon nanotubes. *Carbon*. **39**(12):1913–1917.

2681. Shiv Charan P.S. et al. (2009). Carbon nanotubes: synthesis and application. *Transactions of the Indian Ceramic Society*. **68**(4):163–172.

2682. Sinnott S.B. and Andrews R. (2001). Carbon nanotubes: synthesis, properties, and applications. *Critical Reviews in Solid State and Materials Sciences*. **26**(3):145–249.

2683. Slater S.G. (2002). Nanotechnology—exploring a new horizon with buckyballs and fullerenes. *Home Health Care Management & Practice.* **14**(6):482–483.

2684. Smalley R.E. (1997). Discovering the fullerenes. *Reviews of Modern Physics.* **69**(3):723.

2685. Smalley R.E. and Yakobson B.I. (1998). The future of the fullerenes. *Solid State Communications.* **107**(11):597–606.

2686. Sokolov V.I. and Stankevich I.V. (1993). The fullerenes—new allotropic forms of carbon: molecular and electronic structure, and chemical properties. *Russian Chemical Reviews.* **62**(5):419.

2687. Srinivasan C. (2008). Toxicity of carbon nanotubes—some recent studies. *Current Science.* **95**(3):307–308.

2688. Steinmetz N.F. et al. (2009). Buckyballs meet viral nanoparticles: candidates for biomedicine. *Journal of the American Chemical Society.* **131**(47):17093–17095.

2689. Subramoney S. (1999). Carbon nanotubes: a status report. *Interface-Electrochemical Society.* **8**(4):34–41.

2690. Sun L.F. and Xie S.S. (2000). Creating the narrowest carbon nanotubes. *Nature.* **403**(6768):384.

2691. Sun Y.P. et al. (2002). Functionalized carbon nanotubes: properties and applications. *Accounts of Chemical Research.* **35**(12):1096–1104.

2692. Szwacki N.G. (2008). Boron fullerenes: a first-principles study. *Nanoscale Research Letters.* **3**(2):49–54.

2693. Tagmatarchis N. and Shinohara H. (2001). Fullerenes in medicinal chemistry and their biological applications. *Mini Reviews in Medicinal Chemistry.* **1**(4):339–348.

2694. Tanaka K. et al. (eds.) (1999). *The Science and Technology of Carbon Nanotubes.* Oxford: Elsevier.

2695. Tasis D. et al. (2006). Chemistry of carbon nanotubes. *Chemical Reviews.* **106**(3):1105–1136.

2696. Taubes G. (1991). The disputed birth of buckyballs. *Science.* **253**:1476–1479.

2697. Taylor R. and Walton D.R.M. (1993). The chemistry of fullerenes. *Nature.* **363**:685–693.

2698. Tersoff J. (1992). Energies of fullerenes. *Physical Review B.* **46**(23):15546.

2699. Thiel W. (1994). Recent theoretical fullerene research. *CHIMIA International Journal for Chemistry.* **48**(9):447–448.

2700. Thostenson E.T. et al. (2001). Advances in the science and technology of carbon nanotubes and their composites: a review. *Composites Science and Technology.* **61**(13):1899–1912.

2701. Tomanek D. (1995). Building a better world with buckyballs. *Journal-Korean Physical Society.* **28**:S609–S616.

2702. Trojanowicz M. (2006). Analytical applications of carbon nanotubes: a review. *Trends in Analytical Chemistry.* **25**(5):480–489.

2703. Upadhyayula V.K. et al. (2012). Life cycle assessment as a tool to enhance the environmental performance of carbon nanotube products: a review. *Journal of Cleaner Production.* **26**:37–47.

2704. Van Orden A. and Saykally R.J. (1998). Small carbon clusters: spectroscopy, structure, and energetics. *Chemical Reviews.* **98**(6):2313–2358.

2705. Walther J.H. et al. (2001). Carbon nanotubes in water: structural characteristics and energetics. *The Journal of Physical Chemistry B.* **105**(41):9980–9987.

2706. Wang X.K. et al. (1995). Properties of buckytubes and derivatives. *Carbon.* **33**(7):949–958.

2707. Warheit D.B. (2006). What is currently known about the health risks related to carbon nanotube exposures? *Carbon.* **44**(6):1064–1069.

2708. Wilson L.J. (1999). Medical applications of fullerenes and metallofullerenes. *The Electrochemical Society Interface.* **8**(4):24–28.

2709. Wilson M.A. et al. (1992). Fullerenes—preparation, properties, and carbon chemistry. *Carbon.* **30**(4):675–693.

2710. Withers J.C. et al. (1997). Fullerene commercial vision. *Fullerenes, Nanotubes, and Carbon Nanostructures.* **5**(1):1–31.

2711. Wörle-Knirsch J.M. et al. (2006). Oops they did it again! Carbon nanotubes hoax scientists in viability assays. *Nano Letters.* **6**(6):1261–1268.

2712. Wudl F. (1992). The chemical properties of buckminsterfullerene (C60) and the birth and infancy of fulleroids. *Accounts of Chemical Research.* **25**(3):157–161.

2713. Yam P. (2009). Buckyballs and nanotubes. *Scientific American.* **301**(3):82.

2714. Yamabe T. et al. (1999). *The Science and Technology of Carbon Nanotubes.* Berlin: Elsevier.

2715. Yang W. et al. (2007). Carbon nanotubes for biological and biomedical applications. *Nanotechnology.* **18**(41):412001.

2716. Zangi R. (2014). Are buckyballs hydrophobic? *The Journal of Physical Chemistry B.* **118**(42):12263–12270.

2717. Zhao Q. et al. (2002). Ultimate strength of carbon nanotubes: a theoretical study. *Physical Review B.* **65**(14):144105.

2718. Zhao Y. et al. (2008). Nanotoxicology: are carbon nanotubes safe? *Nature Nanotechnology.* **3**(4):191–192.

2719. Zhennan G. et al. (1991). Buckminsterfullerene C60: synthesis, spectroscopic characterization and structure analysis. *The Journal of Physical Chemistry.* **95**(24):9615–9618.

2720. Zhu Y. and Li W.X. (2008). Cytotoxicity of carbon nanotubes. *Science in China (Series B-Chemistry).* **11**.

Nanowindmills

2721. Yu L. et al. (2011). Solution synthesis and optimization of ZnO nanowindmills. *Applied Surface Science.* **257**(17):7432–7435.

Nanowires

2722. Appenzeller J. et al. (2008). Toward nanowire electronics. *IEEE Transactions on Electron Devices.* **55**(11):2827–2845.

2723. Brambilla G. et al. (2005). Compound-glass optical nanowires. *Electronics Letters.* **41**(7):400–402.

2724. Brambilla G. (2010). Optical fibre nanowires and microwires: a review. *Journal of Optics.* **12**(4):043001.

2725. Briseno A.L. et al. (2008). Introducing organic nanowire transistors. *Materials Today.* **11**(4):38–47.

2726. Cademartiri L. and Ozin G.A. (2009). Ultrathin nanowires— a materials chemistry perspective. *Advanced Materials.* **21**(9):1013–1020.

2727. Chung S.W. et al. (2000). Silicon nanowire devices. *Applied Physics Letters.* **76**(15):2068–2070.

2728. Cobden D.H. (2001). Molecular electronics: nanowires begin to shine. *Nature.* **409**(6816):32–33.

2729. Du G.H. et al. (2003). Potassium titanate nanowires: structure, growth, and optical properties. *Physical Review B.* **67**(3): 035323.

2730. Duan X. et al. (2000). Synthesis and optical properties of gallium arsenide nanowires. *Applied Physics Letters.* **76**(9):1116–1118.

2731. Fasol G. (1998). Nanowires: small is beautiful. *Science.* **280**(5363): 545.

2732. Gall K. et al. (2004). The strength of gold nanowires. *Nano Letters.* **4**(12):2431–2436.

2733. Garnett E.C. et al. (2011). Nanowire solar cells. *Annual Review of Materials Research.* **41**:269–295.

2734. Fert A. and Piraux L. (1999). Magnetic nanowires. *Journal of Magnetism and Magnetic Materials.* **200**(1):338–358.

2735. Häkkinen H. et al. (2000). Nanowire gold chains: formation mechanisms and conductance. *The Journal of Physical Chemistry B.* **104**(39):9063–9066.

2736. Hu J. et al. (1999). Chemistry and physics in one dimension: synthesis and properties of nanowires and nanotubes. *Accounts of Chemical Research.* **32**(5):435–445.

2737. Law M. et al. (2004). Semiconductor nanowires and nanotubes. *Annual Review of Materials Research.* **34**:83–122.

2738. Lee S.T. et al. (2000). Semiconductor nanowires: synthesis, structure and properties. *Materials Science and Engineering: A.* **286**(1):16–23.

2739. Lieber C.M. and Wang Z.L. (2007). Functional nanowires. *MRS Bulletin.* **32**(02):99–108.

2740. Mehrez H. and Ciraci S. (1997). Yielding and fracture mechanisms of nanowires. *Physical Review B.* **56**(19):12632.

2741. Nair P.R. and Alam M. (2007). Design considerations of silicon nanowire biosensors. *IEEE Transactions on Electron Devices.* **54**(12):3400–3408.

2742. Otten C.J. et al. (2002). Crystalline boron nanowires. *Journal of the American Chemical Society.* **124**(17):4564–4565.

2743. Patolsky F. and Lieber C.M. (2005). Nanowire nanosensors. *Materials Today.* **8**(4):20–28.

2744. Patolsky F. et al. (2007). Nanowire-based nanoelectronic devices in the life sciences. *MRS Bulletin.* **32**(02):142–149.

2745. Rao C.N.R. et al. (2003). Inorganic nanowires. *Progress in Solid State Chemistry.* **31**(1):5–147.

2746. Schwarz K.W. and Tersoff J. (2010). Elementary processes in nanowire growth. *Nano Letters.* **11**(2):316–320.

2747. Thelander C. et al. (2006). Nanowire-based one-dimensional electronics. *Materials Today.* **9**(10):28–35.

2748. Wang B. et al. (2001). Novel structures and properties of gold nanowires. *Physical Review Letters.* **86**(10):2046.

2749. Wen B. et al. (2008). Mechanical properties of ZnO nanowires. *Physical Review Letters.* **101**(17):175502.

2750. Xu S. et al. (2010). Self-powered nanowire devices. *Nature Nanotechnology.* **5**(5):366–373.

2751. Yan R. et al. (2009). Nanowire photonics. *Nature Photonics.* **3**(10):569–576.

2752. Yang P. et al. (2002). Inorganic semiconductor nanowires. *International Journal of Nanoscience.* **1**(01):1–39.

2753. Yang P. et al. (2010). Semiconductor nanowire: what's next? *Nano Letters.* **10**(5):1529–1536.

2754. Yu J.Y. et al. (2000). Silicon nanowires: preparation, device fabrication, and transport properties. *The Journal of Physical Chemistry B.* **104**(50):11864–11870.

2755. Zhang Y. et al. (2002). A simple method to synthesize nanowires. *Chemistry of Materials.* **14**(8):3564–3568.

2756. Zimmler M.A. et al. (2010). Optically pumped nanowire lasers: invited review. *Semiconductor Science and Technology.* **25**(2): 024001.

Quantum Bits, Boxes, Dots, Wells, and Wires

2757. Alhassid Y. (2000). The statistical theory of quantum dots. *Reviews of Modern Physics.* **72**(4):895.

2758. Alivisatos A.P. (1996). Semiconductor clusters, nanocrystals, and quantum dots. *Science.* **271**(5251):933–937.

2759. Averin D.V. et al. (1991). Theory of single-electron charging of quantum wells and dots. *Physical Review B.* **44**(12):6199.

2760. Azzazy H.M. et al. (2007). From diagnostics to therapy: prospects of quantum dots. *Clinical Biochemistry.* **40**(13): 917–927.

2761. Bailey R.E. et al. (2004). Quantum dots in biology and medicine. *Physica E: Low-dimensional Systems and Nanostructures.* **25**(1):1–12.

2762. Bastard G. et al. (1982). Exciton binding energy in quantum wells. *Physical Review B.* **26**(4):1974.

2763. Bawendi M.G. et al. (1990). The quantum mechanics of larger semiconductor clusters ("quantum dots"). *Annual Review of Physical Chemistry.* **41**(1):477–496.

2764. Bimberg D. (2005). Quantum dots for lasers, amplifiers and computing. *Journal of Physics D: Applied Physics.* **38**(13): 2055.

2765. Broido D.A. et al. (1992). Theory of holes in quantum dots. *Physical Review B.* **45**(19):11395.

2766. Chakraborty T. (1992). Physics of the artificial atoms: quantum dots in a magnetic field. *Comments on Condensed Matter Physics.* **16**:35–68.

2767. Chakraborty T. (1999). *Quantum Dots: A Survey of the Properties of Artificial Atoms*. Amsterdam: Elsevier.

2768. Chou L.Y. and Chan W.C. (2012). Nanotoxicology: no signs of illness. *Nature Nanotechnology*. **7**(7):416–417.

2769. Clarke J. and Wilhelm F.K. (2008). Superconducting quantum bits. *Nature*. **453**(7198):1031–1042.

2770. Derfus A.M. et al. (2004). Probing the cytotoxicity of semiconductor quantum dots. *Nano Letters*. **4**(1):11–18.

2771. DiVincenzo D. (2010). Quantum bits: better than excellent. *Nature Materials*. **9**(6):468–469.

2772. Hardman R. (2006). A toxicologic review of quantum dots: toxicity depends on physicochemical and environmental factors. *Environmental Health Perspectives*. **114**:165–172.

2773. Harrison P. (2005). *Quantum Wells, Wires and Dots: Theoretical and Computational Physics of Semiconductor Nanostructures*. New York: John Wiley & Sons.

2774. Hauck T.S. et al. (2010). In vivo quantum-dot toxicity assessment. *Small*. **6**(1):138–144.

2775. Henneberger F. and Benson O. (eds.) (2008). *Semiconductor Quantum Bits*. Boca Raton, FL: CRC Press.

2776. Jacak L. et al. (1998). *Quantum Dots*. Berlin: Springer-Verlag.

2777. Jamieson T. et al. (2007). Biological applications of quantum dots. *Biomaterials*. **28**(31):4717–4732.

2778. Jovin T.M. (2003). Quantum dots finally come of age. *Nature Biotechnology*. **21**(1):32–33.

2779. Klostranec J. M. and Chan W.C. (2006). Quantum dots in biological and biomedical research: recent progress and present challenges. *Advanced Materials*. **18**(15):1953–1964.

2780. Kostrykin V. and Schrader R. (1999). Kirchhoff's rule for quantum wires. *Journal of Physics A: Mathematical and General*. **32**(4):595.

2781. Kouwenhoven L. and Marcus C. (1998). Quantum dots. *Physics World*. **11**(6):35–39.

2782. Kuno M. (2013). Colloidal quantum dots: a model nanoscience system. *The Journal of Physical Chemistry Letters.* **4**(4): 680–680.

2783. Loss D. and DiVincenzo D.P. (1998). Quantum computation with quantum dots. *Physical Review A.* **57**(1):120.

2784. Makhlin Y. et al. (2000). Nano-electronic realizations of quantum bits. *Journal of Low Temperature Physics.* **118**(5–6): 751–763.

2785. Miller D.A.B. (2016). Optical physics of quantum wells. Holmdel, NJ: AT&T Bell Laboratories. Available at http://www-ee.stanford.edu/~dabm/181.pdf

2786. Ozkan M. (2004). Quantum dots and other nanoparticles: what can they offer to drug discovery? *Drug Discovery Today.* **9**(24):1065–1071.

2787. Reed M.A. (1993). Quantum dots. *Scientific American.* **268**(1): 118–123.

2788. Sakaki H. (1992). Quantum wires, quantum boxes and related structures: physics, device potentials and structural requirements. *Surface Science.* **267**(1):623–629.

2789. Shiohara A. et al. (2004). On the cytotoxicity caused by quantum dots. *Microbiology & Immunology.* **48**(9):669–675.

2790. Smith A.M. and Nie S. (2004). Chemical analysis and cellular imaging with quantum dots. *Analyst.* **129**(8):672–677.

2791. Smith A.M. and Nie S. (2009). Next-generation quantum dots. *Nature Biotechnology.* **27**(8):732–733.

2792. Stier O. (2001). Electronic and optical properties of quantum dots and wires. PhD Dissertation: Berlin, Technische Universität of Berlin.

2793. Valizadeh A. et al. (2012). Quantum dots: synthesis, bioapplications, and toxicity. *Nanoscale Research Letters.* **7**(1):1–14.

2794. Wendin G. and Shumeiko V.S. (2007). Quantum bits with Josephson junctions (review article). *Low Temperature Physics.* **33**(9): 724–744.

2795. Yang S. and Sham L.J. (1987). Theory of magnetoexcitations in quantum wells. *Physical Review Letters.* **58**(24):2598.

Section 9

Nanotechnology Applications

This section of the book may be the most pertinent for readers because it cites literature related to real-world applications of nanotechnology. These are applications that impact the lives of millions across the globe. However, the majority of these millions are unaware that the very small impacts such large populations through nanotechnology's use in industrial, military, and commercial products, materials, and substances.

General Purpose Technology

2796. Fleischer T. et al. (2005). Assessing emerging technologies: methodological challenges and the case of nanotechnologies. *Technological Forecasting and Social Change.* **72**(9):1112–1121.

2797. Gambardella A. and McGahan A.M. (2010). Business-model innovation: general purpose technologies and their implications for industry structure. *Long Range Planning.* **43**(2):262–271.

2798. Graham S.J and Iacopetta M. (2014). Nanotechnology and the emergence of a general-purpose technology. *Annals of*

The Nanotechnology Revolution: A Global Bibliographic Perspective
Dale A. Stirling
Copyright © 2018 Pan Stanford Publishing Pte. Ltd.
ISBN 978-981-4774-19-2 (Hardcover), 978-1-315-11083-7 (eBook)
www.panstanford.com

Economics and Statistics/Annales d'Économie et de Statistique. **115–116**:25–55.

2799. Kreuchauff F. and Teichert N. (2014). Nanotechnology as general purpose technology (Working Paper Series in Economics No. 53). Karlsruher, Germany: Karlsruher Institute of Technology.

2800. Menz N. and Ott I. (2011). On the role of general purpose technologies within the Marshall-Jacobs controversy: the case of nanotechnologies (Working Paper Series in Economics No. 18). Karlsruher, Germany: Karlsruhe Institute of Technology, Department of Economics & Business Engineering.

2801. Nikulainen T. and Palmberg C. (2010). Transferring science-based technologies to industry: does nanotechnology make a difference? *Technovation.* **30**(1):3–11.

2802. Schaper-Rinkel P. (2013). The role of future-oriented technology analysis in the governance of emerging technologies: the example of nanotechnology. *Technological Forecasting and Social Change.* **80**(3):444–452.

2803. Schultz L. and Joutz F. (2010). Methods for identifying emerging general purpose technologies: a case study of nanotechnologies. *Scientometrics.* **85**(1):155–170.

2804. Shea C.M. et al. (2011). Nanotechnology as general-purpose technology: empirical evidence and implications. *Technology Analysis & Strategic Management.* **23**(2):175–192.

2805. Youtie J. et al. (2008). Assessing the nature of nanotechnology: can we uncover an emerging general purpose technology? *The Journal of Technology Transfer.* **33**(3):315–329.

Aerospace and Defense

2806. Altmann J. (2004). Military uses of nanotechnology: perspectives and concerns. *Security Dialogue.* **35**(1):61–79.

2807. Altmann J. (2006). *Military Nanotechnology.* New York: Routledge.

2808. Altmann J. (2007). *Military Nanotechnology: Potential Applications and Preventive Arms Control.* New York: Routledge.

2809. Altmann J. (2008). Military uses of nanotechnology—too much complexity for international security? *Complexity.* **14**(1):62–70.

2810. Altmann J. and Gubrud M. (2004). Anticipating military nanotechnology. *IEEE Technology and Society Magazine.* **23**(4): 33–40.

2811. Baur J. and Silverman E. (2007). Challenges and opportunities in multifunctional nanocomposite structures for aerospace applications. *MRS Bulletin.* **32**(04):328–334.

2812. Bellucci S. et al. (2007). CNT composites for aerospace applications. *Journal of Experimental Nanoscience.* **2**(3):193–206.

2813. Bielawski M. et al. (2015). Evaluation of doped amorphous carbon coatings for hydrophobic applications in aerospace. *Canadian Aeronautics and Space Journal.* **60**(3):1–12.

2814. Bilhaut L. and Duraffourg L. (2009). Assessment of nanosystems for space applications. *ACTA Astronautica.* 65(9–10): 1272–1283.

2815. Bradley L.D. (2012). Regulating weaponized nanotechnology: how the international criminal court offers a way forward. *Georgia Journal of International and Comparative Law.* **41**:723.

2816. De Neve A. Military uses of nanotechnology and converging technologies: trends and future impacts (Focus Paper No. 8). Brussels: Royal High Institute for Defence, Centre for Security and Defence Studies.

2817. Eichhorn B. et al. (2012). Smart functional nanoenergetic materials. Purdue University, Lafayette, IN. Available at http://www. dtic.mil/get-tr-doc/pdf?AD=ADA568241

2818. Ramaswamy A.L. et al. (2003). Weaponization and characterization of nanoenergetics (Report-51670). Berkeley, CA: Lawrence Berkeley National Laboratory.

2819. Son S.F. et al. (2006). Overview of nanoenergetic materials research at Los Alamos. *MRS Symposium Proceedings.* **896**: 87.

Agriculture and Allied Sciences

General Overviews

2820. Agrawal S. and Rathore P. (2014). Nanotechnology pros and cons to agriculture: a review. *International Journal of Current Microbiology and Applied Sciences.* **3**(3):43–55.

2821. Ali M.A. et al. (2014). Nanotechnology, a new frontier in agriculture. *Advancements in Life Sciences.* **1**(3):129–138.

2822. Bhagat Y. et al. (2015). Nanotechnology in agriculture: a review. *Journal of Pure & Applied Microbiology.* **9**(1).

2823. Card J.W. et al. (2011). An appraisal of the published literature on the safety and toxicity of food-related nanomaterials. *Critical Reviews in Toxicology.* **41**(1):20–49.

2824. Chen H. and Yada R. (2011). Nanotechnologies in agriculture: new tools for sustainable development. *Trends in Food Science & Technology.* **22**(11):585–594.

2825. Chinnamuthu C.R. and Boopathi P.M. (2009). Nanotechnology and agroecosystems. *Madras Agricultural Journal.* **96**:17–31.

2826. Chowdappa P. and Gowda S. (2013). Nanotechnology in crop protection: status and scope. *Pest Management in Horticultural Ecosystems.* **19**(2):131–151.

2827. Croney C.C. et al. (2012). The ethical food movement: what does it mean for the role of science and scientists in current debates about animal agriculture? *Journal of Animal Science.* **90**(5):1570–1582.

2828. Dhewa T. (2015). Nanotechnology applications in agriculture: an update. *Octa Journal of Environmental Research.* **3**(2).

2829. Ditta A. (2012). How helpful is nanotechnology in agriculture? *Advances in Natural Sciences: Nanoscience and Nanotechnology.* **3**(3):033002.

2830. Duncan T.V. (2011). The communication challenges presented by nanofoods. *Nature Nanotechnology.* **6**(11):683.

2831. García M. et al. (2010). Potential applications of nanotechnology in the agro-food sector. *Ciência e Tecnologia de Alimentos.* **30**(3):573–581.

2832. Grover M. et al. (2012). Nanotechnology: scope and limitations in agriculture. *International Journal of Nanotechnology and Applications.* **2**:10–38.

2833. Gul H.T. et al. (2014). Potential of nanotechnology in agriculture and crop protection: a review. *Applied Sciences & Business Economics.* **1**(2):23–28.

2834. Handford C.E. et al. (2014). Implications of nanotechnology for the agri-food industry: opportunities, benefits and risks. *Trends in Food Science & Technology.* **40**(2):226–241.

2835. Helal N.A.S. (2013). Nanotechnology in agriculture: a review. *Agriculture & Forestry.* **59**(1):117–142.

2836. Jahanban L. and Davari M. (2014). Organic agriculture and nanotechnology. *Building Organic Bridges.* **3**:679–682.

2837. Jha Z. et al. (2011). Nanotechnology: prospects of agricultural advancement. *Nano Vision.* **1**(2):88–100.

2838. Kanjana D. (2015). Potential applications of nanotechnology in major agriculture divisions: a review. *International Journal of Agriculture, Environment and Biotechnology.* **8**(3):699–714.

2839. Kuzma J. et al. (2008). Upstream oversight assessment for agrifood nanotechnology: a case studies approach. *Risk Analysis.* **28** (4):1081–1098.

2840. Lopes C.M. et al. (2013). Application of nanotechnology in the agro-food sector. *Food Technology and Biotechnology.* **51**(2):183.

2841. Lu J. and Bowles M. (2013). How will nanotechnology affect agricultural supply chains? *International Food and Agribusiness.* **16**:21–42.

2842. Mishra V.K. et al. (2013). Emerging consequence of nanotechnology in agriculture: an outline. *Trends in Biosciences.* **6**(5):503–506.

2843. Mukhopadhyay S.S. (2014). Nanotechnology in agriculture: prospects and constraints. *Nanotechnology, Science and Applications.* **7**:63.

2844. Mura S. et al. (2013. Advances of nanotechnology in agro-environmental studies. *Italian Journal of Agronomy.* **8**(3):18.

2845. Myczko A. (2006). The application of nanotechnology to the agricultural practice. *Inzynieria Rolnicza.* **2**(77):45–50.

2846. Opara L. (2004). Emerging technological innovation triad for smart agriculture in the 21st century. Part I. Prospects and impacts of nanotechnology in agriculture. *CIGR Journal of Scientific Research & Development.* **6**:1–27.

2847. Prasad R., Kumar V. and Prasad K.S. (2014). Nanotechnology in sustainable agriculture: present concerns and future aspects. *African Journal of Biotechnology.* **13**(6):705–713.

2848. Radha K. et al. (2014). Application of nano technology in dairy industry: prospects and challenges: a review. *Indian Journal of Dairy Science.* **67**:5.

2849. Rai M. and Ingle A. (2012). Role of nanotechnology in agriculture with special reference to management of insect pests. *Applied Microbiology and Biotechnology.* **94**(2):287–293.

2850. Rai V. et al. (2012). Implications of nanobiosensors in agriculture. *Journal of Biomaterials & Nanobiotechnology.* **3**:315–324.

2851. Samantarai S.K. and Achakzai A. (2014). Application of nanotechnology in agriculture and food production: opportunity and challenges. *Middle-East Journal of Scientific Research.* **22**(4):499–501.

2852. Sanjog T. (2013). Nanotechnology in agroecosystem: implications on plant productivity and its soil environment. *Expert Opinion on Environmental Biology.* **2**:1.

2853. Scrinis G. and Lyons K. (2007). The emerging nano-corporate paradigm: nanotechnology and the transformation of nature, food and agri-food systems. *International Journal of Sociology of Food and Agriculture.* **15**(2):22–44.

2854. Sekhon B.S. (2014). Nanotechnology in agri-food production: an overview. *Nanotechnology, Science and Applications.* **7**: 31.

2855. Sharon M. et al. (2010). Nanotechnology in agricultural diseases. *Journal of Phytology.* **2**:83–92.

2856. Siddiqui M.H. et al. (eds.) (2015). *Nanotechnology and Plant Sciences: Nanoparticles and Their Impact on Plants.* Berlin: Springer.

2857. Singh A.K., Lal M., Singh S.P., Khan A.A., Singh S.P. and Tiwari A.K. (2015). Scope of nanotechnology in future agriculture: an overview. *Agrica.* **3**(1 and 2):1–13.

2858. Singh G. and Rattanpal H. (2014). Use of nanotechnology in horticulture: a review. *International Journal of Agricultural Sciences and Veterinary Medicine.* **2**(1):34–42.

2859. Srilatha B. (2011). Nanotechnology in agriculture. *Journal of Nanomedicine & Nanotechnology.* **2**:123.

2860. Sutovsky P. and Kennedy C.E. (2013). Biomarker-based nanotechnology for the improvement of reproductive performance in beef and dairy cattle. *Industrial Biotechnology.* **9**(1):24–30.

2861. Troncarelli M.Z. et al. (2013). Bovine mastitis under nanocontrol: nanostructured propolis as a new perspective of treatment for organic dairy cattle. *Veterinária e Zootecnia.* **20**(Special edition):124–136.

2862. Vandermoere F. et al. (2011). The public understanding of nanotechnology in the food domain: the hidden role of views on science, technology, and nature. *Public Understanding of Science.* **20**(2):195–206.

2863. Verma A.K. et al. (2012). Application of nanotechnology as a tool in animal products processing and marketing: an overview. *American Journal of Food Technology.* **7**(8):445–451.

2864. Vijayalakshmi C. et al. (2014). Trendy usage of nanotechnology in agriculture: a review. In *2014 International Conference on Science Engineering and Management Research (ICSEMR),* Chennai, pp. 1–4.

2865. Vijayalakshmi C. et al. (2015). Modern approaches of nanotechnology in agriculture: a review. *Biosciences Biotechnology Research Asia.* **12**(1):327–331.

2866. Yawson R.M. and Kuzma J. (2010). Systems mapping of consumer acceptance of agrifood nanotechnology. *Journal of Consumer Policy.* **33**(4):299–322.

Animal Production

2867. Guo T. et al. (2004). Application prospects of nanotechnology in animal husbandry. *Chinese Journal of Animal Science.* **1:** 019.

2868. Kuzma J. (2010). Nanotechnology in animal production: up-stream assessment of applications. *Livestock Science.* **130**(1): 14–24.

2869. Ramírez Mella M. and Hernández Mendo O. (2010). Nano-technology on animal production. *Tropical and Subtropical Agroecosystems.* **12**(3):423–429.

2870. Thornton P.K. (2010). Livestock production: recent trends, future prospects. *Philosophical Transactions of the Royal Society of London B: Biological Sciences.* **365**(1554):2853–2867.

Aquaculture and Fisheries

2871. Alishahi A. (2015). Application of nanotechnology in marine-based products: a review. *Journal of Aquatic Food Product Technology.* **24**(5):533–543.

2872. Ashraf M. et al. (2011). Nanotechnology as a novel tool in fisheries and aquaculture development: a review. *Iranian Journal of Energy & Environment.* **2**(3):258–261.

2873. Bhattacharyya A. et al. (2015). Nanotechnology-a unique future technology in aquaculture for the food security. *International Journal of Bioassays.* **4**(07):4115–4126.

2874. Birkelund K. et al. (2011). Enhanced polymeric encapsulation for MEMS based multi sensors for fisheries research. *Sensors and Actuators A: Physical.* **170**(1):196–201.

2875. Myhr A.I. et al. (2011). Precaution or integrated responsibility approach to nanovaccines in fish farming? A critical appraisal of the UNESCO precautionary principle. *NanoEthics.* **5**(1):73–86.

2876. Myskja B.K. et al. (2010). Nanotechnology and trust in aquaculture. In *Global Food Security: Ethical and Legal Challenges.* EurSafe 2010, Bilbao, Spain: Wageningen Academic Publishers, September 16–18, pp. 110–115.

2877. Rather M.A. et al. (2011). Nanotechnology: a novel tool for aquaculture and fisheries development. A prospective mini-review. *Fish Aquaculture.* **2011**:1–5.

2878. Royl S. et al. (2012). Nanotechnology and its application in fisheries and aquaculture. *Aqua International.* **20**(6):26–27.

2879. Yong-liang D. (2002). Prospective for application of nano science and technology on aquaculture in the 21th century. *Modern Fisheries Information.* **2**:002.

2880. Yong-liang D. (2003). NANOST new technology used in fish farming. *Modern Fisheries Information.* **12**:000.

2881. Yong-liang D. (2005). New trends of nano-ST fish farming. *Nanoscience & Technology.* **2**:003.

Crop Production

2882. Kalpana S. (2010). Nanotechnology-based precision farming technologies: an assessment based on R&D indicators. *Society for Technology Management-Newsletter.* Available at http://eprints.naarm.org.in/id/eprint/156

2883. Khot L.R. et al. (2012). Applications of nanomaterials in agricultural production and crop protection: a review. *Crop Protection.* **35**:64–70.

2884. Kole C. et al. (2013). Nanobiotechnology can boost crop production and quality: first evidence from increased plant biomass, fruit yield and phytomedicine content in bitter melon (Momordica charantia). *BMC Biotechnology.* **13**(1): 37.

2885. Laing H.Y. et al. (2008). Application of nanotechnology for preservation of fruit and vegetable. *Storage & Process.* **5**:021.

2886. Lee W.S. et al. (2010). Sensing technologies for precision specialty crop production. *Computers and Electronics in Agriculture.* **74**(1):2–33.

2887. Miller G. (2007). Nanotechnology in agriculture and food production-which food future? *Chain Reaction.* **100**:17–19.

2888. Mousavi S.R. and Rezaei M. (2011). Nanotechnology in agriculture and food production. *Journal of Applied Environmental and Biological Sciences.* **1**(10):414–419.

2889. Ouzounidou G. and Gaitis F. (2011). The use of nano-technology in shelf life extension of green vegetables. *Journal of Innovation Economics & Management.* **8**(2):163–171.

2890. Priester J.H. et al. (2012). Soybean susceptibility to manufac-tured nanomaterials with evidence for food quality and soil fertility interruption. *Proceedings of the National Academy of Sciences.* **109**(37):E2451–E2456.

2891. Sastry R.K. et al. (2013). Nanotechnology in food processing sector: an assessment of emerging trends. *Journal of Food Science and Technology.* **50**(5):831–841.

2892. Sekhon B.S. (2014). Nanotechnology in agri-food production: an overview. *Nanotechnology, Science and Applications.* **7**:31.

2893. Wang H.J. et al. (2011). Application of nanotechnology in post-harvest fruits and vegetables. *Food Science and Technology.* **6**:023.

Forestry and Logging

2894. Beecher J.F. (2007). Organic materials: wood, trees and nanotechnology. *Nature Nanotechnology.* **2**(8):466–467.

2895. Evans P. et al. (2008). Large-scale application of nanotechnol-ogy for wood protection. *Nature Nanotechnology.* **3**(10):577–577.

2896. Howe D.J. et al. (2006). Nanotechnology and the forest products industry: exciting new possibilities. Minneapolis: Dovetail Partners, Incorporated.

2897. Liu Y. et al. (2004). Nanotechnology and the development tendency of nano-wood science. *Journal of Central South Forestry University.* **2004**:5.

2898. Moon R.J. et al. (2006). Nanotechnology applications in the forest products industry. *Forest Products Journal.* **56**(5):4–10.

2899. Wegner T. and Jones P. (2005). Nanotechnology for the forest products industry. *Wood and Fiber Science.* **37**(4):549–551.

2900. Wegner T.H. and Jones P.E. (2006). Advancing cellulose-based nanotechnology. *Cellulose.* **13**(2):115–118.

2901. Wipf T. et al. (2012). Literature review and assessment of nanotechnology for sensing of timber transportation structures final report (General technical report FPL-GTR-210). Madison, WI: U.S. Department of Agriculture, Forest Service, Forest Products Laboratory.

Biology

2902. Fox R. (2004). Nanobiology: the interplay between biology and nanoscience. In *71st Annual Meeting of the Southeastern Section, Oak Ridge, TN.*

2903. Kabanov A.V. et al. (2009). Nanobiology for the pharmacology of cellular ion channels. *Journal of Neuroimmune Pharmacology.* **4**(1):7–9.

2904. Kelly T.F. et al. (2009). Prospects for nanobiology with atom-probe tomography. *MRS Bulletin.* **34**(10):744–750.

2905. Kim S.K. et al. (2011). Editorial on nanobiology. *Physical Chemistry Chemical Physics.* **13**(21):9916–7.

2906. Kroll A. (2012). Nanobiology—convergence of disciplines inspires great applications. *Cellular and Molecular Life Sciences.* **69**(3):335–336.

2907. Mishustina I.E. (2004). Nanobiology of the ocean. *Biology Bulletin of the Russian Academy of Sciences.* **31**(5):495–497.

2908. Naka T. and Tada K. (2003). Trends in nanobiology. *Science and Technology Trends: Quarterly Review.* **5**:68–75.

2909. Nussinov R. and Alemán C. (2006). Nanobiology: from physics and engineering to biology. *Physical Biology.* **3**(1).

2910. Pettitt B.M. (2006). Editorial: nanobiology. *Molecular Simulation.* **31**(10–11):773.

2911. Nussinov R. and Alemán C. (2006). Nanobiology: from physics and engineering to biology. *Physical Biology.* **3**(1).

2912. Spence C.F. (2002). Nanobiology: halting steps into a portion of Richard Feynman's vision. PhD Dissertation: California Institute of Technology.

2913. Thangavel H. (2014). Nanobiology in medicine. In *Nanomedicine.* New York: Springer, pp. 15–31.

2914. Xia T. et al. (2008). Nanobiology: particles slip cell security. *Nature Materials.* **7**(7):519–520.

2915. Xie S.S. (2006). [Importance of nanobiology and nano-medicine]. Zhongguo yi xue ke xue yuan xue bao. *Acta Academiae Medicinae Sinicae.* **28**(4):469–471.

Catalysis

2916. An K. and Somorjai G. (2015). Nanocatalysis I: synthesis of metal and bimetallic nanoparticles and porous oxides and their catalytic reaction studies. *Catalysis Letters.* **145**(1):233–248.

2917. Alayoglu S. and Somorjai G. (2015). Nanocatalysis II: in situ surface probes of nano-catalysts and correlative structure-reactivity studies. *Catalysis Letters.* **145**(1):249–271.

2918. Balasanthiran C. and Hoefelmeyer J.D. (2014). Nanocatalysis: definition and case studies. *Metal Nanoparticles for Catalysis.* (17):6–29.

2919. Gai P.L. et al. (2002). Recent advances in nanocatalysis research. *Current Opinion in Solid State and Materials Science.* **6**(5):401–406.

2920. Gellman A.J. and Shukla N. (2009). Nanocatalysis: more than speed. *Nature Materials.* **8**(2):87–88.

2921. Grabow L.C. and Mavrikakis M. (2008). Nanocatalysis beyond the gold-rush era. *Angewandte Chemie International Edition.* **47**(39):7390–7392.

2922. John N.T. and Francis R.R. (2013). Applications of nanocatalysis in petroleum industry. *Journal of NanoScience, NanoEngineering & Applications.* **3**(2).

2923. Jiang C. et al. (2014). Mechanistic understanding of toxicity from nanocatalysts. *International Journal of Molecular Sciences.* **15**(8):13967–13992.

2924. Kalidindi S.B. and Jagirdar B.R. (2012). Nanocatalysis and prospects of green chemistry. *ChemSusChem.* **5**(1):65–75.

2925. Mohamed R.M. et al. (2012). Enhanced nanocatalysts. *Materials Science & Engineering.* **73**(1):1–13.

2926. Olveira S. et al. (2014). Nanocatalysis: academic discipline and industrial realities. *Journal of Nanotechnology.* **2014**: Article ID 324089.

2927. Pacchioni G. (2009). Nanocatalysis: staying put. *Nature Materials.* **8**(3):167–168.

2928. Polshettiwar V. and Asefa T. (eds.) (2003). *Nanocatalysis: Synthesis and Applications.* Hoboken, NJ: John Wiley & Sons.

2929. Schlögl R. and Abd Hamid S.B. (2004). Nanocatalysis: mature science revisited or something really new? *Angewandte Chemie International Edition.* **43**(13):1628–1637.

2930. Somorjai G.A. et al. (2009). Advancing the frontiers in nanocatalysis, biointerfaces, and renewable energy conversion by innovations of surface techniques. *Journal of the American Chemical Society.* **131**(46):16589–16605.

2931. Zhong Z. et al. (2014). A special section on nanocatalysis and their applications. *Journal of Nanoscience and Nanotechnology.* **14**(9):6789.

2932. Zhu Y. and Hosmane N.S. (2015). Applications of nanocatalysis in boron chemistry. *Coordination Chemistry Reviews.* **293**:357–367.

2933. Zibareva I.V. et al. (2014). Nanocatalysis: a bibliometric analysis. *Kinetics & Catalysis.* **55**(1):1–11.

Chemistry

2934. Ambrogi P. et al. (2008). Make sense of nanochemistry and nanotechnology. *Chemistry Education Research and Practice.* **9**(1):5–10.

2935. Antonietti M. et al. (2001). The vision of "nanochemistry", or is there a promise for specific chemical reactions in nano-restricted environments? *Israel Journal of Chemistry.* **41**(1):1–6.

2936. Arsenault A. and Ozin G.A. (2005). *Nanochemistry: A Chemical Approach to Nanomaterials.* Cambridge: Royal Society of Chemistry.

2937. Ball P. (2005). Putting the nano into nanochemistry. *Chemistry World.* **2**(12):28–33.

2938. Bal'makov M.D. (2002). Information basis of nanochemistry. *Russian Journal of General Chemistry.* **72**(7):1023–1030.

2939. Bao-gui H. (2004). Developments of nanochemistry. *Journal of Wenshan Teachers' College.* **3**:015.

2940. Buchachenko A.L. (2003). Nanochemistry as a direct route to high technology of the new century. *Uspekhi Khimii.* **72**(5):419–437.

2941. Buchachenko A.L. (2009). Nanochemistry and magnetism. *Russian Journal of Physical Chemistry A.* **83**(10):1637–1642.

2942. Cademartiri L. and Ozin G.A. (2009). *Concepts of Nanochemistry.* New York: John Wiley & Sons.

2943. Chasteauneuf A. (2008). Keeping it clean with nanochemical technology. *Nutraceutical Business & Technology.* **4**(6):40–42.

2944. DePalma A. (2003). Nanochemistry brings big ideas to drug development. *Drug Discovery & Development.* **6**(10):47.

2945. Gedanken A. (2003). Sonochemistry and its application to nanochemistry. *Current Science-Bangalore.* **85**(12):1720–1722.

2946. Han B.H. (2015). Nanochemistry. *Chinese Journal of Chemistry.* **33**(1):5–5.

2947. Kharissova O.V. et al. (2011). Ultrasound in nanochemistry: recent advances. *Synthesis and Reactivity in Inorganic, Metal-Organic, and Nano-Metal Chemistry.* **41**(5):429–448.

2948. Masuhara H. et al. (2006). Laser nanochemistry. *Pure and Applied Chemistry.* **78**(12):2205–2226.

2949. Nishide H. et al. (2006). Practical nano-chemistry. *Science and Technology of Advanced Materials.* **7**:395–396.

2950. Ozin G.A. (1992). Nanochemistry: synthesis in diminishing dimensions. *Advanced Materials.* **4**(10):612–649.

2951. Ozin G.A. and Arsenault A.C. (2008). *Nanochemistry: A Chemical Approach to Nanomaterials.* London: Royal Society of Chemistry.

2952. Ozin G.A. and Cademartiri L. (2009). Nanochemistry: what is next? *Small.* **5**(11):1240–1244.

2953. Ozin G.A. and Cademartiri L. (2011). From ideas to innovation: nanochemistry as a case study. *Small.* **7**(1):49–54.

2954. Pagliaro M. (2015). Advancing nanochemistry education. *Chemistry-A European Journal.* **21**(34):11931–11936.

2955. Sergeev G. B. (2001). Nanochemistry of metals. *Russian Chemical Reviews.* **70**(10):809–825.

2956. Shiozawa H. et al. (2009). Screening the missing electron: nanochemistry in action. *Physical Review Letters.* **102**(4): 046804.

2957. Whiteside G.M. et al. (1991). Molecular self-assembly and nanochemistry: a chemical strategy for the synthesis of nanostructures. *Science.* **254**:1312–1319.

Computing

2958. Anderson N.G. and Bhanja S. (eds.) (2014). *Field-Coupled Nanocomputing: Paradigms, Progress, and Perspectives,* vol. 8280. New York: Springer.

2959. Arlat J. et al. (2012). Nanocomputing: Small devices, large dependability challenges. *IEEE Security & Privacy.* **10**(1):69– 72.

2960. Beckett P. and Jennings A. (2002). Towards nanocomputer architecture. *Australian Computer Science Communications.* **24**(3):141–150.

2961. Bourianoff G. (2003). The future of nanocomputing. *Computer.* **36**(8):44–53.

2962. Chen J. et al. (2003). A probabilistic approach to nanocomputing. In *IEEE Non-Silicon Computer Workshop.* San Diego: IEEE Press.

2963. Durbeck L.J. and Macias N.J. (2001). The cell matrix: an architecture for nanocomputing. *Nanotechnology.* **12**(3): 217.

2964. Frank M.P. and Knight Jr, T.F. (1998). Ultimate theoretical models of nanocomputers. *Nanotechnology.* **9**(3):162.

2965. Gopal T.V. (2005). Nanocomputing–trends, directions and applications. Lecture series "next generation information technology—a quantum leap." Chennai, India: Loyola College.

2966. Graham P. and Gokhale M. (2004). Nanocomputing in the presence of defects and faults: a survey. In *Nano, Quantum and Molecular Computing*. US: Springer, pp. 39–72.

2967. Hall J.S. (1994). Nanocomputers and reversible logic. *Nanotechnology*. **5**(3):157.

2968. Hirwani D. and Sharma A. (2012). Nano computing revolution and future prospects. *Recent Research in Science and Technology*. **4**(3):16–17.

2969. Hsu J.J. and Hsu J.Y. (2009). *Nanocomputing: Computational Physics for Nanoscience and Nanotechnology*. Singapore: Pan Stanford Publishing.

2970. Lee R.K. and Hill J.M. (2011). Nanocomputing memory devices formed from carbon nanotubes and metallofullerenes. *World Academy of Science, Engineering and Technology*. **60**: 1413–1416.

2971. Lee T.H. and Dickson R.M. (2004). Nanocomputing with nanoclusters. *Optics and Photonics News*. **15**(6):22–27.

2972. Lieber C.M. (2003). Nanoscience and the pathway towards nanocomputing. Available at http://evideo.lib

2973. Nowak S. (2001). Nanotechnology and nanocomputers. *Archiwum Informatyki Teoretycznej i Stosowanej*. **13**(2):195–204.

2974. Robinson A.L. (1983). Nanocomputers from organic molecules? *Science*. **220**:940–942.

2975. Sager J. et al. (2008). Nanocomputing. In *NanoBioTechnology*. New York: Humana Press, pp. 215–265.

2976. Sahni V. (2008). *Nanocomputing*. New York: Tata McGraw-Hill Education.

2977. Service R.F. (2001). Nanocomputing. Assembling nanocircuits from the bottom up. *Science*. **293**(5531):782.

2978. Service R.F. (2001). Nanocomputing. World's smallest transistor. *Science*. **293**(5531):786.

2979. Uusitalo M.A. et al. (2011). Machine learning: how it can help nanocomputing. *Journal of Computational and Theoretical Nanoscience.* **8**(8):1347–1363.

2980. Waldner J.B. (2013). *Nanocomputers and Swarm Intelligence.* New York: John Wiley & Sons.

Construction

General Overviews

2981. Akhnoukh A.K. (2013). Overview of nanotechnology applications in construction industry in the United States. *Micro and Nanosystems.* **5**(2):147–153.

2982. Alsaffar K.A. (2014). Review of the use of nanotechnology in construction industry. *International Journal of Engineering Research & Development.* **10**(8):67–70.

2983. Amaya D. (2012). Nanotechnology solutions for road infrastructure. *Routes/Roads.* Issue No. 355.

2984. Bartos P.J.M. et al. (eds.) (2004). *Nanotechnology in Construction* (Special Publication No. 292). Cambridge, England: Royal Society of Chemistry.

2985. Davies A. and Gann D. (2002). A review of nanotechnology and its potential applications for construction. Science and Policy Research Unit (SPRU). Brighton: University of Sussex.

2986. Halicioglu F.H. (2009). The potential benefits of nanotechnology for innovative solutions in the construction sector. In *Nanotechnology in Construction 3.* Berlin Heidelberg: Springer, pp. 209–214.

2987. Hanus M.J. and Harris A.T. (2013). Nanotechnology innovations for the construction industry. *Progress in Materials Science.* **58**(7):1056–1102.

2988. Khitab A. (2014). Nano construction materials: a review. *Reviews on Advanced Materials Science.* 181–189.

2989. Lee J. et al. (2010). Nanomaterials in the construction industry: a review of their applications and environmental health and safety considerations. *ACS Nano.* **4**(7):3580–3590.

2990. Pacheco-Torgal F. and Jalali S. (2011). Nanotechnology: advantages and drawbacks in the field of construction and building materials. *Construction and Building Materials.* **25**(2):582–590.

2991. Porwal Y. (2014). Nanotechnology in construction and civil engineering. *International Journal of Emerging Trends in Science and Technology.* **1**(06):845–856.

2992. Rana A.K. et al. (2009). Significance of nanotechnology in construction engineering. *International Journal of Recent Trends in Engineering.* **1**(4):46–48.

2993. Sinha M. (2014). Nanotechnology in civil engineering and construction: a review on state of the art and future prospects. *International Journal of Global Technology Initiatives.* **3**(1):A31–A36.

2994. Srivastava A. and Singh K. (2011). Nanotechnology in civil engineering and construction: a review on state of the art and future prospects. In *Proceedings of Indian Geotechnical Conference* (Paper No. R-024), Kochi, December 15–17.

2995. Teizer J. et al. (2011). Nanotechnology and its impact on construction: bridging the gap between researchers and industry professionals. *Journal of Construction Engineering and Management.* **138**(5):594–604.

2996. Uzoegbo H. (2014). A review of current and future developments in the application of nanotechnology to construction materials. In *Proceedings of the 1st African International Conference/Workshop on Applications of Nanotechnology to Energy, Health and Environment*, March 23–29.

2997. Zhu W. et al. (2004). Application of nanotechnology in construction. *Materials and Structures.* **37**(9):649–658.

Construction Materials

2998. Apicella A. et al. (2010). Application of nanostructured smart materials in sustainable buildings. *International Journal of Sustainable Manufacturing.* **2**(1):66–79.

2999. Arora S.K. et al. (2014). Drivers of technology adoption—the case of nanomaterials in building construction. *Technological Forecasting and Social Change.* **87**:232–244.

3000. Azahar W. et al. (2015). Application of nanotechnology in asphalt binder: a conspectus and overview. *Jurnal Teknologi.* **76**(14).

3001. Chandrika R. et al. (2015). Application of nanotechnology in building materials. *International Journal of Applied Engineering Research.* **10**(33):2015.

3002. Feldman D. (2014). Polymer nanocomposites in building, construction. *Journal of Macromolecular Science, Part A.* **51**(3): 203–209.

3003. Jones W. et al. (2015). Nanomaterials in construction and demolition-how can we assess the risk if we don't know where they are? *Journal of Physics: Conference Series.* **617**(1): 012031.

3004. Khitab A. and Arshad M.T. (2014). Nano construction materials: review. *Reviews on Advanced Materials Science.* **38**:181–189.

3005. Lee J. et al. (2010). Nanomaterials in the construction industry: a review of their applications and environmental health and safety considerations. *ACS Nano.* **4**:3580–3590.

3006. Pacheco-Torgal F. and Jalali S. (2011). Nanotechnology: advantages and drawbacks in the field of construction and building materials. *Construction and Building Materials.* **25**(2):582–590.

3007. Sanchez F. and Sobolev K. (2010). Nanotechnology in concrete: a review. *Construction and Building Materials.* **24**(11):2060–2071.

3008. Sobolev K. and Gutiérrez M.F. (2005). How nanotechnology can change the concrete world. *American Ceramic Society Bulletin.* **84**(10):14.

3009. Sobolev K. et al. (2009). Bibliography on application of nanotechnology and nanomaterials in concrete. Skokie, IL: Portland Cement Association.

3010. Wong S. (2014). An overview of nanotechnology in building materials. *Canadian Young Scientist Journal.* **2014**(2):18–21.

Green and Sustainable

3011. Andersen M.M. and Geiker M.R. (2009). Nanotechnologies for climate friendly construction–key issues and challenges. In *Nanotechnology in Construction 3*. Berlin Heidelberg: Springer, pp. 199–207.

3012. Daryoush B. and Darvish A. (2013). A case study and review of nanotechnology and nanomaterials in green architecture. *Research Journal of Environmental and Earth Sciences.* **5**(2):78–84.

3013. Elvin G. (2007). Nanotechnology for green building, green technology forum. Available online at http://esonn.fr/esonn2010/xlectures/mangematin/Nano_Green_Building 55ex.pdf

3014. Karimimoshaver M. et al. (2014). A study of the place of nanotechnology in green roof design for reducing energy consumption. *Research Journal of Recent Sciences.* **4**(5):95–102.

3015. Kibert C.J. (2012). *Sustainable Construction: Green Building Design and Delivery.* New York: John Wiley & Sons.

3016. Samer M. (2013). Towards the implementation of the green building concept in agricultural buildings: a literature review. *Agricultural Engineering International: CIGR Journal.* **15**(2):25–46.

3017. Sev A. and Ezel M. (2014). Nanotechnology innovations for the sustainable buildings of the future. *International Journal of Civil, Environmental, Structural Construction & Architectural Engineering.* **8**(8):872–882.

3018. Smith G.B. and Claes-Goran S.G. (2010). *Green Nanotechnology: Solutions for Sustainability and Energy in The Built Environment.* Boca Raton, FL: CRC Press.

Heavy and Civil Engineering Construction

3019. Faruqi M. et al. (2015). State-of-the-art review of the applications of nanotechnology in pavement materials. *Journal of Civil Engineering Research.* **5**(2):21–27.

3020. Gopalakrishnan K. et al. (2011). *Nanotechnology in Civil Infrastructure*. Berlin Heidelberg: Springer.

3021. Ugwu O.O. (2013). Nanotechnology for sustainable infrastructure in 21st century civil engineering. *Journal of Innovative Engineering.* **1**(1):2.

3022. Zheng W. et al. (2010). Impact of nanotechnology on future civil engineering practice and its reflection in current civil engineering education. *Journal of Professional Issues in Engineering Education and Practice.* **137**(3):162–173.

3023. Yan J. et al. (2012). The comprehensive utilizing method for urban rain-flood based on nanotechnology. *Applied Mechanics and Materials.* **110**:3867–3872.

3024. Steyn W.J. (2009). Potential applications of nanotechnology in pavement engineering. *Journal of Transportation Engineering.* **135**(10):764–772.

3025. Ugwu O.O. et al. (2012). Nanotechnology as a preventive engineering solution to highway infrastructure failures. *Journal of Construction Engineering and Management.* **139**(8):987–993.

Cosmetics and Personal Care Products

3026. Mihranyan A. et al. (2012). Current status and future prospects of nanotechnology in cosmetics. *Progress in Materials Science.* **57**(5):875–910.

3027. Mu L. and Sprando R.L. (2010). Application of nanotechnology in cosmetics. *Pharmaceutical Research.* **27**(8):1746–1749.

3028. Silpa R. et al. (2012). Nanotechnology in cosmetics: opportunities and challenges. *Journal of Pharmacy and Bioallied Sciences.* **4**(3):186–193.

3029. Starzyk E. et al. (2008). Nanotechnology: does it have a future in cosmetics? *SÖFW Journal.* **134**(6).

Cultural and Historic Preservation

3030. Baglioni P. et al. (2003). Nanotechnology for wall paintings conservation. In Robinson B.H. (ed.), *Self-Assembly*. Amsterdam: IOS Press, pp. 32–41.

3031. Baglioni P. et al. (2006). The maya site of Calakmul: in situ preservation of wall paintings and limestone using nanotechnology. *Studies in Conservation.* **51**(Suppl 2):162–169.

3032. Baglioni M. et al. (2012). Smart cleaning of cultural heritage: a new challenge for soft nanoscience. *Nanoscale.* **4**(1):42–53.

3033. Casini M. (2014). Smart materials and nanotechnology for energy retrofit of historic buildings. *International Journal of Civil and Structural Engineering.* **1**(3):88–97.

3034. Dei L. and Salvadori B. (2006). Nanotechnology in cultural heritage conservation: nanometric slaked lime saves architectonic and artistic surfaces from decay. *Journal of Cultural Heritage.* **7**(2):110–115.

3035. Di Salvo S. (2014). Nanotechnology for cultural heritage. *International Journal of Science, Technology & Society.* **2**(2):28–32.

3036. Giorgi R. et al. (2002). Nanotechnologies for conservation of cultural heritage: paper and canvas deacidification. *Langmuir.* **18**(21):8198–8203.

3037. Giorgi R. et al. (2005). Nanoparticles of calcium hydroxide for wood conservation. The deacidification of the Vasa warship. *Langmuir.* **21**(23):10743–10748.

3038. Giorgi R. et al. (2010). Nanoparticles for cultural heritage conservation: calcium and barium hydroxide nanoparticles for wall painting consolidation. *Chemistry-A European Journal.* **16**(31):9374–9382.

3039. La Russa M.F. et al. (2012). Multifunctional TiO2 coatings for cultural heritage. *Progress in Organic Coatings.* **74**(1):186–191.

Earth Sciences

3040. Banfield J.F. et al. (2010). A perspective on nanominerals and their roles in microbial ecosystems. *Geochimica et Cosmochimica Acta.* **74**(12):A48.

3041. Harris P. (2003). it's a small world: nanominerals' growing influence. *Industrial Minerals.* (433):60–63.

3042. Hochella M.F. (2002). Nanoscience and technology: the next revolution in the Earth sciences. *Earth and Planetary Science Letters.* **203**(2):593–605.

3043. Hochella M.F. (2002). There's plenty of room at the bottom: nanoscience in geochemistry. *Geochimica et Cosmochimica Acta.* **66**(5):735–743.

3044. Hochella M.F. et al. (2008). Nanominerals, mineral nanoparticles, and earth systems. *Science.* **319**(5870):1631–1635.

3045. Jones R. (2007). Are natural resources a curse? *Nature Nanotechnology.* **2**(11):665–666.

3046. Patil G. et al. (2011). Toxicological concerns of nanominerals: old problem with new challenges. *Journal of Bionanoscience.* **5**(1):33–40.

3047. Wang Y. et al. (2003). Nanogeochemistry: geochemical reactions and mass transfers in nanopores. *Geology.* **31**(5):387–390.

Electronics

3048. Akinwande D. et al. (2014). Two-dimensional flexible nanoelectronics. *Nature Communications.* **2014**: Article ID 5678.

3049. Allen R. (2003). Nano-technology: the next revolution to redefine electronics. *Electronic Design.* Available online at http://electronicdesign.com/components/nanotechnology-next-revolution-redefine-electronics.

3050. Allsop M. et al. (2007). Nanotechnologies and nanomaterials in electrical and electronic goods: a review of uses and health concerns (Greenpeace Research Laboratories Technical Note 09-2007). Available at http://www.greenpeace.to/publications.asp#2007. Accessed online August 25, 2015.

3051. Bandyopadhyay S. and Roychowdhury V.P. (1996). Granular nanoelectronics. *IEEE Potentials.* **15**(2):8–11.

3052. Bate R.T. et al. (1987). Nanoelectronics (Final Technical Report). A report prepared for the U.S. Army Research Office. Dallas, TX: Texas Instruments Incorporated.

3053. Beaumont S.P. (1996). III–V nanoelectronics. *Microelectronic Engineering.* **32**(1):283–295.

3054. Brosseau C. (2011). Emerging technologies of plastic carbon nanoelectronics: a review. *Surface and Coatings Technology.* **206**(4):753–758.

3055. Cerofolini G.F. and Ferla G. (2002). Toward a hybrid micro-nanoelectronics. *Journal of Nanoparticle Research.* **4**:185–191.

3056. Cerofolini G.F. et al. (2005a). A hybrid approach to nanoelectronics. *Nanotechnology.* **16**(8):1040.

3057. Cerofolini G.F. et al. (2005b). Strategies for nanoelectronics. *Microelectronic Engineering.* **81**(2):405–419.

3058. Chau R. et al. (2007). Integrated nanoelectronics for the future. *Nature Materials.* **6**(11):810–812.

3059. Che Y. et al. (2014). Review of carbon nanotube nano-electronics and macroelectronics. *Semiconductor Science and Technology.* **29**(7):073001.

3060. Cohen-Karni T. et al. (2012). The smartest materials: the future of nanoelectronics in medicine. *ACS Nano.* **6**(8):6541–6545.

3061. Compañó R. (2001). Trends in nanoelectronics. *Nanotechnology.* **12**(2):85.

3062. Declerck G. (2005). A look into the future of nanoelectronics. *VLSI Technology Digest of Technical Papers.* 6–10.

3063. Duan X. et al. (2013). Nanoelectronics-biology frontier: from nanoscopic probes for action potential recording in live cells to three-dimensional cyborg tissues. *Nano Today.* **8**(4):351–373.

3064. Durkan C. (2007). *Current at The Nanoscale: An Introduction to Nanoelectronics.* London: Imperial College Press.

3065. Fahrner W.R. (2005). *Nanotechnology and Nanoelectronics.* Berlin: Springer.

3066. Feiner L.F. (2006). Nanoelectronics: crossing boundaries and borders. *Nature Nanotechnology.* **1**(2):91–92.

3067. Fink D. et al. (2005). Ion track-based nanoelectronics. *International Journal of Nanoscience.* **4**(05–06):965–973.

3068. Forshaw M. et al. (2004). A short review of nanoelectronic architectures. *Nanotechnology.* **15**(4):S220.

3069. Fountain T.J. et al. (1998). The use of nanoelectronic devices in highly parallel computing systems. *IEEE Transactions on Very Large Scale Integration (VLSI) Systems.* **6**(1):31–38.

3070. Frazier G. (1988). An ideology for nanoelectronics. In *Concurrent Computations.* New York: Springer, pp. 3–21.

3071. Gargini P.A. (2004). Silicon nanoelectronics and beyond. *Journal of Nanoparticle Research.* **6**(1):11–26.

3072. Hanson G.W. (2008). *Fundamentals of Nanoelectronics.* Upper Saddle River, NJ: Pearson/Prentice Hall.

3073. Haselman M. and Hauck S. (2010). The future of integrated circuits: a survey of nanoelectronics. *Proceedings of the IEEE.* **98**(1):11–38.

3074. Heikkilä T.T. (2013). *The Physics of Nanoelectronics: Transport and Fluctuation Phenomena at Low Temperatures.* London: Oxford University Press.

3075. Hoenlein W. et al. (2006). Nanoelectronics beyond silicon. *Microelectronic Engineering.* **83**(4):619–623.

3076. Hu C. (1999). Silicon nanoelectronics for the 21st century. *Nanotechnology.* **10**(2):113.

3077. Iniewski K. (2010). *Nanoelectronics: Nanowires, Molecular Electronics, and Nanodevices.* New York: McGraw Hill Professional.

3078. Ionescu A.M. (2012). Nanoelectronics: ferroelectric devices show potential. *Nature Nanotechnology.* **7**(2):83–85.

3079. Ismail R. et al. (eds.) (2012). *Advanced Nanoelectronics.* Boca Raton, FL: CRC Press.

3080. Korkin A. and Rosei F. (eds.) (2008). *Nanoelectronics and Photonics: From Atoms to Materials, Devices, and Architectures.* New York: Springer Science & Business Media.

3081. Liebau M. et al. (2005). Nanoelectronics based on carbon nanotubes: technological challenges and recent developments. *Fullerenes, Nanotubes, and Carbon Nanostructures.* **13**(S1):255–258.

3082. Lu W. and Lieber C.M. (2007). Nanoelectronics from the bottom up. *Nature Materials.* **6**(11):841–850.

3083. Mizuta H. and Oda S. (2008). Bottom-up approach to silicon nanoelectronics. *Microelectronics Journal.* **39**(2):171–176.

3084. Murali R. (2012). *Graphene Nanoelectronics: From Materials to Circuits.* New York: Springer Science & Business Media.

3085. Murawski F. (2004). Market for nanoelectronics. *Nanotechnology Law and Business.* **1**:364.

3086. Nikolic K. and Forshaw M. (2003). The current status of nanoelectronic devices. *International Journal of Nanoscience.* **2**(1/2):7.

3087. Nyberg T. et al. (2002). Macromolecular nanoelectronics. *Current Applied Physics.* **2**(1):27–31.

3088. Rabe K.M. (2006). Nanoelectronics: new life for the 'dead layer'. *Nature Nanotechnology.* **1**(3):171–172.

3089. Rae A. (2013). Real life applications of nanotechnology in electronics. *OnBoard Technologies.* 36–39.

3090. Rao W. et al. (2007). Towards nanoelectronics processor architectures. *Journal of Electronic Testing.* **23**(2–3):235–254.

3091. Rittner M.N. (2004). Nanomaterials in nanoelectronics: who's who and what's next. *JOM.* **56**(6):22.

3092. Roche S. (2011). Nanoelectronics: graphene gets a better gap. *Nature Nanotechnology.* **6**(1):8–9.

3093. Russer P. and Fichtner N. (2010). Nanoelectronics in radio-frequency technology. *IEEE Microwave Magazine.* **11**(3):119–135.

3094. Schwierz F. (2011). Nanoelectronics: flat transistors get off the ground. *Nature Nanotechnology.* **6**(3):135–136.

3095. Shen T.C. (2000). Role of scanning probes in nanoelectronics: a critical review. *Surface Review and Letters.* **7**(05–06):683–688.

3096. Snider G. et al. (2005). Nanoelectronic architectures. *Applied Physics A.* **80**(6):1183–1195.

3097. Tan S.G. and Mansoor B.A.J. (2012). *Introduction to the Physics of Nanoelectronics.* Philadelphia: Woodhead Publishing.

3098. Tian B. and Lieber C.M. (2013). Synthetic nanoelectronic probes for biological cells and tissue. *Annual Review of Analytical Chemistry.* **6**:31.

3099. Tsu R. (2001). Challenges in nanoelectronics. *Nanotechnology.* **12**(4):625.

3100. Vaidyanathan M. (2011). Electronics from the bottom up: strategies for teaching nanoelectronics at the undergraduate level. *IEEE Transactions on Education.* **54**(1):77–86.

3101. Vanmaekelbergh D. (2009). Nanoelectronics: from droplets to devices. *Nature Nanotechnology.* **4**(8):475–476.

3102. van Roermund A. and Hoekstra J. (2000). Design philosophy for nanoelectronic systems, from sets to neural nets. *International Journal of Circuit Theory and Applications.* **28**(6):563–584.

3103. Van Roosmalen A.J. and Zhang G.Q. (2006). Reliability challenges in the nanoelectronics era. *Microelectronics Reliability.* **46**(9):1403–1414.

3104. Van Rossum M. (1993). From micro-to nanoelectronics: new technology requirements. *Materials Science and Engineering: B.* **20**(1):128–133.

3105. Wang K.L. (2002). Issues of nanoelectronics: a possible roadmap. *Journal of Nanoscience and Nanotechnology.* **2**(3–4):235–266.

3106. Wang W. et al. (2006). Hybrid nanoelectronics: future of computer technology. *Journal of Computer Science and Technology.* **21**(6):871–886.

3107. Wolf E.L. (2009). *Quantum Nanoelectronics: An Introduction to Electronic Nanotechnology and Quantum Computing.* New York: Wiley-VCH.

3108. Xue Q.K. (2011). Nanoelectronics: a topological twist for transistors. *Nature Nanotechnology.* **6**(4):197–198.

3109. Yu B. and Meyyappan M. (2006). Nanotechnology: role in emerging nanoelectronics. *Solid-State Electronics.* **50**(4): 536–544.

3110. Zhirnov V.V. and Herr D.J. (2001). New frontiers: self-assembly and nanoelectronics. *Computer.* (1):34–43.

Energy and Power Generation, Transmission and Distribution

Green and Sustainable Energy

3111. Baxter J. et al. (2009). Nanoscale design to enable the revolution in renewable energy. *Energy & Environmental Science.* **2**(6):559–588.

3112. Brinker C.J. and Ginger D. (2011). Nanotechnology for sustainability: energy conversion, storage, and conservation. In *Nanotechnology Research Directions for Societal Needs in 2020.* Netherlands: Springer, pp. 261–303.

3113. Fromer N.A. and Diallo M.S. (2013). Nanotechnology and clean energy: sustainable utilization and supply of critical materials. *Journal of Nanoparticle Research.* **15**(11):1–15.

3114. Guo K.W. (2012). Green nanotechnology of trends in future energy: a review. *International Journal of Energy Research.* **36**(1):1–17.

3115. Kapilashrami M. et al. (2014). Probing the optical property and electronic structure of TiO2 nanomaterials for renewable energy applications. *Chemical Reviews.* **114**(19):9662–9707.

3116. Liu C.J. et al. (2010). Preparation and characterization of nanomaterials for sustainable energy production. *ACS Nano.* **4**(10):5517–5526.

3117. Mao S.S. and Chen X. (2007). Selected nanotechnologies for renewable energy applications. *International Journal of Energy Research.* **31**(6–7):619–636.

3118. Reddy K.G. et al. (2014). On global energy scenario, dye-sensitized solar cells and the promise of nanotechnology. *Physical Chemistry Chemical Physics.* **16**(15):6838–6858.

3119. Saunders J.R. et al. (2007). Nanotechnology's implications for select systems of renewable energy. *International Journal of Green Energy.* **4**(5):483–503.

3120. Serrano E. et al. (2009). Nanotechnology for sustainable energy. *Renewable and Sustainable Energy Reviews.* **13**(9):2373–2384.

3121. Shrair J. (2009). Advances in nanotechnology can provide clean energy resources and sustainable development (Technical report). Budapest University of Technology and Economics, Department of Electronic Devices.

3122. Shi L. et al. (2012). A review on sustainable design of renewable energy systems. *Renewable and Sustainable Energy Reviews.* **16**(1):192–207.

3123. Siril P.F. (2003). Nanotechnology and its application in renewable energy. Himachal Pradesh, India: IIT Mandi, School of Basic Sciences. Available online at http://www.iitmandi. ac.in/ciare/images/Nanotechnology%20and%20its%20 application%20in%20renewable%20energy.pdf

3124. Zang L. (2011). *Energy Efficiency and Renewable Energy Through Nanotechnology.* Berlin: Springer.

Biomass Power

3125. Chen Y.X. et al. (2014). Nanotechnology makes biomass electrolysis more energy efficient than water electrolysis. *Nature Communications.* **5**:4036.

3126. Kramb J. (2011). Potential applications of nanotechnology in bioenergy. Master's Thesis: University of Jyväskylä, Department of Physics.

3127. Luque R. and Balu A.M. (eds.) (2013). *Producing Fuels and Fine Chemicals from Biomass Using Nanomaterials.* Boca Raton, FL: CRC Press.

Electric Power

3128. Cao Y. et al. (2004). The future of nanodielectrics in the electrical power industry. *IEEE Transactions on Dielectrics and Electrical Insulation.* **11**(5):797–807.

3129. Elcock D. (2007). Potential impacts of nanotechnology on energy transmission applications and needs (ANL/EVS/TM-08-3). Argonne National Laboratory.

3130. Uldrich J. (2006). A cautionary tale: nanotechnology and the changing face of the electric utility industry. *Management Quarterly.* **47**(2):16.

Fossil Fuel

3131. Nassar N.N. et al. (2011). Application of nanotechnology for heavy oil upgrading: catalytic steam gasification/cracking of asphaltenes. *Energy & Fuels.* **25**(4):1566–1570.

Geothermal

3132. Ames M. et al. (2015). The utility of threshold reactive tracers for characterizing temperature distributions in geothermal reservoirs. *Mathematical Geosciences.* **47**(1):51–62.

Natural Gas

3133. Mokhatab S. et al. (2006). Applications of nanotechnology in oil and gas E&P. *Journal of Petroleum Technology.* **58**(04):48–51.
3134. Mokhatab S. and Towler B.F. (2007). Nanomaterials hold promise in natural gas industry. *International Journal of Nanotechnology.* **4**(6):680–690.

Nuclear Energy

3135. Bang I.C. and Jeong J.H. (2011). Nanotechnology for advanced nuclear thermal-hydraulics and safety. *Nuclear Engineering and Technology.* **43**(3):217–242.
3136. Besley J.C. and McComas K.A. (2015). Something old and something new: comparing views about nanotechnology and nuclear energy. *Journal of Risk Research.* **18**(2):215–231.
3137. Federation of American Scientists (2015). Nuclear power and nanomaterials: big potential for small particles. *Public Interest Report.* **68**(1).
3138. Shi W.-Q. et al. (2012). Nanomaterials and nanotechnologies in nuclear energy chemistry. *Radiochimica Acta.* **100**:727–736.

Solar Power

3139. Abdin Z. et al. (2013). Solar energy harvesting with the application of nanotechnology. *Renewable & Sustainable Energy Reviews.* **26**:837–852.

3140. International Electrotechnical Commission (2013). Nanotechnology in the sectors of solar energy and energy storage. Geneva.

3141. Jadhav M.V. et al. (2011). Nanotechnology for powerful solar energy. *International Journal of Advanced Biotechnology Research.* **2**:208–212.

3142. Kamat P. (2007). Meeting the clean energy demand: nanostructure architectures for solar conversion. *Journal of Physics and Chemistry.* **111**(7):2834–2860.

3143. Kamat P. (2008). Quantum dot solar cells. Semiconductor nanocrystals as light harvesters. *The Journal of Physical Chemistry C.* **112**(48):18737–18753.

3144. Mahian O. et al. (2013). A review of the applications of nanofluids in solar energy. *International Journal of Heat and Mass Transfer.* **57**(2):582–594.

3145. Oelhafen P. and Schüler A. (2005). Nanostructured materials for solar energy conversion. *Solar Energy.* **79**(2):110–121.

3146. Ramsurn H. and Gupta R.B. (2013). Nanotechnology in solar and biofuels. *ACS Sustainable Chemistry & Engineering.* **1**(7):779–797.

3147. Soga T. (ed.) (2006). *Nanostructured Materials for Solar Energy Conversion.* Elsevier, New York.

Environmental Engineering

General Overviews

3148. Chuanping H.W.F. (2006). Application of nanometer materials & nanotechnology in environmental protection. *New Chemical Materials.* **11**:004.

3149. Tratnyek P.G. and Johnson R.L. (2006). Nanotechnologies for environmental cleanup. *Nano Today.* **1**(2):44–48.

Air

3150. Baraton M.I. and Merhari L. (2004). Advances in air quality monitoring via nanotechnology. *Journal of Nanoparticle Research.* **6**(1):107–117.

3151. Pummakarnchana O. et al. (2005). Air pollution monitoring and GIS modeling: a new use of nanotechnology based solid state gas sensors. *Science and Technology of Advanced Materials.* **6**(3):251–255.

3152. Senthilnathan T. (2010). Ambient air investigation with nanotechnology. *Journal of Environmental Research & Development.* **5**(2):359–363.

3153. Zhao J. and Yang X. (2003). hotocatalytic oxidation for indoor air purification: a literature review. *Built Environment.* **38**:645–654.

Remediation

3154. Bhawana P. and Fulekar M.H. (2012). Nanotechnology: remediation technologies to clean up the environmental pollutants. *Research Journal of Chemical Sciences.* **2**(2):90–96.

3155. Duran N. (2008). Use of nanoparticles in soil-water bioremediation processes. *Revista De La Ciencia Del Suelo Y Nutrición Vegetal.* **8**(Especial):33–38.

3156. Karn B. et al. (2009). Nanotechnology and in situ remediation: a review of the benefits and potential risks. *Environmental Health Perspectives.* **117**(12):1823–1831.

3157. Li X.Q. et al. (2006). Zero-valent iron nanoparticles for abatement of environmental pollutants: materials and engineering aspects. *Critical Reviews in Solid State and Materials Sciences.* **31**(4):111–122.

3158. Macé C. et al. (2006). Nanotechnology and groundwater remediation: a step forward in technology understanding. *Remediation Journal.* **16**(2):23–33.

3159. Obare S.O. and G. J. Meyer. (2005). Nanostructured materials for environmental remediation of organic contaminants in water. *Journal of Environmental Science and Health, Part A.* **39**(10):2549–2582.

3160. Otto M. et al. (2008). Nanotechnology for site remediation. *Remediation.* **19**(1):99–108.

3161. Prabhakar V., Bibi T., Vishnu P. and Bibi T. (2013). Nanotechnology, future tools for water remediation. *International Journal of Emerging Technology and Advanced Engineering.* **3**(7):54–59.

3162. Rajan C.S. (2011). Nanotechnology in groundwater remediation. *International Journal of Environmental Science & Development.* **2**(3):182–187.

3163. Ren X. et al. (2011). Carbon nanotubes as adsorbents in environmental pollution management: a review. *Chemical Engineering Journal.* **170**(2):395–410.

3164. U.S. EPA (2007). Proceedings of the nanotechnology site remediation workshop, September 6–7, 2006 (EPA/905/K-07/001). Chicago: EPA Region 5.

3165. Watlington K. (2005). Emerging nanotechnologies for site remediation and wastewater treatment. A report prepared for the U.S. EPA. Raleigh: North Carolina State University.

3166. Zhang W.-S. (2003). Nanoscale iron particles for environmental remediation: an overview. *Journal of Nanoparticle Research.* **5**:323–332.

Water and Wastewater

3167. Ahmed T. et al. (2014). Emerging nanotechnology-based methods for water purification: a review. *Desalination and Water Treatment.* **52**(22–24):4089–4101.

3168. Brame J. et al. (2011). Nanotechnology enabled water treatment and reuse: emerging opportunities and challenges for developing countries. *Trends in Food Science & Technology.* **22**(11):618–624.

3169. Brunetti G. et al. (2015). Fate of zinc and silver engineered nanoparticles in sewerage networks. *Water Research.* **77**:72–84.

3170. Cloete T.E. et al. (eds.) (2010). *Nanotechnology in Water Treatment Applications*. Norfolk, UK: Caister Academic Press.

3171. Ghasemzadeh G. et al. (2014). Applications of nanomaterials in water treatment and environmental remediation. *Frontiers of Environmental Science & Engineering*. **8**(4):471–482.

3172. Hillie T. and Hlophe M. (2007). Nanotechnology and the challenge of clean water. *Nature Nanotechnology*. **2**(11):663–664.

3173. Kumar S. et al. (2014). Nanotechnology-based water treatment strategies. *Journal of Nanoscience & Nanotechnology*. **14**(2): 1838–1858.

3174. Loncto J. et al. (2007). Nanotechnology in the water industry. *Nanotechnology Law and Business*. **4**:157.

3175. Mohmood I. et al. (2013). Nanoscale materials and their use in water contaminants removal: a review. *Environmental Science and Pollution Research*. **20**(3):1239–1260.

3176. Qu X. et al. (2012). Nanotechnology for a safe and sustainable water supply: enabling integrated water treatment and reuse. *Accounts of Chemical Research*. **46**(3):834–843.

3177. Qu X. et al. (2013). Applications of nanotechnology in water and wastewater treatment. *Water Research*. **47**(12):3931–3946.

3178. Savage N. and Diallo M.S. (2005). Nanomaterials and water purification: opportunities and challenges. *Journal of Nanoparticle Research*. **7**(4–5):331–342.

3179. Shannon M.A. et al. (2008). Science and technology for water purification in the coming decades. *Nature*. **452**(7185):301–310.

3180. Tansel B. (2008). New technologies for water and wastewater treatment: a survey of recent patents. *Recent Patents on Chemical Engineering*. **1**(1):17–26.

3181. Theron J. et al. (2008). Nanotechnology and water treatment: applications and emerging opportunities. *Critical Reviews in Microbiology*. **34**(1):43–69.

Forensic Science

3182. Chen Y.F. (2011). Forensic applications of nanotechnology. *Journal of the Chinese Chemical Society.* **58**(6):828–835.

3183. Hallikeri V.R. et al. (2012). Nanotechnology–the future armour of forensics: a short review. *Journal of the Scientific Society.* **39**(1):10.

3184. Pitkethly M. (2009). Nanotechnology and forensics. *Materials Today.* **12**(6):6.

3185. Topal Z. and Kaya-Akyüzlü D. (2015). Nanotechnology in forensic sciences. *Turkish Journal of Occupational/ Environmental Medicine and Safety.* **1**(S1):88.

3186. Valle F. et al. (2012). Nanotechnology for forensic sciences: analysis of PDMS replica of the case head of spent cartridges by optical microscopy, SEM and AFM for the ballistic identification of individual characteristic features of firearms. *Forensic Science International.* **222**(1):288–297.

Life Cycle Assessment

3187. Alencar M.S.M. et al. (2007). Nanopatenting patterns in relation to product life cycle. *Technological Forecasting and Social Change.* **74**(9):1661–1680.

3188. Asmatulu E. et al. (2012). Life cycle and nano-products: end-of-life assessment. *Journal of Nanoparticle Research.* **14**(3):1–8.

3189. Bauer C. et al. (2008). Towards a framework for life cycle thinking in the assessment of nanotechnology. *Journal of Cleaner Production.* **16**(8):910–926.

3190. Beaudrie C.E.H. (2013). From cradle-to-grave at the nano-scale: expert risk perceptions, decision-analysis, and life cycle regulation for emerging nanotechnologies. PhD Dissertation: University of British Columbia.

3191. Gavankar S. et al. (2012). Life cycle assessment at nanoscale: review and recommendations. *International Journal of Life Cycle Assessment.* **17**(3):295–303.

3192. Gavankar S. et al. (2015). The role of scale and technology maturity in life cycle assessment of emerging technologies: a case study on carbon nanotubes. *Journal of Industrial Ecology.* **19**(1):51–60.

3193. Grieger K.D. et al. (2012). Analysis of current research addressing complementary use of life-cycle assessment and risk assessment for engineered nanomaterials: have lessons been learned from previous experience with chemicals? *Journal of Nanoparticle Research.* **14**(7):1–23.

3194. Hetherington A.C. et al. (2014). Use of LCA as a development tool within early research: challenges and issues across different sectors. *The International Journal of Life Cycle Assessment.* **19**(1):130–143.

3195. Hischier R. (2014). Life cycle assessment of manufactured nanomaterials: inventory modelling rules and application example. *The International Journal of Life Cycle Assessment.* **19**(4):941–943.

3196. Hischier R. and Walser T. (2012). Life cycle assessment of engineered nanomaterials: state of the art and strategies to overcome existing gaps. *Science of the Total Environment.* **425**:271–282.

3197. Huber J. (2008). Technological environmental innovations (TEIs) in a chain-analytical and life-cycle-analytical perspective. *Journal of Cleaner Production.* **16**(18):1980–1986.

3198. Meyer D.E. and Upadhyayula V.K. (2014). The use of life cycle tools to support decision making for sustainable nano-technologies. *Clean Technologies and Environmental Policy.* **16**(4):757–772.

3199. Meyer D.E. et al. (2009). An examination of existing data for the industrial manufacture and use of nanocomponents and their role in the life cycle impact of nanoproducts. *Environmental Science & Technology.* **43**(5):1256–1263.

3200. Olapiriyakul S. and Caudill R.J. (2009). Thermodynamic analysis to assess the environmental impact of end-of-life recovery processing for nanotechnology products. *Environmental Science & Technology.* **43**(21):8140–8146.

3201. Olsen S. and Jørgensen M.S. (2005). Environmental assessment of micro/nano production in a life cycle perspective. *MRS Proceedings.* **895**:G01.

3202. Olsen S.I. and Miseljic M. (2013). Life cycle assessment in nanotechnology–issues in impact assessment and case studies. In *Safety Issues of Nanomaterials along their Life Cycle*, Barcelona, Spain.

3203. Rivera J.L. and Sutherland J.W. (2015). A design of experiments (DOE) approach to data uncertainty in LCA: application to nanotechnology evaluation. *Clean Technologies and Environmental Policy.* 1–11.

3204. Russell A.J. (2006). Human and ecological risk assessment and life cycle assessment: intersections, collisions, and future directions. *Human and Ecological Risk Assessment.* **12**(3):427–430.

3205. Seager T.P. and Linkov I. (2008). Coupling multicriteria decision analysis and life cycle assessment for nanomaterials. *Journal of Industrial Ecology.* **12**(3):282–285.

3206. Som C. et al. (2010). Life cycle concepts for the development of safe nanoproducts: state of the art, existing gaps and challenges. *Toxicology.* **269**(2–3):160–169.

3207. Som C. et al. (2010). The importance of life cycle concepts for the development of safe nanoproducts. *Toxicology.* **269**(2):160–169.

3208. Sweet L. and Strohm B. (2006). Nanotechnology—life-cycle risk management. *Human and Ecological Risk Assessment.* **12**(3):528–551.

3209. Theis T.L. et al. (2011). A life cycle framework for the investigation of environmentally benign nanoparticles and products. *Rapid Research Letters.* **5**(9):312–317.

3210. Tsuzuki T. (2014). Life cycle thinking and green nanotechnology. *Austin Journal of Nanomedicine & Nanotechnology.* **2**(1):1.

3211. Wardak A. et al. (2008). Identification of risks in the life cycle of nanotechnology-based products. *Journal of Industrial Ecology.* **12**(3):435–448.

3212. Wender B. (2013). LCA and responsible innovation of nanotechnology. M.S. Thesis: Arizona State University.

Manufacturing

General Overviews

3213. Chryssolouris G. et al. (2004). Nanomanufacturing processes: a critical review. *International Journal of Materials and Product Technology.* **21**(4):331–348.

3214. Narayan J. (2012). Nanoscience to nanotechnology to manufacturing transition. *International Journal of Nanotechnology.* **9**(10–12):914–941.

3215. Phoenix C. and Drexler E. (2004). Safe exponential manufacturing. *Nanotechnology.* **15**(8):869.

3216. Vaseashta A. et al. (2008). Green nanotechnologies for responsible manufacturing. *MRS Proceedings.* **1106**:3.

Apparel and Textiles

3217. Antczak T. (2012). Nanotechnology-methods of manufacturing cellulose nanofibers. *Fibres & Textiles in Eastern Europe.* **20**(2):91.

3218. Chen J. et al. (2010). Perspective on development of nanotechnology in textiles. *Advanced Materials Research.* **113**:670–673.

3219. Coyle S. et al. (2007). Smart nanotextiles: a review of materials and applications. *MRS Bulletin.* **32**(5):4434–4442.

3220. El-Khatib E.M. (2012). Antimicrobial and self-cleaning textiles using nanotechnology. *Research Journal of Textile & Apparel.* **16**(3):156–174.

3221. Gupta D. (2011). Functional clothing: definition and classification. *Indian Journal of Fibre and Textile Research.* **36**(4):321.

3222. Hinestroza J.P. (2007). Can nanotechnology be fashionable? *Materials Today.* **10**(9):64.

3223. Kaounides L. et al. (2007). Nanotechnology innovation and applications in textiles industry: current markets and future growth trends. *Materials Science and Technology.* **22**(4):209–237.

3224. Patanaik A. et al. (2007). Nanotechnology in fibrous materials: a new perspective. *Textile Progress.* **39**(2):67–120.

3225. Patra J.K. and Gouda S. (2013). Application of nanotechnology in textile engineering: an overview. *Journal of Engineering and Technology Research.* **5**(5):104–111.

3226. Qian L. (2004). Nanotechnology in textiles: recent developments and future prospects. *AATCC Review.* **4**(5):14–16.

3227. Ren Y.Z. et al. (2006). Application of nanotechnology to sport costumes. *Sports Science Research.* **27**(3):36.

3228. Yue-ling L. (2007). The properties of fine wool after nanotechnology finish. *Wool Textile Journal.* **6**:011.

Ceramics and Glass

3229. Arora A. (2004). Ceramics in nanotech revolution. *Advanced Engineering Materials.* **6**(4):244–247.

3230. Cain M. and Morrell R. (2001). Nanostructured ceramics: a review of their potential. *Applied Organometallic Chemistry.* **15**(5):321–330.

3231. Hirao K. (2003). Nanotechnology and ceramics. Development of nanoglasses, toward practical applications. *Bulletin of the Ceramic Society of Japan.* **38**(5):323–330.

3232. Mauro J.C. et al. (2013). Glass: the nanotechnology connection. *International Journal of Applied Glass Science.* **4**(2):64–75.

3233. Mauro J.C. et al. (2014). Glass science in the United States: current status and future directions. *International Journal of Applied Glass Science.* **5**(1):2–15.

3234. Min'ko N.I. and Nartsev V.M. (2008). Nanotechnology in glass materials (review). *Glass and Ceramics.* **65**(5–6):148–153.

3235. Pivinskii Y.E. (2008). Nanodispersed silica and some aspects of nanotechnology in the field of silicate materials science. Part 4. *Refractories and Industrial Ceramics.* **49**(1):67–74.

3236. Pivinskii Y.E. (2011). HCBS ceramic concretes in the XXI century—problems and prospects for applying technology in the field of silicate materials science. Part 1. *Refractories and Industrial Ceramics*. **52**(2):107–115.

3237. Scalisi F. (2014). Nanotechnology and earth construction: the mechanical properties of adobe brick stabilized by laponite nanoparticles. *Advanced Materials Research*. **983**: 63–66.

3238. Shevchenko V.Y. (2005). From ancient inorganic chemistry and alchemy of ceramics to modern nanotechnology. *Glass Physics and Chemistry*. **31**(1):11–26.

3239. Singh L.P. et al. (2013). Beneficial role of nanosilica in cement based materials–A review. *Construction and Building Materials*. **47**:1069–1077.

Chemicals

3240. Aegerter M.A. et al. (2008). Coatings made by sol–gel and chemical nanotechnology. *Journal of Sol-Gel Science and Technology*. **47**(2):203–236.

3241. Baer D.R. et al. (2003). Enhancing coating functionality using nanoscience and nanotechnology. *Progress in Organic Coatings*. **47**(3):342–356.

3242. Butler L. et al. (2004). Use of nanotechnology in producing protective wood coatings. *Surface Coatings Australia*. **41**(1–2): 14–21.

3243. Castano V.M. and Rodriguez R. (2004). A nanotechnology approach to high performance anti-graffiti coatings'. *International Journal of Applied Management and Technology*. **2**(2):53–58.

3244. Kaiser J.P. et al. (2013). Is nanotechnology revolutionizing the paint and lacquer industry? A critical opinion. *Science of the Total Environment*. **442**:282–289.

3245. Lee C.M. et al. (2010). Nanotech for vehicle paint and flame-retardant coatings. In *IEEE 2010 International Forum on Strategic Technology (IFOST)*, Ulsan, pp. 306–310.

3246. Mathiazhagan A. and Joseph R. (2011). Nanotechnology-a new prospective in organic coating. *International Journal of Chemical Engineering Applications.* **2**(4):225–237.

3247. Nikolic M. et al. (2015). Use of nanofillers in wood coatings: a scientific review. *Journal of Coatings Technology & Research.* **12**(3):445–461.

3248. Stange R. (2003). From nanotechnology to stir-in pigments: raw material developments driven by environmental requirements. *European Coatings Journal.* (5):28–30.

3249. Szewczyk P. (2010). The role of nanotechnology in improving marine antifouling coatings. *Zeszyty Naukowe/Akademia Morskaw Szczecinie.* **24**:118–123.

3250. Zhao Q.Q. et al. (2003). Nanotechnology in the chemical industry–opportunities and challenges. *Journal of Nanoparticle Research.* **5**(5–6):567–572.

3251. Zhou S. and Wu L. (2009). Development of nanotechnology-based organic coatings. *Composite Interfaces.* **16**(4–6):281–292.

Food Products

3252. Akbari Z. et al. (2007). Improvement in food packaging industry with biobased nanocomposites. *International Journal of Food Engineering.* **3**(4):24.

3253. Arora A. and Padua G.W. (2010). Review: nanocomposites in food packaging. *Journal of Food Science.* **75**(1):R43–R49.

3254. Bin W.U. et al. (2008). Potential applications of nanotechnology in fruit wine industry. *China Brewing.* **5**:004.

3255. Blasco C. and Picó Y. (2011). Determining nanomaterials in food. *Trends in Analytical Chemistry.* **30**(1):84–99.

3256. Brody A.L. (2003). Nano, nano food packaging technology. *Food Technology.* **57**(12):52–54.

3257. Chau C.F. et al. (2007). The development of regulations for food nanotechnology. *Trends in Food Science & Technology.* **18**(5):269–280.

3258. Chaudhry Q. et al. (2008). Applications and implications of nanotechnologies for the food sector. *Food Additives*. **25**:241–258.

3259. Cushen M. et al. (2012). Nanotechnologies in the food industry–recent developments, risks and regulation. *Trends in Food Science & Technology*. **24**(1):30–46.

3260. Das M., Saxena N. and Dwivedi P.D. (2009). Emerging trends of nanoparticles application in food technology: safety paradigms. *Nanotoxicology*. **3**(1):10–18.

3261. Dudo A. et al. (2011). Food nanotechnology in the news. Coverage patterns and thematic emphases during the last decade. *Appetite*. **56**(1):78–89.

3262. Duran N. and Marcato P.D. (2013). Nanobiotechnology perspectives. Role of nanotechnology in the food industry: a review. *International Journal of Food Science & Technology*. **48**(6):1127–1134.

3263. Hiregoudar S. et al. (2011). Recent trends in application of nanotechnology in food processing. *Food Science Research Journal*. **2**(2):219–225.

3264. Liu Q. et al. (2012). Evaluation of antioxidant activity of chrysanthemum extracts and tea beverages by gold nanoparticles-based assay. *Colloids and Surfaces B: Biointerfaces*. **92**:348–352.

3265. Magnuson B. et al. (2011). A brief review of the occurrence, use, and safety of food-related nanomaterials. *Journal of Food Science*. **76**(6):R126–R133.

3266. Markman G. and Livney Y.D. (2012). Maillard-conjugate based core–shell co-assemblies for nanoencapsulation of hydrophobic nutraceuticals in clear beverages. *Food & Function*. **3**(3):262–270.

3267. Meetoo D. (2011). Nanotechnology and the food sector: from the farm to the table. *Emirates Journal of Food and Agriculture*. **23**(5):387–403.

3268. Momin J.K. and Joshi B.H. (2015). Nanotechnology in foods. In *Nanotechnologies in Food and Agriculture*. Berlin: Springer International Publishing, pp. 3–24.

3269. Moraru C. et al. (2009). Food nanotechnology: current developments and future prospects. *Global Issues in Food Science and Technology.* **21**:369–399.

3270. Morris V.J. (2008). Nanotechnology in the food industry. *New Food Magazine.* **4**:53–55.

3271. Nana E. et al. (2013). The efficiency of application nanosilver in technological processes of making red wine. *Journal of Food Science and Engineering.* **3**(8).

3272. Neethirajan S. and Jayas D.S. (2011). Nanotechnology for the food and bioprocessing industries. *Food and Bioprocess Technology.* **4**(1):39–47.

3273. Ozimek L. et al. (2010). Nanotechnologies in food and meat processing. *Acta Scientiarum Polonorum Technologia Alimentaria.* **9**(4):401–412.

3274. Popov K.I. et al. (2010). Food nanotechnologies. *Russian Journal of General Chemistry.* **80**(3):630–642.

3275. Rashidi L. and Khosravi-Darani K. (2011). The applications of nanotechnology in food industry. *Critical Reviews in Food Science and Nutrition.* **51**(8):723–730.

3276. Ravichandran R. (2010). Nanotechnology applications in food and food processing: innovative green approaches, opportunities and uncertainties for global market. *International Journal of Green Nanotechnology: Physics and Chemistry.* **1**(2):72–96.

3277. Reza Mozafari M. et al. (2008). Nanoliposomes and their applications in food nanotechnology. *Journal of Liposome Research.* **18**(4):309–327.

3278. Rossi M., Cubadda F., Dini L., Terranova M.L., Aureli F., Sorbo A. and Passeri D. (2014). Scientific basis of nanotechnology, implications for the food sector and future trends. *Trends in Food Science & Technology.* **40**(2):127–148.

3279. Sanguansri P. and Augustin M.A. (2006). Nanoscale materials development–a food industry perspective. *Trends in Food Science & Technology.* **17**(10):547–556.

3280. Sastry R.K. et al. (2013). Nanotechnology in food processing sector: an assessment of emerging trends. *Journal of Food Science & Technology.* **50**(5):831–841.

3281. Sekhon B.S. (2010). Food Nanotechnology: an overview. *Nanotechnology, Science and Application.* **3**:1–15.

3282. Siegrist M. et al. (2008). Perceived risks and perceived benefits of different nanotechnology foods and nanotechnology food packaging. *Appetite.* **51**(2):283–290.

3283. Song Z.P. et al. (2009). Application of nanometer paint during bulk curing of the flue-cured tobacco. *Journal of Northwest A & F University (Natural Science Edition).* **8**:019.

3284. Sozer N. and Kokini J.L. (2009). Nanotechnology and its applications in the food sector. *Trends in Biotechnology.* **27**(2):82–89.

3285. Sumit G. (2012). Nanotechnology in food packaging: a critical review. *Russian Journal of Agricultural and Socio-Economic Sciences.* **10**(10):14–24.

3286. Tkac J. et al. (2007). Nanotechnology gets into winemaking. *Nano Today.* **2**(4):48.

3287. Takhistova K. (2008). Food nanotechnology: in search of a regulatory framework. *Rutgers Computer & Technology Law Journal.* **35**:255.

3288. van Rijn C.J. and Raspe O. (2008). Membrane technology, nanotechnology and beer filtration. *Food Manufacturing Efficiency.* **1**(3):29.

3289. Weiss J. et al. (2006). Functional materials in food nanotechnology. *Journal of Food Science.* **719**:R107–R116.

3290. Yu J. et al. (2009). Application effect of nano-coating to curing barns in tobacco curing. *Chinese Tobacco Science.* **5**:012.

Paper Manufacturing

3291. Brodin F.W. et al. (2014). Cellulose nanofibrils: challenges and possibilities as a paper additive or coating material—a review. *Nordic Pulp & Paper Research Journal.* **29**(1):156–166.

3292. Chauhan V.S. and Chakrabarti S.K. (2012). Use of nanotechnology for high performance cellulosic and papermaking products. *Cellulose Chemistry and Technology*. **46**(5):389.

3293. Johnston J.H. et al. (2004). Nano-structured silicas and silicates—new materials and their applications in paper. *Current Applied Physics*. **4**(2):411–414.

3294. Jones P. and Wegner T.H. (2007). Small world, big results: nanotechnology and forest products are teaming up to create the material of the 21st century? Coated paper and paperboard. *Paper Age*. July/August 22–26.

3295. Kharisov B.I. and Kharissova O.V. (2010). Advances in nanotechnology in paper processing. *International Journal of Green Nanotechnology: Materials Science & Engineering*. **2**(1):M1–M8.

3296. Mohieldin S.D. et al. (2011). Nanotechnology in pulp and paper industries: a review. *Key Engineering Materials*. **471**: 251–256.

3297. Pätäri S. et al. (2011). Opening up new strategic options in the pulp and paper industry: case biorefineries. *Forest Policy and Economics*. **13**(6):456–464.

3298. Sishun C. et al. (2006). Achievements on the application of nanotechnology in papermaking industry. *Shanghai Paper Making*. **1**:006.

3299. Xue-ren A. (2008). Some application areas of nanotechnology in paper industry. *Paper and Paper Making*. **3**:039.

Plastic and Rubber

3300. Qiongzhi G. et al. (2003). New advances on applications of nanotechnology in rubber industry. *Synthetic Rubber Industry*. **4**.

3301. Ratnayake U.N. (2009). Nanotechnology for upgrading the rubber industry. *Bulletin of the Rubber Research Institute of Sri Lanka*. **50**:125–134.

3302. Thomas S. and Stephen R. (2010). *Rubber Nanocomposites: Preparation, Properties and Applications*. New York: John Wiley & Sons.

3303. Tunnicliffe L.B. and Busfield J.J. N.D. Nanotechnology in tires. White Paper. London: Queen Mary University of London, Soft Matter Group.

Primary Metals (Ferrous and Nonferrous)

3304. Kolpakov S.V. et al. (2007). Nanotechnology in the metallurgy of steel. *Steel in Translation.* **37**(8):716–721.

Printing and Related Support Activities

3305. Torres C.M.S. (ed.) (2012). *Alternative Lithography: Unleashing the Potentials of Nanotechnology.* Berlin: Springer Science & Business Media.

Wood and Forest Products

3306. Candan Z. and Akbulut T. (2013). Developing environmentally friendly wood composite panels by nanotechnology. *BioResources.* **8**(3):3590–3598.
3307. Candan Z. and Akbulut T. (2014). Nano-engineered plywood panels: performance properties. *Composites Part B: Engineering.* **64**:155–161.
3308. Civardi C. et al. (2015). Micronized copper wood preservatives: an efficiency and potential health risk assessment for copper-based nanoparticles. *Environmental Pollution.* **200**:126–132.
3309. Clausen C.A. (2007). Nanotechnology: implications for the wood preservation industry (IRG/WP 07-30415). Stockholm, Sweden: IRG Secretariat.
3310. Evans P. et al. (2008). Large-scale application of nanotechnology for wood protection. *Nature Nanotechnology.* **3**(10):577–577.
3311. Gardner D.J. and Han Y. (2013). Nanotechnology applications in forest products: current trends. *Proceedings of the 55th International Convention of Society of Wood Science and Technology.* August 27–31, 2012, Beijing, China.

3312. Gomes S.I. et al. (2015). Cu-nanoparticles ecotoxicity–explored and explained? *Chemosphere.* **139**:240–245.

3313. Howe D.J. et al. (2006). Nanotechnology and the forest products industry: exciting new possibilities. Minneapolis, MN: Dovetail Partners, Incorporated.

3314. Mahltig B. et al. (2008). Functionalising wood by nanosol application. *Journal of Materials Chemistry.* **18**(27):3180–3192.

3315. Moon R.J. et al. (2006). Nanotechnology applications in the forest products industry. *Forest Products Journal.* **56**(5):4–10.

3316. Papadopoulou E. (2013). Nanocellulose in wood-based panels: a review. Available at http://www.costfp1205.com/en/events/documents/papadopoulou_nanocelluloseinwood-basedcomposites.pdf

3317. Salari A. et al. (2012). Effect of nanoclay on some applied properties of oriented strand board (OSB) made from underutilized low quality paulownia (Paulownia fortunei) wood. *Journal of Wood Science.* **58**(6):513–524.

3318. Taghiyari H.R. et al. (2015). Effects of nanotechnology on fluid flow in agricultural and wood-based composite materials. In *Agricultural Biomass Based Potential Materials.* Switzerland: Springer International Publishing, pp. 73–89.

3319. Wegner T. and Jones P. (2005). Nanotechnology for forest products, part 1-Nano-sized particles may be small, but for our industry they offer huge potential. *Solutions-for People Processes and Paper.* **88**(7):44–46.

3320. Wegner T. and Jones P. (2005). Nanotechnology for forest products, part 2. *Solutions-Norcross.* **50**:44.

3321. Zheng X.G. et al. (2008). Application of new nano-biocide in wood preservation. *Forestry Machinery & Woodworking Equipment.* **7**:004.

Microscopes

General Overviews

3322. Bunk S. (2001). Better microscopes will be instrumental in nanotechnology development. *Nature.* **410**(6824):127–129.

Field Ion

3323. Chen Y.C. and Seidman D.N. (1971). On the atomic resolution of a field ion microscope. *Surface Science.* **26**(1):61–84.

3324. Forbes R.G. (1985). Seeing atoms: the origins of local contrast in field-ion images. *Journal of Physics D: Applied Physics.* **18**(6):973.

3325. Forbes R.G. (2003). Field electron and ion emission from charged surfaces: a strategic historical review of theoretical concepts. *Ultramicroscopy.* **95**:1–18.

3326. Grivet P. and Septier A. (1978). Ion microscopy: history and actual trends. *Annals of the New York Academy of Sciences.* **306**(1):158–182.

3327. McMullan D. (1990). The prehistory of scanned image microscopy. Part 2. The scanning electron microscope. *Proceedings of the Royal Microscopical Society.* **25**:189–194.

3328. McMullan D. (2004). Appendix II a history of the scanning electron microscope, 1928–1965. *Advances in Imaging and Electron Physics.* **133**:523–545.

3329. Miller M.K. (2000). The development of atom probe field-ion microscopy. *Materials Characterization.* **44**(1):11–27.

3330. Müller E.W. and Bahadurt K. (1956). Field ion microscope. *Physical Review.* **102**(3).

3331. Müller E.W. (1961). The field ion microscope. *American Scientist.* **49**(1):88–98.

3332. Müller E.W. et al. (1968). The atom-probe field ion microscope. *Review of Scientific Instruments.* **39**(1):83–86.

3333. Müller E.W. (1970). The atom-probe field ion microscope. *Naturwissenschaften.* **57**(5):222–230.

3334. Müller E.W. (1974). Advances in atom-probe field ion microscopy. *Journal of Microscopy.* **100**(2):121–131.

3335. Oatley C.W. (1972). *The Scanning Electron Microscope, Part 1, The Instrument.* Cambridge, England: Cambridge University Press.

3336. Panitz J.A. (1982). Field-ion microscopy-a review of basic principles and selected applications. *Journal of Physics E: Scientific Instruments.* **15**:1281–1294.

3337. Pawley J.B. (1997). The development of field-emission scanning electron microscopy for imaging biological surfaces. *Scanning.* **19**:324–336.

3338. Ralph B. (1970). The metallurgical applications of the field-ion microscope. *Surface Science.* **23**(1):130–143.

3339. Ralph B. and Brandon D.G. (1964a). The field-ion microscope. 1. Design and development. *Journal of the Royal Microscopical Society.* **82**(3):179–184.

3340. Ralph B. and Brandon D.G. (1964b). The field-ion microscope. 2. Applications. *Journal of the Royal Microscopical Society.* **82**(3):185–188.

3341. Ralph B. and Southon M.J. (1965). Field-ion microscope. *Journal of Scientific Instruments.* **42**(8):543.

3342. Ryan H.F. and Suiter J. (1965). An all-metal field ion microscope. *Journal of Scientific Instruments.* **42**(8):645.

3343. Sakurai T. et al. (1990). Field ion-scanning tunneling microscopy. *Progress in Surface Science.* **33**(1):3–89.

3344. Thurstans R.E. and Walls J.M. (1980). *Field-Ion Microscopy and Related Techniques: A Bibliography 1951–1978.* Birmingham: Warwick Publishing.

3345. Tsong T.T. (1994). Atom-probe field ion microscopy and applications to surface science. *Surface Science.* **299**:153–169.

3346. Vurpillot F. et al. (2007). Towards the three-dimensional field ion microscope. *Surface and Interface Analysis.* **39**(2–3):273–277.

Helium Ion

3347. Economou N.P. et al. (2012). The history and development of the helium ion microscope. *Scanning.* 34(2):83–89.

3348. Joy D.C. (2013). *Helium Ion Microscopy: Principles and Applications.* New York: Springer.

Scanning Electron

3349. Adams J.D. et al. (2004). Microtechnology, nanotechnology, and the scanning-probe microscope: an innovative course. *IEEE Transactions on Education.* **47**(1):51–56.

3350. Biel S.S. and Gelderblom H.R. (1999). Diagnostic electron microscopy is still a timely and rewarding method. *Journal of Clinical Virology.* **13**(1):105–119.

3351. Big E.J. (1956). A short history of the electron microscope. *Bios.* **27**(1):33–37.

3352. Breton P. J. (1999). From microns to nanometers: early landmarks in the science of scanning electron microscope imaging. *Scanning Microscopy.* **13**(1):1–6.

3353. Coffey T. et al. (2015). Exploring nanoscience and scanning electron microscopy in K–12 classrooms. *Microscopy Today.* **23**(01):44–47.

3354. Cosslett V.E. (1967). The future of the electron microscope. *Journal of the Royal Microscopical Society.* **87**(1):53–76.

3355. Freundlich M.M. (1963). Origin of the electron microscope. *Science.* **142**(3589):185–188.

3356. Ghadially F.N. (1999). As you like it, part 2: a critique and historical review of the electron microscopy literature. *Ultrastructural Pathology.* **23**(1):1–17.

3357. Griffith O.H. et al. (1991). Bibliography on emission microscopy, mirror electron microscopy, low-energy electron microscopy and related techniques: 1985–1991. *Ultramicroscopy.* **36**(1):262–274.

3358. Haguenau F. et al. (2003). Key events in the history of electron microscopy. *Microscopy and Microanalysis.* **9**(02):96–138.

3359. Haine M.E. (1947). The design and construction of a new electron microscope. *Journal of the Institution of Electrical Engineers-Part I: General.* **94**(82):447–462.

3360. Katterwe H. et al. (1981). The comparison scanning electron microscope within the field of forensic science. *Scanning Electron Microscopy.* (Pt 2):499–504.

3361. Mulvey T. (1962). Origins and historical development of the electron microscope. *British Journal of Applied Physics.* **13**:197–207.

3362. Nakayama T. et al. (2012). Development and application of multiple-probe scanning probe microscopes. *Advanced Materials.* **24**(13):1675–1692.

3363. Newbury D.E. and Williams D.B. (2000). The electron microscope: the materials characterization tool of the millennium. *Acta Materialia.* **48**(1):323–346.

3364. Oatley C.W. (1982). The early history of the scanning electron microscope. *Journal of Applied Physics.* **53**(2):R1–R13.

3365. Rasmussen N. (1999). Picture control: the electron microscope and the transformation of biology in America, 1940–1960. Palo Alton, CA: Stanford University Press.

3366. Rhodin T. (2001). Scanning probe microscopies, nanoscience and nanotechnology. *Applied Physics A.* **72**(1):S141–S143.

3367. Ruska E. (1987). The development of the electron microscope and of electron microscopy. *Reviews of Modern Physics.* **59**(3):627.

3368. Shedd G.M. and Russell P. (1990). The scanning tunneling microscope as a tool for nanofabrication. *Nanotechnology.* **1**(1):67.

3369. Smith K.C.A. (1956). The scanning electronic microscope and its field of application. PhD Dissertation: Cambridge University.

3370. Smith L.T. (2014). Project NANO: will allowing high school students to use research grade scanning electron microscopes increase their interest in science? Master's Thesis (MST): Portland State University.

3371. Souza W.D. (2008). Electron microscopy of trypanosomes: a historical view. *Memórias do Instituto Oswaldo Cruz.* **103**(4):313–325.

3372. Spence J.C. (1999). The future of atomic resolution electron microscopy for materials science. *Materials Science and Engineering: R: Reports.* **26**(1):1–49.

3373. Taylor T.N. (1968). Application of the scanning electron microscope in paleobotany. *Transactions of the American Microscopical Society.* **87**(4):510–515.

3374. Weiss P.S. (2007). A conversation with Dr. Heinrich Rohrer: STM co-inventor and one of the Founding Fathers of nanoscience. *ACS Nano.* **1**(1):3–5.

3375. Zaluzec N.J. (2006). The scanning confocal electron microscope: a new tool for defect studies in semiconductor devices. In *2006 13th International Symposium on the Physical and Failure Analysis of Integrated Circuits*, Singapore, pp. 49–53.

Mining, Quarrying, and Oil and Gas Extraction

3376. Buckingham D. (2007). Nanotechnology: an emerging technology. *Mining Engineering.* **59**(12):23–29.

3377. Cocuzza M. et al. (2012). Current and future nanotech applications in the oil industry. *American Journal of Applied Sciences.* **9**(6):784–793.

3378. Diallo M.S. et al. (2015). Mining critical metals and elements from seawater: opportunities and challenges. *Environmental Science & Technology.* **49**(16):9390–9399.

3379. Hu C. et al. (2005). Application of nanoscale science and technology in mineral materials. *Mining and Metallurgical Engineering.* **3**:019.

3380. Kotova O.B. and Ponaryadov A.V. (2009). Nanotechnological mineralogy. *Journal of Mining Science.* **45**(1):93–98.

3381. Krishnamoorti R. (2006). Extracting the benefits of nanotechnology for the oil industry. *Journal of Petroleum Technology.* **58**(11):24–26.

3382. Mokhatab S. et al. (2006). Applications of nanotechnology in oil and gas E&P. *Journal of Petroleum Technology.* **58**(4):48–51.

3383. Oleynikova G.A. and Panova E.G. (2011). Geochemistry of nanoparticles in the rocks, ores and waste. *Journal of Earth Science and Engineering.* **1**(3).

3384. Popel S.I. et al. (2014). Nanoscale particles in technological processes of beneficiation. *Beilstein journal of nanotechnology.* **5**(1):458–465.

3385. Rytwo G. (2008). Clay minerals as an ancient nanotechnology: historical uses of clay organic interactions, and future possible perspectives. *Macla.* **9**:15–17.

3386. van Loon A.J. (2002). From the benefits of micro to the threats of nano for the ore-mining and ore-refining sectors. *Earth-Science Reviews.* **58**(1):233–241.

3387. Williams P.E. (2005). Study on the improvement of mining explosives by nanotechnology. 含能材料. **13**(5):337–339.

3388. Zhenhua D. (1999). Dilemma and chance in mineralogy: enlightenment to mineralogy from nano-science. *Acta Mineralogica Sinica.* **3**:018.

3389. Zhou C.H. and Keeling J. (2013). Fundamental and applied research on clay minerals: from climate and environment to nanotechnology. *Applied Clay Science.* **74**:3–9.

Nanobiotechnology

3390. Adhikari R. (2005). Nanobiotechnology: will it deliver? *Healthcare Purchasing News.* **29**(1):60.

3391. Bennett D.J. and Schuurbiers D. (2005). Nanobiotechnology: responsible actions on issues in society and ethics. *NSTI Nanotech.* **2**:765–768.

3392. Billingston C. et al. (2014). Prevention of bacterial foodborne disease using nanobiotechnology. *Nanotechnology, Science & Applications.* **7**:73–83.

3393. de Morais M.G. et al. (2014). Biological applications of nanobiotechnology. *Journal of Nanoscience & Nanotechnology.* **14**(1):1007–1017.

3394. Dordick J.S. and Lee K.H. (2014). Editorial overview: nanobiotechnology. *Current Opinion in Biotechnology*. doi: 10.1016/j.copbio.2014.06.015

3395. Duran N. et al. (2009). State of the art of nanobiotechnology applications in neglected diseases. *Current Nanoscience*. **5**(4):396–408.

3396. Emerich D.F. et al. (2007). Role of nanobiotechnology in cell-based nanomedicine: a concise review. *Journal of Biomedical Nanotechnology*. **3**(3):235–244.

3397. Fakruddin M. et al. (2012). Prospects and applications of nanobiotechnology: a medical perspective. *Journal of Nanobiotechnology*. **10**:31.

3398. Fortina P. et al. (2005). Nanobiotechnology: the promise and reality of new approaches to molecular recognition. *Trends in Biotechnology*. **23**(4):168–173.

3399. Galbraith D.W. (2007). Nanobiotechnology: silica breaks through in plants. *Nature Nanotechnology*. **2**(5):272–273.

3400. Gerwin V. (2006). Nanobiotechnology: small talk. *Nature*. **444**(7118):514–515.

3401. Grunwald A. (2004). The case of nanobiotechnology. *EMBO Reports*. **5**(Suppl 1):S32–S36.

3402. Gusić N. et al. (2014). Nanobiotechnology and bone regeneration: a mini-review. *International Orthopaedics*. **38**(9):1877–1884.

3403. Jain K.K. (2005). The role of nanobiotechnology in drug discovery. *Drug Discovery Today* **10**(21):1435–1442.

3404. Jain K.K. (2006). *Nanobiotechnology in Molecular Diagnostics: Current Techniques and Applications*. Norwich (UK): Horizon Bioscience.

3405. Jain K.K. (2007a). Applications of nanobiotechnology in clinical diagnostics. *Clinical Chemistry*. **53**(11):2002–2009.

3406. Jain K.K. (2007b). *Nanobiotechnology: Applications, Markets and Companies*. Basel: PharmaBiotech Publications.

3407. Jain K.K. (2009). The role of nanobiotechnology in drug discovery. *Advances in Experimental Medicine and Biology*. **655**:37–43.

3408. Jain K.K. (2010). Potential of nanobiotechnology in the management of glioblastoma multiforme. In *Glioblastoma*. New York: Springer, pp. 399–419.

3409. Jain K.K. (2011a). Nanobiotechnology and personalized medicine. *Progress in Molecular Biology & Translational Science*. **103**:277–352.

3410. Jain K.K. (2011b). The role of nanobiotechnology in the development of personalized medicine. *Medical Principles and Practice*. **20**(1):1–3.

3411. Jewett M.C. and Patlosky F. (2013). Nanobiotechnology: synthetic biology meets materials science. *Current Opinion in Biotechnology*. **24**(4):551–554.

3412. Kayser O. and Trejo N. (2005). The impact of nanobiotechnology on the development of new drug delivery systems. *Current Pharmaceutical Biotechnology*. **6**(1):3–5.

3413. Klefenz H. (2004). Nanobiotechnology: from molecules to systems. *Engineering in Life Sciences*. **4**(3):211–218.

3414. Koopmans R.J. and Aggeli A. (2010). Nanobiotechnology— quo vadis? *Current Opinion in Microbiology*. **13**(3):327–334.

3415. Lawrence R.N. (2002). James Gimzewski discusses the potential of nanobiotechnology. *Drug Discovery Today*. **7**(1):18–21.

3416. Lowe C.R. (2000). Nanobiotechnology: the fabrication and applications of chemical and biological nanostructures. *Current Opinion in Structural Biology*. **10**(4):428–434.

3417. Maharana B.R. et al. (2010). Nanobiotechnology: a voyage to the future? *Veterinary World*. **3**(3):145–147.

3418. Mahmood T. and Hussain S.T. (2010). Nanobiotechnology for the production of biofuels from spent tea. *African Journal of Biotechnology*. **9**(6):858–868.

3419. Main E. et al. (2014). The emergence of the nanobiotechnology industry. *Nature Nanotechnology*. **9**(1):2–5.

3420. Medvedeva N.V. et al. (2007). Nanobiotechnology and nanomedicine. *Biochemistry (Moscow) Supplement Series B: Biomedical Chemistry*. **1**(2):114–124.

3421. Mirkin C.A. and Niemeyer C.M. (eds.) (2004). *Nanobiotechnology: Concepts, Applications and Perspectives*. New York: John Wiley & Sons.

3422. Mirkin C.A. and Niemeyer C.M. (eds.) (2007). *Nanobiotechnology II: More Concepts and Applications*. New York: John Wiley & Sons.

3423. Mohanty C. et al. (2009). Nanobiotechnology: application of nanotechnology in therapeutics and diagnosis. *International Journal of Green Nanotechnology: Biomedicine*. **1**(1):B24–B38.

3424. Moos W.H. and Barry S. (2006). Nanobiotechnology: It's a small world after all. *Drug Development Research*. **67**(1):1–3.

3425. Morais M.G.D. et al. (2014). Biological applications of nanobiotechnology. *Journal of Nanoscience and Nanotechnology*. **14**(1):1007–1017.

3426. Muraleedharan H. (2010). Nanobiotechnology: bioinspired devices and materials of the future (a review). *Journal of Biosciences Research*. **1**(2):108–117.

3427. Nguyen P. et al. (2010). Applications of nanobiotechnology in ophthalmology. Part 1. *Ophthalmic Research*. **44**(1):1–16.

3428. Nicolini C.A. (2009). *Nanobiotechnology and Nanobiosciences*. Singapore: Pan Stanford Publishing.

3429. Niemeyer C.M. and Mirkin C.A. (eds.) (2006). *Nanobiotechnology: Concepts, Applications and Perspectives*. Weinbheim, Germany: Wiley-VCH.

3430. Paradise J. et al. (2008). Developing oversight frameworks for nanobiotechnology. *Minnesota Journal of Law, Science & Technology*. **9**:399.

3431. Paradise J. et al. (2008). Exploring emerging nanobiotechnology drugs and medical devices. *Food & Drug Law Journal*. **63**:407.

3432. Paradise J. et al. (2009). Challenge of developing oversight approaches to nanobiotechnology. *The Journal of Law, Medicine & Ethics*. **37**:543.

3433. Pavon L.F. and Okamoto O.K. (2007). Applications of nanobiotechnology in cancer. *Einstein*. **5**:74–77.

3434. Plows A. and Reinsborough M. (2008). Nanobiotechnology and ethics: converging civil society discourses. In *Emerging Conceptual, Ethical and Policy Issues in Bionanotechnology.* Netherlands: Springer, pp. 133–156.

3435. Priest S.H. (2009). Risk communication for nanobiotechnology: to whom, about what, and why? *The Journal of Law, Medicine & Ethics.* **37**(4):759–769.

3436. Reshetilov A.N. and Bezborodov A.M. (2008). Nanobiotechnology and biosensor research. *Applied Biochemistry & Microbiology.* **44**(1):1–5.

3437. Rollins K. (2009). Nanobiotechnology regulation: a proposal for self-regulation with limited oversight. *Nanotechnology Law and Business.* **6**:221.

3438. Shoseyov O. and Levy I. (2008). *Nanobiotechnology: Bioinspired Devices and Materials of the Future.* New York: Springer Science & Business Media.

3439. Siep L. (2008). Ethical problems of nanobiotechnology. *Nanobiotechnology, Nanomedicine and Human Enhancement.* **7**:17.

3440. Sikyta B. (2001). [Nanobiotechnology in pharmacology and medicine]. *Ceska a Slovenska farmacie: casopis Ceske farmaceuticke spolecnosti a Slovenske farmaceuticke spolecnosti.* **50**(6):263–266.

3441. Sobha K. et al. (2010). Emerging trends in nanobiotechnology. *Biotechnology and Molecular Biology Reviews.* **5**:1–12.

3442. Takeda Y. et al. (2009). Nanobiotechnology as an emerging research domain from nanotechnology: a bibliometric approach. *Scientometrics.* **80**(1):23–38.

3443. Weber W.L. and Xia Y. (2011). The productivity of Nanobiotechnology research and education in U.S. universities. *American Journal of Agricultural Economics.* **93**:1151–1167.

3444. Whitesides G.M. (2003). The 'right' size in nanobiotechnology. *Nature Biotechnology.* **21**(10):1161–1165.

3445. Willner I. (2007). Nanobiotechnology. *FEBS Journal.* **274**(2): 301–301.

3446. Zhou W. (2002). Ethics of nanobiotechnology at the frontline. *Santa Clara Computer & High Technology Law Journal.* **19**:481.

Nanoengineering

3447. Ahmed H. (2004). Historical preface–three decades of micro and nano engineering. *Microelectronic Engineering.* **73**:1–4.

3448. Anonymous (2000). Magnetic properties key to nanoengineering. *Dr. Dobbs Journal.* **25**(6):18.

3449. Davies A.G. and Thompson J. (2007). *Advances in Nanoengineering: Electronics, Materials, Assembly.* London: Imperial College Press.

3450. Ganesh V.K. (2012). Nanotechnology in civil engineering. *European Scientific Journal.* **8**(27).

3451. Hobæk T.C. et al. (2011). Surface nanoengineering inspired by evolution. *BioNanoScience.* **1**(3):63–77.

3452. Kelkar A.D. et al. (eds). (2014). *Nanoscience and Nanoengineering: Advances and Applications.* Boca Raton, FL: CRC Press.

3453. Mustelin T. (2006). Challenges and optimism for nanoengineering. *Nanomedicine (Lond).* **1**(4):383–385.

3454. Naganathan S. et al. (2014). Nanotechnology in civil engineering: a review. *Advanced Materials Research.* **935**:151–154.

3455. Rawat P. et al. (2015). A review on nanotechnology in civil engineering. *History.* **39**(179):152–158.

3456. Rohrer H. (1998). Nanoengineering beyond nanoelectronics. *Microelectronic Engineering.* **41**:31–36.

3457. Schulz M. J. et al. (eds.) (2005). *Nanoengineering of Structural, Functional and Smart Materials.* Boca Raton, FL: CRC Press.

Nanofabrication

3458. Cabrini S. and Kawata S. (eds.) (2012). *Nanofabrication Handbook.* Boca Raton, FL: CRC Press.

3459. Carter K.R. (2011). Nanofabrication: past, present and future. *Journal of Materials Chemistry.* **21**(37):14095–14096.

3460. Chen W. and Ahmed H. (1997). Nanofabrication for electronics. *Advances in Imaging and Electron Physics*. **102**:87–185.

3461. Chen Y. and Pepin A. (2001). Nanofabrication: conventional and nonconventional methods. *Electrophoresis*. **22**(2):187–207.

3462. Gates B.D., Xu Q., Love J.C., Wolfe D.B. and Whitesides G.M. (2004). Unconventional nanofabrication. *Annual Review of Materials Research*. **34**:339–372.

3463. Gates B.D., Xu Q., Stewart M., Ryan D., Willson C.G. and Whitesides G.M. (2005). New approaches to nanofabrication: molding, printing, and other techniques. *Chemical Reviews*. **105**(4): 1171–1196.

3464. Mailly D. (2009). Nanofabrication techniques. *The European Physical Journal Special Topics*. **172**(1):333–342.

3465. Marrian C.R. and Tennant D.M. (2003). Nanofabrication. *Journal of Vacuum Science & Technology A*. **21**(5):S207–S215.

3466. Natelson D. (2006). Nanofabrication: best of both worlds. *Nature Materials*. **5**(11):853–854.

3467. Smith H.I. and Craighead H.G. (2008). Nanofabrication. *Physics Today*. **43**(2):24–30.

3468. Stepanova M. and Dew S. (2011). *Nanofabrication: Techniques and Principles*. New York: Springer Science & Business Media.

3469. Wang Y., Mirkin C.A. and Park S.J. (2009). Nanofabrication beyond electronics. *ACS Nano*. **3**(5):1049–1056.

3470. Wiederrecht G. (2010). *Handbook of Nanofabrication*. New York: Academic Press.

3471. Wilkinson C.D.W. (1987). Nanofabrication. *Microelectronic Engineering*. **6**(1):155–162.

3472. Wilkinson C.D.W., Curtis A.S.G. and Crossan J. (1998). Nanofabrication in cellular engineering. *Journal of Vacuum Science & Technology B*. **16**(6):3132–3136.

Nanomechanics

3473. Cleland A.N. (2003). *Foundation of Nanomechanics: From Solid-State Theory to Device Applications.* New York: Springer-Verlag.

3474. Feringa B.L. and Browne W.R. (2008). Nanomechanics: macro-molecules flex their muscles. *Nature Nanotechnology.* **3**(7): 383–384.

3475. Fritz J. et al. (2000). Translating biomolecular recognition into nanomechanics. *Science.* **288**(5464):316–318.

3476. Gerber C. (2005). Nanomechanics: opening new frontiers in bio analyses and diagnostics. *NanoBiotechnology.* **1**(3):289–289.

3477. Gerberich W. and Mook W. (2005). Nanomechanics: a new picture of plasticity. *Nature Materials.* **4**(8):577–578.

3478. Guo X.E. (2008). What is nanomechanics of bone and why is it important? *Journal of Muskoskeletal & Neuronal Interactions.* **8**(4):327–328.

3479. Kippenberg T.J. (2008). Photonics: nanomechanics gets the shakes. *Nature.* **456**(7221):458.

3480. Poggio M. (2013). Nanomechanics: sensing from the bottom up. *Nature Nanotechnology.* **8**(7):482–483.

3481. Rodgers P. (2010). Nanomechanics: welcome to the quantum ground state. *Nature Nanotechnology.* **5**(4):245.

3482. Vodnick D. (2006). Nanomechanical characterization of coatings. *Paint & Coatings Industry.* **22**(8):L5–L10.

3483. Wilson-Rae I. (2015). Nanomechanics: rocking at the nanoscale. *Nature Nanotechnology.* **10**(6): 489–490.

Nanomedicine: General Topics

Environmental and Health Risks

3484. Baun A. and Hansen S.F. (2008). Environmental challenges for nanomedicine. *Nanomedicine.* **3**(5):605–608.

3485. Cattaneo A.G. et al. (2010). Nanotechnology and human health: risks and benefits. *Journal of Applied Toxicology.* **30**(8):730–744.

3486. de Jong W. et al. (2005). Nanotechnology in medical applications: possible risks for human health (RIVM rapport 265001002). Bilthoven, NL: Rijksinstituut voor Volksgezondheid en Milieu.

3487. Faunce T.A. and Shats K. (2007). Researching safety and cost-effectiveness in the life cycle of nanomedicine. *Journal of Law, Medicine & Ethics.* **15**:128–135.

3488. Hammed O. et al. (2016). Nanomedicine: overview, problem, solution and future. *International Journal of Computer Science & Software Engineering.* **5**(9):218–222.

3489. Hogle L.F. (2012). Concepts of risk in nanomedicine research. *Journal of Law, Medicine & Ethics.* **40**(4):809–822.

3490. Igarashi E. (2015). *Nanomedicines and Nanoproducts: Applications, Disposition, and Toxicology in the Human Body.* Boca, Raton, FL: CRC Press.

3491. Krug H.F. (2014). Nanosafety research–are we on the right track? *Angewandte Chemie International Edition in English.* **53**:12304–12319.

3492. Mahapatra I. et al. (2013). Potential environmental implications of nano-enabled medical applications: critical review. *Environmental Science: Processes & Impacts.* **15**(1):123–144.

3493. Moore R. (2007). Nanomedicine and risk: further perspectives. *Medical Device Technologies.* **18**(6):28–29.

3494. Murashov V. (2009). Occupational exposure to nanomedical applications. *Wiley Interdisciplinary Reviews: Nanomedicine and Nanobiotechnology.* **1**(2):203–213.

3495. Resnik D.B. (2012). Responsible conduct in nanomedicine research: environmental concerns beyond the common rule. *Journal of Law, Medicine & Ethics.* **40**(4):848–855.

3496. Rucinski T.L. (2013). Searching for the nano-needle in a green haystack: researching the environmental, health, and safety ramifications of nanotechnology. *Pace Environmental Law Review.* **30**(2):397–440.

3497. Shatkin J.A. (2012). *Nanotechnology: Health and Environmental Risks*, 2nd ed. Boca Raton, FL: CRC Press.

3498. Wagner V. et al. (2008). Nanomedicine: drivers for development and possible impacts. Seville, Spain: European Commission, Joint Research Centre, Institute for Prospective Technological Studies.

3499. World Health Organization (2012). Nanotechnology and human health: scientific evidence and risk governance (Report of the WHO expert meeting, December 10–11, 2012, Bonn Germany). Geneva.

Ethics

3500. Aala M. et al. (2008). Bioethical issues of nanotechnology at a glance. *Iranian Journal of Public Health.* **37**(1 Suppl):12–17.

3501. Baumgartner C. (2004). [Ethische Aspekte nanotechnologischer Forschung und Entwicklung in der Medizin]. *Politik und Zeitgeschichte.* **23–24**:39–46 (German).

3502. Bawa R. and Johnson S. (2007). The ethical dimensions of Nanomedicine. *Medical Clinics of North America.* **91**(5):881–887.

3503. Bennett M.G. and Naranja R.J. (2013). Getting nano tattoos right - a checklist of legal and ethical hurdles for an emerging nanomedical technology. *Nanomedicine.* **9**(6):729–731.

3504. Berger F. et al. (2008). Ethical, legal and social aspects of brain-implants using nano-scale materials and techniques. *NanoEthics.* **2**(3):241–249.

3505. Best R. and Khushf G. (2006). The social conditions for nanomedicine: disruption, systems, and lock-in. *The Journal of Law, Medicine & Ethics.* **34**(4):733–740.

3506. Boenink M. (2010). Molecular medicine and concepts of disease: the ethical value of a conceptual analysis of emerging biomedical technologies. *Medicine, Health Care and Philosophy.* **13**(1):11–23.

3507. Bouwman M.T. (2010). Legal and ethical aspects associated with nanomedicine in patient care: a literature review (Student Paper No. 340099). Rotterdam: Erasmus University.

3508. Brownsord R. (2008). Regulating nanomedicine—the smallest of our concerns? *NanoEthics.* **2**(1):73–86.

3509. Costa H.S. et al. (2011). Scientist's perception of ethical issues in nanomedicine: a case study. *Nanomedicine.* **6**(4):681–691.

3510. Dresser R. (2012). Building an ethical foundation for first-in-human nanotrials. *The Journal of Law, Medicine & Ethics.* **40**(4):802–808.

3511. Duncan R. and Gaspar R. (2011). Nanomedicine(s) under the microscope. *Molecular Pharmaceutics.* **8**:2101–2141.

3512. Ebbesen M. and Jensen T.G. (2006). Nanomedicine: techniques, potentials, and ethical implications. *Journal of Biomedicine and Biotechnology.* **2006**:51516.

3513. Erdmann M. (2008). Implications of nanomedicine: applying converging technologies in nanomedicine. Taking stock of challenges and benefits. *European Journal of Nanomedicine.* **1**(1):37–39.

3514. Fatehi L. et al. (2012). Recommendations for nanomedicine human subjects research oversight: an evolutionary approach for an emerging field. *The Journal of Law, Medicine & Ethics.* **40**(4):716–750.

3515. Faunce T.A. (2007). Nanotechnology in global medicine and human biosecurity: private interests, policy dilemmas and the calibration of public health law. *The Journal of Law, Medicine & Ethics (US).* **35**(4):629–642.

3516. Gordijn B. (2007). Ethical issues in nanomedicine. In ten Have Henk A.M.J. (ed.), *Nanotechnologies, Ethics and Politics.* Paris, France: UNESCO Pub., pp. 99–123.

3517. Hall R.M. et al. (2012). A portrait of nanomedicine and its bioethical implications. *The Journal of Law, Medicine & Ethics.* **40**(4):763–779.

3518. Hermerén G. (2007). [Nanomedicine challenges European ethicians, lawyers and toxicologists]. *Lakartidningen.* **104**(17):1326–1330 (Swedish).

3519. Hock S.C. et al. (2011). A review of the current scientific and regulatory status of nanomedicines and the challenges ahead. *Journal of Pharmaceutical Science and Technology.* **65**(2):177–195.

3520. Hogle L.F. (2012). Concepts of risk in nanomedicine research. *The Journal of Law, Medicine & Ethics.* **40**(4):809–822.

3521. Johnson S. (2009). The era of Nanomedicine and nanoethics: has it come, is still coming, or will it pass us by? *American Journal of Bioethics.* **9**(10):1–2.

3522. Kazemi A. et al. (2014). The question of ethics in nanomedicine. *Journal of Clinical Research & Bioethics.* **5**:193. http://dx.doi.org/10.4172/2155-9627.1000193

3523. Khushf G. (2007). Upstream ethics in nanomedicine: a call for research. *Nanomedicine (Lond).* **2**(4):511–521.

3524. Kuiken T. (2011). Nanomedicine and ethics: is there anything new or unique? *Wiley Interdisciplinary Reviews: Nanomedicine and Nanobiotechnology.* **3**(2):111–118.

3525. Lenk C. et al. (2007). Nanomedicine—emerging or re-emerging ethical issues? A discussion of four ethical themes. *Medicine, Health Care & Philosophy.* **10**(2):173–184.

3526. Leontis V.L. and Agich G.J. (2010). Freitas on disease in nanomedicine: implications for ethics. *NanoEthics.* **4**:205–214.

3527. Lupton M. (2011). The social, moral & ethical issues raised by nanotechnology in the field of medicine. *Medicine and Law.* **30**(2):187–200.

3528. MacDonald C. and Williams-Jones B. (2012). Nothing new (ethically) under the sun: policy & clinical implications of nanomedicine. *Bioéthique Online*, 1/11.

3529. McGee E.M. (2009). Nanomedicine: ethical concerns beyond diagnostics, drugs, and techniques. *American Journal of Bioethics.* **9**(10):14–15.

3530. Oftedal G. (2014). The role of philosophy of science in Responsible Research and Innovation (RRI): the case of nanomedicine. *Life Sciences, Society and Policy.* **10**(1):1–12.

3531. Pelle S. and Nurock V. (2012). Of nanochips and persons: towards an ethics of diagnostic technology in personalized medicine. *NanoEthics.* **6**:155–165.

3532. Pérez Alvarez S. (2012). [Meta-legal paradigms of nanomedicine]. *Rev Derecho Genoma Hum.* **37**:61–91 (Spanish).

3533. Poirot-Mazères I. (2011). Legal aspects of the risks raised by nanotechnologies in the field of medicine. *Journal International de Bioethique.* **22**(1):99–118, 212.

3534. Ponsaran M. (2012). Nanomedicine: ethical and societal challenges to major stakeholders. *Philippiniana Sacra.* **47**(140): 490–494.

3535. Prabhala B. and Dharmendra J. (2012). Ethical issues in nanomedicine. *The Holistic Approach to Medicine.* **2**(4):171–175.

3536. Resnik D.B. and Tinkle S.S. (2007a). Ethical issues in clinical trials involving nanomedicine. *Contemp Clin Trails.* **28**(4):433–441.

3537. Resnik D.B. and Tinkle S.S. (2007b). Ethics in nanomedicine. *Nanomedicine.* **2**(3):345–350.

3538. Riehemann K. et al. (2009). Nanomedicine: challenge and perspectives. *Angewandte Chemie International Edition.* **48**(5):872–897.

3539. Sandler R. (2009). Nanomedicine and nanomedical ethics. *American Journal of Bioethics* **9**(10):16–17.

3540. Silva Costa H. et al. (2011). Scientist's perception of ethical issues in nanomedicine: a case study. *Nanomedicine.* **6**(4):681–691.

3541. Slade C.P. (2011). Public value mapping of equity in emerging nanomedicine. *Minerva: A Review of Science, Learning & Policy.* **49**(1):71–86.

3542. Spagnolo A.G. and Daloiso V. (2009). Outlining ethical issues in nanotechnologies. *Bioethics.* **23**(7):394–402.

3543. Tiefenauer L.X. (2006). Ethics of nanotechnology in medicine. *Nanobiotechnology.* **2**(1–2):1–3.

3544. Trisolino A. (2014). Nanomedicine: building a bridge between science and law. *NanoEthics.* **8**:141–163.

3545. Virdi J. (2009). Bridging the knowledge gap: examining potential limits in nanomedicine. *Spontaneous Generations: A Journal for the History and Philosophy of Science.* **2**(1): 25.

3546. White G.B. (2009). Missing the boat on nanoethics. *American Journal of Bioethics.* **9**(10):18–19.

3547. Wiesing U. and Clausen J. (2014). The clinical research of nanomedicine: a new ethical challenge? *NanoEthics.* **8**(1):19–28.

3548. Wolf S.M. and Jones C.M. (2011). Designing oversight for nanomedicine research in human subjects. *Journal of Nanoparticle Research.* **13**:1449–1465.

History, Trends, and Future Directions

3549. Abeer S. (2012). Future medicine: nanomedicine. *Journal of International Medical Sciences Academy.* **25**(3):187–192.

3550. Alakhova D.Y. and Kabanov A.V. (2015). Nanomedicine and nanotechnology are rapidly developing fields across the nation and worldwide. *Journal of Controlled Release.* **201**:1

3551. Allhoff F. (2009). The coming era of Nanomedicine. *American Journal of Bioethics.* **9**(10):3–11.

3552. Anonymous (2003). Nanomedicine: grounds for optimism, and a call for papers. *Lancet.* **362**(9385):673.

3553. Baba Y. (2006). [Nanotechnology in medicine]. *Nihon Rinsho.* **64**(2):189–198 (Japanese).

3554. Bellare J.R. (2011). Nanotechnology and nanomedicine for healthcare: challenges in translating innovations from bench to bedside. *Journal of Biomedical Nanotechnology.* **7**(1):36–37.

3555. Berger J. (2011). The age of biomedicine: current trends in traditional subjects. *Journal of Applied Biomedicine.* **9**(2):57–61.

3556. Berube D.M. (2009). The public acceptance of Nanomedicine: a personal perspective. *Wiley Interdisciplinary Reviews: Nanomedicine and Nanobiotechnology.* **1**(1):2–5.

3557. Bhowmik D. et al. (2010). Nanomedicine: an overview. *International Journal of PharmTech Research.* **2**(4):2143–2151.

3558. Bishop C.J. et al. (2014). Highlights from the latest articles in nanomedicine. *Nanomedicine.* **9**(7):945–947.

3559. Bogunia-Kubik K. and Sugisaka M. (2002). From molecular biology to nanotechnology and nanomedicine. *Biosystems.* **65**(2–3):123–138.

3560. Boisseau P. and Loubaton B. (2011). Nanomedicine, nano-technology in medicine. *Comptes Rendus Physique.* **12**(7): 620–636.

3561. Bosetti R. and Vereeck L. (2011). Future of Nanomedicine: obstacles and remedies. *Nanomedicine.* **6**(4):747–755.

3562. Bottini M. et al. (2011). Public optimism towards nano-medicine. *International Journal of Nanomedicine.* **6**:3473–3485.

3563. Boulaiz H. et al. (2011). Nanomedicine: application areas and development prospects. *International Journal of Molecular Sciences.* **12**(5):3303–3321.

3564. Bounia-Kubik K. et al. (2002). From molecular biology to nanotechnology and nanomedicine. *Biosystems.* **65**(2/3):123.

3565. Bullis K. (2006). Nanomedicine. *Technology Review.* **109**(1): 58–59.

3566. Caruso F. et al. (2012). Nanomedicine. *Chemical Society Reviews.* **41**(7):2537–2538.

3567. Caruthers S.D. et al. (2007). Nanotechnological applications in medicine. *Current Opinion in Biotechnology.* **18**(1):26–30.

3568. Ciutan M. et al. (2010). Nanomedicine: the future medicine. *Management in Health.* **14**(1).

3569. Coccia M. and Finardi U. (2012). Emerging nanotechnological research for future pathways of biomedicine. *International Journal of Biomedical Nanoscience and Nanotechnology.* **2**(3–4):299–317.

3570. Conti P.S. et al. (2008). Molecular imaging: the future of mod-ern medicine. *The Journal of Nuclear Medicine.* **49**(6):16N.

3571. Corabian P. and Chojecki D. (2012). Exploratory brief on nanomedicine or the application of nanotechnology in human healthcare. Alberta, Canada: Institute of Health Economics.

3572. de Silva M.N. (2007). Nanotechnology and nanomedicine: a new horizon for medical diagnostics and treatment. *Archivos de la Sociedad Española de Oftalmología.* **82**(6):331–334.

3573. Dab W. (2010). [Nanomedicine: small size, big stakes]. *La Revue du Praticien.* **60**(7):896–897.

3574. D'Aquino R. (2006). Fulfilling the promise to nanomedicine. *Chemical Engineering Progress.* **102**(2):35–37.

3575. Darshan S. and Tyshenko M.G. (2010). Identifying recent trends in nanomedicine development. *International Journal of Nanotechnology.* **7**(2–3):173–186.

3576. Datta R. and Jaitawat S.S. (2006). Nanotechnology: the new frontier of medicine. *Medical Journal Armed Forces India.* **62**(3):263–268.

3577. Diwan P. and Bharadwaj A. (2006). *Nanomedicine.* New Delhi, India: Pentagon Press.

3578. Donnor A. (2010). Nanotechnology in molecular medicine. *Trends in Molecular Medicine.* **16**(12):551–552.

3579. Ebbesen M. and Jensen T.G. (2006). Nanomedicine: techniques, potentials, and ethical implications. *Journal of Biomedicine and Biotechnology.* **2006**:51516.

3580. Emerich D.F. and Thanos C.G. (2003). Nanotechnology and medicine. *Expert Opinion on Biological Therapy.* **3**(4):655–663.

3581. Erdmann M. (2009). Implications of nanomedicine: the spiritualization of science, technology, and education in a one-world society. *European Journal of Nanomedicine.* **2**(1):31.

3582. Escoffier L. et al. (2015). *Commercializing Nanomedicine: Industrial Applications, Patents, and Ethics.* Singapore: Pan Stanford Publishing.

3583. Etheridge M.L. et al. (2012). The big picture on nanomedicine: the state of investigational and approved nanomedicine products. *Nanomedicine.* **9**(1):1–14.

3584. European Commission (2008). Nanomedicine: drivers for development and possible impacts (EUR-23494-EN). Seville, Spain: Joint Research Centre, Institute for Prospective Technological Studies.

3585. European Science Foundation (2005). Nanomedicine. Strausburg, France.

3586. European Commission (2005). Vision paper and basis for a strategic research agenda for nanomedicine. Luxembourg.

3587. European Commission (2009). Roadmaps in nanomedicine towards 2020. Luxembourg.

3588. European Science Commission (2005). Nanomedicine: an ESF-European medical research councils (EMRC) forward look report.

3589. Ferrari M. (2008). The mathematical engines of nano-medicine. *Small.* **4**(1):20–25.

3590. Ferrari M. et al. (2009). Nanomedicine and society. *Clinical Pharmacology & Therapeutics.* **85**(5):466–467.

3591. Filipponi L. and Sutherland D. (2007). Applications of nanotechnology: medicine (Part 2). Denmark: University of Aarhus, Interdisciplinary Nanoscience Center (iNANO).

3592. Flynn T. and Wei C. (2005). The pathway to commercialization for nanomedicine. *Nanomedicine.* **1**(1):47–51.

3593. Formoso P. et al. (2015). Nanotechnology for the environment and medicine. *Mini-Reviews in Medicinal Chemistry.* [Epub ahead of print].

3594. Freitas R.A. Jr. (2005). Nanotechnology, nanomedicine and nanosurgery. *International Journal of Surgery.* **3**(4):243–246.

3595. Freitas R.A. Jr. (2005). What is nanomedicine? *Nanomedicine.* **1**(1):2–9.

3596. Freitas R.A. Jr. (2006). What is nanomedicine? *Disease-a-Month.* **51**(6):325–341.

3597. Freitas R.A. Jr. (2010). The future of nanomedicine. *Futurist.* **44**(1):21–22.

3598. Gaur A. et al. (2008). Significance of nanotechnology in medical sciences. *Asian Journal of Pharmaceutics.* **2**(2):80–85.

3599. Ge Y. et al. (eds.) (2014). *Nanomedicine.* New York: Springer.

3600. Gendelman H.E. et al. (2014). The promise of nanoeuromedicine. *Nanomedicine.* **9**(2):171–176.

3601. Grenha A. (2011). The era of nanomedicine. *Journal of Pharmacy and Bioallied Sciences.* **3**(2):181.

3602. Gupta J. (2011). Nanotechnology applications in medicine and dentistry. *Journal of Investigative and Clinical Dentistry.* **2**(2):81–s88.

3603. Haberzettl C.A. (2002). Nanomedicine: destination or journey? *Nanotechnology.* **13**(4).

3604. Hammond P.T. (2014). A growing place for nano in medicine. *ACS Nano.* **8**(8):7551–7552.

3605. Hehenberger M. (2012). *Nanomedicine: Science, Business and Impact.* Boca Raton, FL: CRC Press.

3606. Hermes C. et al. (2014). [Nanotechnology: scientific, material, global and ethic progress]. *Pers Bioet.* **18**(2):107–118 (Portuguese).

3607. Herzog A. (2002). Of genomics, cyborgs, and nanotechnology: a look into the future of medicine. *Connecticut Medicine.* **66**(1):53–54.

3608. Holm B.A. et al. (2002). Nanotechnology in biomedical applications. *Molecular Crystals and Liquid Crystals.* **374**(1):589–598.

3609. Hunter R.J. and Preedey V.R. (eds.) (2011). *Nanomedicine in Health and Disease.* Enfield, NH: Science Publishers.

3610. Hunziker P. (2010). Nanomedicine: shaping the future of medicine in a context of academia, industry and politics. *European Journal of Nanomedicine.* **3**(1):6.

3611. Hunziker P.R. et al. (2002). Nanotechnology in medicine: moving from bench to the bedside. *Chimia.* **56**:520–526.

3612. Jain K.K. (2008). *The Handbook of Nanomedicine.* Totowa, NJ: Humana Press.

3613. Jain K.K. (2008). Nanomedicine: application of nanobiotechnology in medical practice. *Medical Principles and Practice.* **17**(2):89–101.

3614. Jain K.K. and Jain V. (2006). Impact of nanotechnology on healthcare. *Nanotechnology Law & Business.* **3**:411–418.

3615. Jena M., Mishra S., Jena S. and Mishra S.S. (2013). Nanotechnology-future prospect in recent medicine: a review. *International Journal of Basic & Clinical Pharmacology.* **2**(4): 353–359.

3616. JiMin W.U. and ZiJian L.I. (2013). Applications of nano-medicine in biomedicine. *Science Bulletin.* **58**(35).

3617. Jokanović V. (2014). The deep scientific and philosophic approach to the future nanomedicine, given on the base of author introduction in the monograph nanomedicine, the greatest challenge of the 21st century. *Drug Designing.* **3**(2). doi:10.4172/2169-0138.1000113

3618. Juliano R.L. (2012). The future of Nanomedicine: promises and limitations. *Science & Public Policy.* **39**:99–104.

3619. Juliano R.L. (2013). Nanomedicine: is the wave cresting? *Nature Reviews Drug Discovery.* **12**(3):171–172.

3620. Kapoor D.N. et al. (2013). Advanced nanomedicine: present contributions and future expectations. *American Journal of Phytomedicine and Clinical Therapeutics.* **1**(2):124–139.

3621. Katayama Y. (2005). [Development of nano-diagnosis and nano-medicine]. *Fukuoka Igaku Zasshi.* **96**(6): 281–283 (Japanese).

3622. Kato K. (2012). [Development trend of nanomedicines]. *Yakugaku Zasshi: Journal of the Pharmaceutical Society of Japan.* **133**(1):43–51.

3623. Kaur A. et al. (2012). How nanotechnology works in medicine. *International Journal of Electronics and Computer Science Engineering.* **1**(4):2452–2459.

3624. Khan A.U. (2012). Medicine at nanoscale: a new horizon. *International Journal of Nanomedicine.* **7**:2997–2998.

3625. Khushf G. and Siegel R.A. (2012). What is unique about nano-medicine? The significance of the mesoscale? *Journal of Law, Medicine & Ethics.* **40**(4):780–794.

3626. Kim B.Y. et al. (2010). Nanomedicine. *New England Journal of Medicine.* **363**(25):2434–2443.

3627. Kostarelos K. (2006). The emergence of nanomedicine: a field in the making. *Nanomedicine.* **1**(1):1–3.

3628. Kostarelos K. (2006). Establishing nanomedicine. *Nano-medicine.* **1**(3):259–260.

3629. Kostarelos K. (2009). Nanomedicine: transcending from embryonic to adolescent. *Nanomedicine.* **4**(2):123–124.

3630. Kranz C. et al. (2011). Analytical challenges in nanomedicine. *Analytical and Bioanalytical Chemistry.* **399**(7):2309–2311.

3631. Labhasetwar V. and Leslie-Pelecky D.L. (eds.) (2007). *Biomedical Applications of Nanotechnology.* Hoboken, NJ: John Wiley & Sons.

3632. Lim J.M. (2004). [The present and future of nanotechnology in medicine]. *Korean Journal of Hepatology.* **10**(3):185–190.

3633. Liu Y. and Wang H. (2007). Nanomedicine: nanotechnology tackles tumors. *Nature Nanotechnology.* **2**(1):20–21.

3634. Logothetidis S. (2006). Nanotechnology in medicine: the medicine of tomorrow and nanomedicine. *Hippokratia.* **10**:7–21.

3635. Logothetidis S. (ed.) (2012). *Nanomedicine and Nanobiotechnology.* New York: Springer-Verlag.

3636. Lokesh P. and Ashish D. (2014). Nanomedicines: present scenario and future challenges. doi:10.13140/2.1.1507.6648

3637. Lord R. (2013). Explore the world of nanomedicine. *American Biolology Teacher.* **75**(8):595–596.

3638. Luxenhofer R. et al. (2014). Quo vadis nanomedicine? *Nanomedicine.* **9**(14):2071–2074.

3639. Maebius S. and Jamison D. (2008). Realizing the opportunities of nanomedicine. *Nanotechnology Law & Business.* 121.

3640. Mahadevan V. and Sethuraman S. (2003). Nanomaterials and nanosensors for medical applications. In *Trends in Nanoscale Mechanics.* Amsterdam, Netherlands: Springer, pp. 207–228.

3641. Malhotra P. et al. (2010). Nano medicine: a futuristic approach. *JK Science.* **12**(1):3–5.

3642. Malinoski F.J. (2014). The nanomedicines alliance: an industry perspective on nanomedicines. *Nanomedicine.* **10**(8): 1819–1820.

3643. Malsch N.H. (ed.) (2005). *Biomedical Nanotechnology.* Boca Raton, FL: CRC Press.

3644. Manzar N. and Mujeeb E. (2012). Nanomedicine. *Journal of College of Physicians and Surgeons Pakistan.* **22**(8):481–483.

3645. Marchesan S. and Prato M. (2012). Nanomaterials for (nano) medicine. *ACS Medicinal Chemistry Letters.* **4**(2):147–149.

3646. Meetoo D. (2009). Nanotechnology: the revolution of the big future with tiny medicine. *British Journal of Nursing.* **18**(19):1201–1206.

3647. Mehlich J. and Thiele F. (2014). Nanomedicine: visions, risk, potential. *European Journal of Nanomedicine.* **6**(1):47.

3648. Mehta M.D. (2004). The future of Nanomedicine looks promising, but only if we learn from the past. *Health Law Review.* **13**(1):16–18.

3649. Miksanek T. (2001). Microscopic doctors and molecular black bags: science fiction's prescription for nanotechnology and medicine. *Literature and Medicine.* **20**(1):55–70.

3650. Moghimi S.M. et al. (2005). Nanomedicine: current status and future prospects. *FASEB Journal.* **19**(3):311–330.

3651. Morigi V. et al. (2012). Nanotechnology in medicine: from inception to marked domination. *Journal of Drug Delivery.* **2012**: Article ID 389485.

3652. Müller B. (2012). [What is nanomedicine? Nanotechnology for most of the patients!]. *Revue Médicale Suisse.* **8**(325):152–153 (German).

3653. Navalakhe R.M. and Nandedkar T.D. (2007). Application of nanotechnology in biomedicine. *Indian Journal of Experimental Biology.* **45**(2):160.

3654. Oftedal G. (2014). The role of philosophy of science in responsible research and innovation: the case of nanomedicine. *Life Sciences, Society and Policy.* **10**(1):5.

3655. Omanović-Mikličanin E. et al. (2015). The future of healthcare: nanomedicine and internet of nano things. *Folia Medica Facultatis Medicinae Universitatis Saraeviensis.* **50**(1):23–28.

3656. Ornelas-Megiatto C. et al. (2015). Interlocked systems in nanomedicine. *Current Topics in Medicinal Chemistry.* **15**(13):1236–1256.

3657. Pan D. (ed.) (2015). *Nanomedicine: A Soft Matter Perspective.* Boca Raton, FL: CRC Press.

3658. Owen A. et al. (2014). The application of nanotechnology in medicine: treatment and diagnostics. *Nanomedicine.* **9**(9): 1291–1294.

3659. Pal'tsev M.A. et al. (2009). Nanotechnology in medicine. *Herald of the Russian Academy of Sciences.* **79**:369–377.

3660. Park H.H. et al. (2007). Rise of the nanomachine: the evolution of a revolution in medicine. *Nanomedicine.* **2**(4):425–439.

3661. Park H.H. et al. (2007). Rise of the nanomachine: the evolution of a revolution in medicine. *Nanomedicine.* 2(4):415–423.

3662. Patil M. et al. (2008). Future impact of nanotechnology on medicine and dentistry. *Journal of Indian Society of Periodontology.* **12**(2):34–40.

3663. Pautler M. and Brenner S. (2010). Nanomedicine: promises and challenges for the future of public health. *International Journal of Nanomedicine.* **5**:803–809.

3664. Peiris P.M. and Karathanasis E. (2011). Is nanomedicine still promising? *Oncotarget.* **2**(6):430–432.

3665. Peplow M. (2015). A smarter bandage. *Scientific American.* **312**(4):47–49.

3666. Perry K-M.E. (2010). Nanotechnology and health: from boundary object to bodily intervention. Thesis: Department of Sociology & Anthropology, Simon Fraser University.

3667. Peters R. (2006). Nanoscopic medicine: the next frontier. *Small.* **2**(4):452–456.

3668. Petherick A. (2008). Advances in nanomedicine. *New Statesman.* **137**:4928–4930.

3669. Pilarski L.M. et al. (2004). Microsystems and nanoscience for biomedical applications: a view to the future. *Bulletin of Science, Technology & Society.* **24**(1):40–45.

3670. Piotrovskiĭ L.B. (2010). [Nanomedicine as a part of nanotechnology]. Vestn Ross Akad Med Nauk. (3):41–6 (Russian).

3671. Raghvendra S.T. et al. (2010). Clinical applications of nanomedicines: a review. *International Journal of Applied Biology and Pharmaceutical Technology.* **1**(2):660–665.

3672. Rannard S. and Owen A. (2009). Nanomedicine: not a case of "one size fits all." *Nano Today.* **4**(5):382–384.

3673. Riehemann K. et al. (2009). Nanomedicine: challenge and perspectives. *Angewandte Chemie International Edition.* **48**(5):872–897.

3674. Roco M. (2003). Nanotechnology: convergence with modern biology and medicine. *Current Opinion in Biotechnology.* **14**(3):337–346.

3675. Roszek B. et al. (2005). Nanotechnology in medical applications: state of the art materials and devices (RIVM Report 265001001). Bilthoven: RIVM.

3676. Rzigalinski B.A. et al. (2006). Radical nanomedicine. *Nanomedicine.* **1**(4):399–412.

3677. Saji V.S. et al. (2010). Nanotechnology in biomedical applications: a review. *International Journal of Nano and Biomaterials.* **3**(2):119–139.

3678. Salieb-Beugelaar G.B. (2014). What's up in nanomedicine? *European Journal of Nanomedicine.* **6**(1):5–7.

3679. Sadanandan N. (2011). Nanomedicine: the basics. *Western Journal of Medicine.* **3**(3):11–14.

3680. Sahoo S.K., Parveen S. and Panda J.J. (2007). The present and future of nanotechnology in human health care. *Nanomedicine: Nanotechnology, Biology and Medicine.* **3**(1), 20–31.

3681. Saini R. et al. (2010). Nanotechnology: the future medicine. *Journal of Cutaneous and Aesthetic Surgery.* **3**(1):32–33.

3682. Sandhiya S. et al. (2009). Emerging trends of nanomedicine— an overview. *Fundamental & Clinical Pharmacology.* **23**(3): 263–269.

3683. Sanhai W.R. et al. (2008). Seven challenges for nanomedicine. *Nature Nanotechnology.* **3**(5):242–244.

3684. Saniotis A. (2008). Mythogenesis and nanotechnology: future medical directions. *Journal of Futures Studies.* **12**(3):71–82.

3685. Saniotis A. (2012). Nanomedicine and future body enhancement. *Nanotechnology Perceptions.* **8**(1):76.

3686. Satvekar R.K. et al. (2014). Emerging trends in medical diagnosis: a thrust on nanotechnology. *Medicinal Chemistry.* **4**(4):407–416.

3687. Seaton A. (2006). Nanotechnology and the occupational physician. *Occupational Medicine.* **56**(5):312–316.

3688. Sechi G. et al. (2014). The perception of nanotechnology and nanomedicine: a worldwide social media study. *Nanomedicine.* **9**(10):1475–1486.

3689. Sekhon B.S. (2012). Current scenario on impact of nanomedicine. *Journal of Pharmaceutical Education & Research.* **3**(1):71–76.

3690. Shelley T. (2006). *Nanotechnology: New Promises, New Dangers.* New York: Zed Books, Ltd.

3691. Sheikh F.A. (2014). Highlights from recent advances in nanomedicine. *Nanomedicine.* **9**(9): 1287–1289.

3692. Shrivastava S. and Dash D. (2009). Applying nanotechnology to human health: revolution in biomedical sciences. *Journal of Nanotechnology.* **2009**: Article ID 184702.

3693. Singh Y. (2014). Trends in biomedical nanotechnology. *Journal of Nanomedicine Biotherapeutic Discovery.* **4**(2): 1000e130.

3694. Solano-Umana V. et al. (2015). The new field of nanomedicine. *International Journal of Applied Science & Technology.* **5**(1): 79–88.

3695. Sparks S. (2012). *Nanotechnology: Business Applications and Commercialization.* Boca Raton, FL: CRC Press.

3696. Stylios G.K. et al. (2005). Applications of nanotechnologies in medical practice. *Injury.* **36**(4 Suppl 1):S6–S13.

3697. Surendiran A. et al. (2009). Novel applications of nanotechnology in medicine. *The Indian Journal of Medical Research.* **130**(6):689–701.

3698. Teli M.K. et al. (2010). Nanotechnology and nanomedicine: going small means aiming big. *Current Pharmaceutical Design.* **16**(16):1882–1892.

3699. Terlega K. and Latocha M. (2012). [Nanotechnology future of medicine]. *Polski Merkuriusz Lekarski.* (196):229–32 (Polish).

3700. Thacker E. (2004). Nanomedicine: molecules that matter. *Biomedia.* **11**:115–140.

3701. Thierry B. and Textor M. (2012). Nanomedicine in focus: opportunities and challenges ahead. *Biointerphases.* **7**(1–4):19.

3702. Thorley A.J. and Tetley T.D. (2013). New perspectives in nanomedicine. *Pharmacology and Therapeutics.* **140**(2):176–185.

3703. Thrall J.H. (2004). Nanotechnology and medicine. *Radiology.* **230**(2):315–318.

3704. Tibbals H.F. (2010). *Medical Nanotechnology and Nanomedicine.* Boca Raton, FL: CRC Press.

3705. Tong R. and Kohane D.S. (2014). Shedding light on nanomedicine. *Wiley Interdisciplinary Reviews: Nanomedicine and Nanobiotechnology.* **4**(6):638–662.

3706. Tong S. et al. (2014). Nanomedicine: tiny particles and machines give huge gains. *Annals of Biomedical Engineering.* **42**(2):243–259.

3707. Torchilin V. (2013). Many faces of nanomedicine. *Drug Delivery and Translational Research.* **3**(5):382–383.

3708. Tosi G. et al. (2012). Nanomedicine: the future for advancing medicine and neuroscience. *Nanomedicine.* **7**(8):1113–1116.

3709. Trisolino A. (2010). Nanomedicine: governing uncertainties. Thesis (LL.M.): University of Toronto, Canada.

3710. Tsai N. et al. (2014). Nanomedicine for global health. *Journal of Laboratory Automation.* **19**(6):511–516.

3711. Ventola C.L. (2012). The Nanomedicine revolution. Part 1. Emerging concepts. *P&T: A Peer-Reviewed Journal for Formulary Management.* **37**(9):512–525.

3712. Urban G.A. (2005). [Nanotechnology in medicine]. *Praxis (Bern 1994).* **94**(41):1591–1593 (German).

3713. Venkatraman S. (2014). Has nanomedicine lived up to its promise? *Nanotechnology.* **25**(37):372501.

3714. Virdi J. (2008). Bridging the knowledge gap: examining potential limits in nanomedicine. *Spontaneous Generations: A Journal for the History and Philosophy of Science.* **2**(1):25–44.

3715. Ventola C.L. (2012a). The nanomedicine revolution. Part 1: Emerging concepts. *Pharmacy and Therapeutics.* **37**(9): 512.

3716. Ventola C.L. (2012b). The nanomedicine revolution. Part 2: Current and future clinical applications. *P&T: A Peer-Reviewed Journal for Formulary Management.* **37**(10):582–591.

3717. Ventola C.L. (2012c). The nanomedicine revolution. Part 3: Regulatory and safety challenges. *P&T: A Peer-Reviewed Journal for Formulary Management.* **37**(11):631–639.

3718. Vishwakarma K. et al. (2008). Nanotechnology: a boon for medical science. *International Journal of Nanotechnology and Applications.* **2**(1):69–73.

3719. Wagner V. et al. (2006). The emerging nanomedicine landscape. *Nature Biotechnology* **24**(10):1211–1217.

3720. Walker B. Jr. and Mouton C.P. (2006). Nanotechnology and nanomedicine: a primer. *Journal of the National Medical Association.* **98**(12):1985–1988.

3721. Watson T. (2010). The future of health care: nano-medicine's fantastic voyage. *Canadian Business.* **83**(4/5):42–43.

3722. Weber D.O. (1999). Nanomedicine. *Health Forum Journal.* **42**(4):32, 36–37.

3723. Webster T.J. (2006). Nanomedicine: what's in a definition? *International Journal of Nanomedicine.* **1**(2):115–116.

3724. Webster T.J. (2007). Nanomedicine: real commercial potential or just hype? *International Journal of Nanomedicine.* **1**(4):373–374.

3725. Wei A. et al. (2012). Challenges and opportunities in the advancement of nanomedicine. *Journal of Controlled Release.* **164**(2): 236–246.

3726. Wei C. (2007). *Nanomedicine.* Philadelphia, PA: Saunders.

3727. Wilkinson J.M. (2003). Nanotechnology applications in medicine. *Medical Device Technology.* **14**(5):29–31.

3728. Wong I.Y. et al. (2013). Nanotechnology: emerging tools for biology and medicine. *Genes & Development.* **27**(22):2397–2408.

3729. Woodrow Wilson International Center for Scholars (2007). Nanofrontiers: on the horizons of medicine and healthcare. Issue 1, May.

3730. Zhao W. et al. (2002). Application of nanotechnology in biomedical sciences. *Di Yi Jun Yi Da Xue Xue Bao* [*Academic Journal of the First Medical College of PLA*]. **22**(5):461–463.

3731. Zuo L. et al. (2007). New technology and clinical applications of nanomedicine. *Medical Clinics of North America*. **91**(5):845–862.

3732. Xie S-S. (2006). Importance of nanobiology and nanomedicine. *Acta Academiae Medicinae Sinicae*. **28**(4):469–471.

Informatics

3733. Barirani A. et al. (2013). Discovering and assessing fields of expertise in nanomedicine: a patent co-citation network perspective. *Scientometrics*. **94**(3):1111–1136.

3734. Chemical Heritage Foundation (2010). Nanomedicine terminology and standards workshop report. Philadelphia.

3735. Chen K. and Guan J. (2011). A bibliometric investigation of research performance in emerging nanobiopharmaceuticals. *Journal of Informetrics*. **5**(2):233–247.

3736. Chiesa S. et al. (2008). Building an index of nanomedical resources: an automatic approach based on text mining. *KES 2008, Part 2*. 50–57.

3737. Chiesa S. et al. (2009). European efforts in nanoinformatics research applied to nanomedicine. *Studies in Health Technology and Informatics*. **150**:757–761.

3738. Decker M. (2003). Definitions of nanotechnology – who needs them? *Newsletter of the European Academy Neuenahr-Ahrweiler*. **41**:1–3.

3739. De la Iglesia D. et al. (2009). Nanoinformatics: new challenges for biomedical informatics at the nano level. *Studies in Health Technology and Informatics*. **150**:987–991.

3740. De la Iglesia D. et al. (2011). International efforts in nanoinformatics research applied to nanomedicine. *Methods of Information in Medicine*. **1**:84–95.

3741. De la Iglesia D. et al. (2013). Nanoinformatics knowledge infrastructures: bringing efficient information management to nanomedical research. *Computational Science & Discovery.* **6**(1):014011.

3742. Gallud A. et al. (2015). Recent nanomedicine articles of outstanding interest. *Nanomedicine.* **10**(12):1859–1861.

3743. Gordon N. and Sagman U. (2003). Nanomedicine taxonomy. Verdun, Quebec: Canadian Institute of Health Research and Canadian NanoBusiness Alliance.

3744. Hasman A. et al. (2011). Biomedical informatics: a confluence of disciplines? *Methods of Information in Medicine.* **50**(6):508.

3745. Jones C. et al. (2015). Highlights from the latest articles in nanomedicine for reproductive oncology. *Nanomedicine.* **10**(9):1375–1377.

3746. Kuhn K.A. et al. (2008). Informatics and medicine – from molecules to populations. *Methods of Information in Medicine.* **47**(4):283–295.

3747. Liu X. and Webster T.J. (2013). Nanoinformatics for biomedicine: emerging approaches and applications. *International Journal of Nanomedicine.* **8**(Suppl 1):1.

3748. Lopez-Alonso V. et al. (2008). Action GRID: assessing the impact of nanotechnology on biomedical informatics. *AMIA Annual Symposium Proceedings.* 1046.

3749. Luby B.M. et al. (2014). Research highlights: highlights from the latest articles in nanomedicine. *Nanomedicine.* **9**(4):385–388.

3750. Maojo V. and Kulikowski C.A. (2006). Reflections on biomedical informatics: from cybernetics to genomic medicine and nanomedicine. *Studies in Health Technology and Informatics.* **124**:19–24.

3751. Maojo V. et al. (2010). Nanoinformatics and DNA-based computing: catalyzing nanomedicine. *Pediatric Research.* **67**(5):481–489.

3752. Maojo V. et al. (2012). Nanoinformatics: a new area of research in nanomedicine. *International Journal of Nanomedicine.* **7**:3867–3890.

3753. Maojo V. et al. (2011). Nanoinformatics: developing advanced informatics applications for nanomedicine. In *Intracellular Delivery*. Amsterdam, Netherlands: Springer, pp. 847–860.

3754. McNamara L.E. et al. (2014). Highlights from the latest articles in nanomedicine. *Nanomedicine*. **9**(6):755–757.

3755. Mian S.H. et al. (2015). Research highlights from the International Journal of Nanomedicine 2014. *International Journal of Nanomedicine*. **10**:2503–2505.

3756. Pires R.A. et al. (2014). Highlights from the latest articles in nanomedicine. *Nanomedicine*. **9**(5):573–576.

3757. Shao S.F. and Lovell J. (2014). Highlights from the latest articles in nanomedicine. *Nanomedicine*. **9**(3):385–386.

3758. Shi D. and Yarmush M.L. (2010). The emerging future at the nexus of nanotechnology and biomedicine: an introduction to Nano Life. *Nano Life*. **1**(1). doi:10.1142/S1793984410000146

3759. Thomas D.G. et al. (2011). Informatics and standards for nanomedicine technology. *Wiley Interdisciplinary Reviews: Nanomedicine and Nanobiotechnology*. **3**(5):511–532.

3760. Wang S.L. et al. (2015). Highlights from the latest research in nanomedicine. *Nanomedicine*. **10**(1):5–8.

3761. Webster T.J. (2007). IJN's second year is now a part of nanomedicine history! *International Journal of Nanomedicineicine*. **2**(1):1–2.

3762. Woodson T. (2012). Research inequality in nanomedicine. *Journal of Business Chemistry*. **9**(3):133–146.

Nanomedicine: Applications and Concepts

General Overviews

3763. Ali S. and Tariq M. (2011). Editorial: nanotechnology and its implication in medical science. *Journal of the Pakistan Medical Association*. **64**(9):984–986.

3764. Angeli E. et al. (2008). Nanotechnology applications in medicine. *Tumori*. **94**(2):206–215.

3765. Arachchige M.C. et al. (2015). Advanced targeted nanomedicine. *Journal of Biotechnology.* **202**: 88–97.

3766. BCC Research LLC (2012). Nanotechnology in medical applications: the global market (Report Code. HLC069B). Wellesley, Massachusetts.

3767. Burgess R. (2012). *Understanding Nanomedicine: An Introductory Textbook.* Boca Raton, FL: CRC Press.

3768. Freitas R.A. Jr. (1999). *Nanomedicine, Volume I: Basic Capabilities.* Georgetown, TX: Landes Bioscience.

3769. Freitas R.A. Jr. (2003). *Nanomedicine, Volume 11A: Biocompatability.* Boca Raton, CRC Press.

3770. Kenwright K. and Pifer L.L. (2010). Nanotechnology: nanomedicine. *Clinical Laboratory Science.* **23**(2): 112–116.

Alternative Medicine

3771. Upadhyay R. and Nayak C. (2011). Homeopathy emerging as nanomedicine. *International Journal of High Dilution Research.* **10**(37).

Aptamers

3772. Sun H. and Zu Y. (2015). Aptamers and their applications in nanomedicine. *Small.* **11**(20):2352–2364.

Biosensors

3773. Ghoshal S. et al. (2010). Biosensors and biochips for nanomedical applications: a review. *Sensors & Transducers.* **113**(2):1.

3774. Vaddiraju S. et al. (2010). Emerging synergy between nanotechnology and implantable biosensors: a review. *Biosensors and Bioelectronics.* **25**(7):1553–1565.

Cardiology

3775. Buxton D.B. (2009). Current status of nanotechnology approaches for cardiovascular disease: a personal perspective.

Wiley Interdisciplinary Reviews: Nanomedicine and Nanobiotechnology. **1**(2):149–155.

3776. Kipshidze N. (2009). Nanotechnology in cardiology. *Bulletin of The Georgian National Academy of Sciences.* **3**:165–177.

3777. Lanza G. et al. (2006). Nanomedicine opportunities in cardiology. *Annals of the New York Academy of Sciences.* **1080**:451–465.

3778. Patel D.N. and Bailey S.R. (2007). Nanotechnology in cardiovascular medicine. *Catheterization and Cardiovascular Interventions.* **69**(5):643–654.

3779. Tillmann C. et al. (2011). Nanotechnology in interventional cardiology. *Wiley Interdisciplinary Reviews: Nanomedicine and Nanobiotechnology.* **4**(1):82–95.

Clinical Nanomedicine

3780. Baker J.R. Jr. (2011). The need to pursue and publish clinical trials in nanomedicine. *Wiley Interdisciplinary Reviews: Nanomedicine and Nanobiotechnology.* **3**(4):341–342.

3781. Bawa R. et al. (eds.) (2015). *Handbook of Clinical Nanomedicine: Law, Business, Regulation, Safety and Risk.* New York: Taylor & Francis.

3782. Brenner S. (2013). *The Clinical Nanomedicine Handbook.* Boca Raton, FL: CRC Press.

3783. Fadeel B. et al. (2010). Nanomedicine: reshaping clinical practice. *Journal of Internal Medicine.* **267**(1):2–8.

3784. Jotterand F. (2007). Nanomedicine: how it could reshape clinical practice. *Nanomedicine.* **2**(4):401–405.

3785. Karagkiozaki V. and Logothetidis S. (eds.) (2014). *Horizons in Clinical Nanomedicine.* Boca Raton, FL: CRC Press.

3786. Morrow K.J. Jr. et al. (2007). Recent advances in basic and clinical nanomedicine. *Medical Clinics of North America.* **91**(5):805–843.

3787. Svenson S. (2012). Clinical translation of nanomedicines. *Current Opinion in Solid State & Materials Science.* **16**(6):287–294.

3788. Wising U. and Clausen J. (2014). The clinical research of nano-medicine: a new ethical challenge? *NanoEthics*. **8**:19–28.

Cosmetics

3789. Beck R. et al. (eds.) (2011). *Nanocosmetics and Nano-medicines: New Approaches for Skin Care*. Berlin: Springer-Verlag.
3790. Mihranyan A. et al. (2012). Current status and future prospects of nanotechnology in cosmetics. *Progress in Materials Science*. **57**(5):875–910.

Cryonics

3791. Charles T. J. (2011). Cryonics with nanotechnology for extending life. In *International Conference on Nanoscience, Engineering and Technology (ICONSET 2011)*, Chennai, pp. 454–459.
3792. Tejaswini V.Y.L. et al. (2014). Implementation of Cryonics using HCI. *International Journal of Advanced Trends in Computer Science & Engineering*. **3**(1):149–153.

Dendrimers

3793. Aixiang L. (2003). The research and applications of den-drimers in biology and medicine. *Shandong Journal of Biomedical Engineering*. **1**:017.
3794. Astruc D. (1996). Research avenues on dendrimers towards molecular biology: from biomimetism to medicinal engi-neering. *Comptes rendus de l'Académie des sciences. Série II, Mécanique, physique, chimie, astronomie*. **322**(10):757–766.
3795. Balogh L. et al. (2002). Dendrimer nanocomposites in medicine (White Paper). Ann Arbor, MI: University of Michigan Center for Biologic Nanotechnology.
3796. Barrett T. et al. (2008). Dendrimers in medical nanotech-nology. *IEEE Engineering in Medicine and Biology Magazine*. **28**(1):12–22.

3797. Boas U., Christensen J.B. and Heegaard P.M. (2006). *Dendrimers in Medicine and Biotechnology: New Molecular Tools.* London: Royal Society of Chemistry.

3798. Denning J. (2003). Dendrimers in medicine and biotechnology. *Topics in Current Chemistry.* **228**:227.

3799. Ghobril C. et al. (2012). Dendrimers in nuclear medical imaging. *New Journal of Chemistry.* **36**(2):310–323.

3800. Hui Z. (2004). On the application of dendrimers in medical science. *Journal of Zhongzhou University.* **1**:037.

3801. Jianhua Z. and Guanghua Y. (2007). New application of dendrimer in medicine. *Science & Technology Information.* **29**:326.

3802. Jun L. et al. (2001). The application of dendrimers in biology and medicine. *Chemical Reagents.* **2**:007.

3803. Klajnert B. and Bryszewska M. (2007). *Dendrimers in Medicine.* Hauppauge, New York: Nova Science Publishers.

3804. Micha-Screttas M. (2007). Biomedical applications of dendrimers. *Current Topics in Medicinal Chemistry.* **8**(14):1159–1160.

3805. Mignani S. et al. (2013). Dendrimer space concept for innovative nanomedicine: a futuristic vision for medicinal chemistry. *Progress in Polymer Science.* **38**(7):993–1008.

3806. Mintzer M.A. and Grinstaff M.W. (2011). Biomedical applications of dendrimers: a tutorial. *Chemical Society Reviews.* **40**(1):173–190.

3807. Mody V. (2010). Dendrimers in medicine. *Chronicles of Young Scientists.* **1**(4):31.

3808. Na M. et al. (2006). Dendrimers as potential drug carriers. Part II. Prolonged delivery of ketoprofen by in vitro and in vivo studies. *European Journal of Medicinal Chemistry,* **41**(5):670–674.

3809. Oliveira J.M. et al. (2010). Dendrimers and derivatives as a potential therapeutic tool in regenerative medicine strategies: a review. *Progress in Polymer Science.* **35**(9):1163–1194.

3810. Oliveira J.M. et al. (2012). Dendrimer-based nanoparticles in tissue engineering and regenerative medicine approaches. *Journal of Tissue Engineering and Regenerative Medicine.* **6**:9.

3811. Rolland O., Turrin C.O., Caminade A.M. and Majoral J.P. (2009). Dendrimers and nanomedicine: multivalency in action. *New Journal of Chemistry.* **33**(9):1809–1824.

3812. Sharma A. and Kakkar A. (2015). Designing dendrimer and miktoarm polymer based multi-tasking nanocarriers for efficient medical therapy. *Molecules.* **20**(9):16987–17015.

3813. Shaunak S. and Brocchini S. (2006). Dendrimer-based drugs as macromolecular medicines. *Biotechnology and Genetic Engineering Reviews.* **23**(1):309–316.

3814. Svenson S. (2015). The dendrimer paradox–high medical expectations but poor clinical translation. *Chemical Society Reviews.* **44**(12):4131–4144.

3815. Svenson S. and Tomalia D.A. (2005). Dendrimers in biomedical applications: reflections on the field. *Advanced Drug Delivery Reviews.* **57**(15):2106–2129.

3816. Tolia G.T. et al. (2008). The role of dendrimers in drug delivery. *Pharmaceutical Technology.* **32**:88–98.

3817. Yang H. and Kao W.J. (2006). Dendrimers for pharmaceutical and biomedical applications. *Journal of Biomaterials Science, Polymer Edition.* **17**(1–2):3–19.

3818. Wu L.P. et al. (2015). Dendrimers in medicine: therapeutic concepts and pharmaceutical challenges. *Bioconjugate Chemistry.* **26**(7):1198–1211.

3819. Yiyun C. and Tongwen X. (2005). Dendrimers as potential drug carriers. Part I. Solubilizing of non-steroidal anti-inflammatory drugs in the presence of polyamidoamine dendrimers. *European Journal of Medicinal Chemistry.* **40**(11):1188–1192.

Dentistry

3820. Babel S. and Mathur S. (2011). Nanorobotics-headway towards dentistry. *International Journal of Research in Science & Technology.* **1**:1–9.

3821. Bhardwaj A., Bhardwaj A., Misuriya A., Maroli S., Manjula S. and Singh A.K. (2014). Nanotechnology in dentistry: present and future. *Journal of International Oral Health: JIOH.* **6**(1):121.

3822. Chandki R. et al. (2012). Nanodentistry: exploring the beauty of miniature. *Journal of Clinical and Experimental Dentistry.* **4**(2):e119–e124.

3823. Dash S. and Kallepalli S. (2015). Influence of nanotechnology in operative dentistry and endodontics—think "big," act "small." *Journal of Advanced Medical and Dental Sciences Research.* **3**(4):51–56.

3824. Freitas R.A. Jr. (2000). Nanodentistry. *Journal of the American Dental Association.* **131**:1559–1566.

3825. Gupta J. (2011). Nanotechnology applications in medicine and dentistry. *Journal of Investigative and Clinical Dentistry.* **2**(2):81–88.

3826. Jhaveri H.M. and Balaji P.R. (2005). Nanotechnology: the future of dentistry. *The Journal of Indian Prosthodontic Society.* **5**(1):15.

3827. Kanaparthy R. and Kanaparthy A. (2011). The changing face of dentistry: nanotechnology. *International Journal of Nanomedicine.* **6**:2799–2804.

3828. Kaur J. et al. (2011). Nanotechnology-the era of molecular dentistry. *Indian Journal of Dental Sciences.* **3**:80–82.

3829. Kumar S.R. and Vijayalakshmi R. (2006). Nanotechnology in dentistry. *Indian Journal of Dental Research.* **17**(2):62–65.

3830. Lainović T., Blažić L. and Potran M. (2012). Nanotechnology in dentistry: current state and future perspectives. *Stomatoloski glasnik Srbije.* **59**(1):44–50.

3831. Nagalaxmi V. (2012). Small wonders paving a great future-nanotechnology. *Annals and Essence of Dentistry*. **4**(1):99–100.

3832. Nagpal A. et al. (2011). Nanotechnology: the era of molecular dentistry. *Indian Journal of Dental Science*. **5**(3):80–82.

3833. Padovani G. et al. (2015). Advances in dental materials through nanotechnology: facts, perspectives and toxicological aspects. *Trends in Biotechnology*. **33**(11):621–636.

3834. Salerno M. and Diaspro A. (2015). Dentistry on the bridge to nanoscience and nanotechnology. *Frontiers in Materials*. 2. 10.3389/fmats.2015.00019

3835. Shetty N.J. et al. (2013). Nanorobots: future in dentistry. *The Saudi Dental Journal*. **25**(2):49–52.

3836. Shiva Manjunath R.G. and Rana A. (2015). Nanotechnology in periodontal management. *Journal of Advanced Oral Research*. **6**(1).

3837. Ure D. and Harris J. (2002). Nanotechnology in dentistry: reduction to practice. *Dental Update*. **30**(1):10–15.

3838. Verma S.K. and Chauhan R. (2014). Nanorobotics in dentistry: a review. *Indian Journal of Dentistry*. **5**:62–70.

Dermatology

3839. Antonio J.R. et al. (2014). Nanotechnology in dermatology. *Anais brasileiros de dermatologia*. **89**(1):126–136.

3840. Basavaraj K.H. (2012). Nanotechnology in medicine and relevance to dermatology. *Indian Journal of Dermatology*. **57**(3):169–174.

3841. Blecher P.K. and Friedman A. (2012). Nanotechnology and the diagnosis of dermatological infectious disease. *Journal of Drugs in Dermatology*. **11**(7):846–851.

3842. Cevc G. and Vierl U. (2010). Nanotechnology and the transdermal route: a state of the art review and critical appraisal. *Journal of Controlled Release*. **141**(3):277–299.

3843. Collins A. and Nasir A. (2011). Nanotechnology and dermatology: benefits and pitfalls. *Giornale italiano di dermatologia e venereologia: organo ufficiale, Societa italiana di dermatologia e sifilografia.* **146**(2):115–126.

3844. DeLouise L. A. (2012). Applications of nanotechnology in dermatology. *Journal of Investigative Dermatology.* **132**:964–975.

3845. Friedman A. (2013). Nanodermatology: the giant role of nanotechnology in diagnosis and treatment of skin disease. *Nanomedicine in Drug Delivery.* 89–130.

3846. Saraceno R. et al. (2013). Emerging applications of Nanomedicine in dermatology. *Skin Research & Technology.* **19**(1): e13–e19.

3847. Nasir A. et al. (eds.) (2012). *Nanotechnology in Dermatology.* New York: Springer Science.

Diagnostics

3848. Alexiou C. (ed.) (2012). *Nanomedicine: Basic and Clinical Applications in Diagnostics and Therapy.* Basel: S. Karger AG.

3849. Alharbi K.K. and Al-Sheikh Y.A. (2014). Role and implications of nanodiagnostics in the changing trends of clinical diagnosis. *Saudi Journal of Biological Sciences.* **21**(2):109–117.

3850. Azzazy H.M., Mansour M.M. and Kazmierczak S.C. (2006). Nanodiagnostics: a new frontier for clinical laboratory medicine. *Clinical Chemistry.* **52**(7):1238–1246.

3851. Cheng M.M. et al. (2006). Nanotechnologies for biomolecular detection and medical diagnostics. *Current Opinion in Chemical Biology.* **10**:11–19.

3852. Menezes G.A. et al. (2011). Nanoscience in diagnostics: a short review. *Internet Journal of Medical Update.* **6**(1):16–23.

3853. Satvekar R.K. et al. (2014). Emerging trends in medical diagnosis: a thrust on nanotechnology. *Medicinal Chemistry.* **4**(4):407–416.

3854. Uchegbu I.F. and Siew A. (2013). Nanomedicines and nanodiagnostics come of age. *Journal of Pharmaceutical Sciences.* **102**(2):305–310.

Diseases

General Overviews

3855. Adhikari R. and Thapa S. (eds.) (2012). *Infectious Diseases and Nanomedicine I. First International Conference (ICIDN-2012), December 15–18, 2012, Kathmandu, Nepal.* Delhi: Springer India.

3856. Blecher K. et al. (2011). The growing role of nanotechnology in combating infectious disease. *Virulence.* **2**(5):395–401.

3857. Boenink M. (2009). Tensions and opportunities in convergence: shifting concepts of disease in emerging molecular medicine. *NanoEthics.* **3**(3):243–255.

3858. Bragg R.R. and Kock L. (2013). Nanomedicine and infectious diseases. *Expert Review of Anti-Infective Therapies.* **11**(4):359–361.

Alzheimer's

3859. Fluri F. (2010). Clinical nanomedicine: nanomedical approaches in Alzheimer's disease. *European Journal of Nanomedicine.* **3**(1):7.

3860. Gregori M. et al. (2015). Nanomedicine for the treatment of Alzheimer's disease. *Nanomedicine.* **10**(7):1203–1218.

Diabetes

3861. Arya A.K. et al. (2008). Applications of nanotechnology in diabetes. *Digest Journal of Nanomaterials and Biostructures.* **3**(4):221–225.

3862. Dannis S. and Dhruba B. (2011). The role of nanotechnology in diabetes treatment: current and future perspectives. *International Journal of Nanotechnology.* **8**(1/2):53–65.

3863. Harsoliya M.S. (2012). Recent advances & applications of nanotechnology in diabetes. *International Journal of Pharmaceutical & Biological Archive.* **3**(2):255–261.

3864. Meeto D. and Lappin M. (2009). Nanotechnology and the future of diabetes management. *Journal of Diabetes Nursing.* **13**:8.

3865. Sung H.W. et al. (2011). Nanomedicine for diabetes treatment. *Nanomedicine.* **6**(8):1297–1300.

HIV/Aids

3866. Kumar L. et al. (2015). Nanotechnology: a magic bullet for HIV AIDS treatment. *Artificial Cells, Nanomedicine, and Biotechnology.* **43**(2):71–86.
3867. Lisziewicz J. and Toke E.R. (2013). Nanomedicine applications towards the cure of HIV. *Nanomedicine.* **9**(1):28–38.

Inflammatory Diseases

3868. Hussain A. et al. (2013). Nanomedicines for the treatment of inflammatory bowel diseases. *European Journal of Nanomedicine.* **5**(1):23.
3869. Khaja F.A. et al. (2012). Nanomedicines for inflammatory diseases. *Methods in Enzymology.* **508**:355–375.

Malaria

3870. Aditya N.P. et al. (2013). Advances in nanomedicines for malaria treatment. *Advances in Colloid and Interface Science.* **201**:1–17.

Parkinson's

3871. Fluri F. (2009). Clinical nanomedicine: nanomedical approaches in Parkinson's disease. *European Journal of Nanomedicine.* **2**(1):48.

Tuberculosis

3872. Dube A. et al. (2013). State of the art and future directions in nanomedicine for tuberculosis. *Expert Opinion on Drug Delivery.* **10**(12):1725–1734.

Drug Delivery (includes nanocarriers)

3873. Agrawal U. et al. (2014). Is nanotechnology a boon for oral drug delivery? *Drug Discovery Today.* **19**(10):1530–1546.

3874. Amiji M.M. (2005). Nanotechnology for drug delivery: an overview. *Nano Science and Technology Institute.* **2**:345–349.

3875. Arias José L. (ed.) (2014). *Nanotechnology and Drug Delivery, Volume One: Nanoplatforms in Drug Delivery.* Boca Raton, FL: CRC Press.

3876. Arruebo M. et al. (2007). Magnetic nanoparticles for drug delivery. *Nano Today.* **2**(3):22–32.

3877. Bae Y.H. and Park K. (2011). Targeted drug delivery to tumors: myths, reality and possibility. *Journal of Controlled Release.* **153**(3):198.

3878. Baviskar D.T. et al. (2012). Carbon nanotubes: an emerging drug delivery tool in nanotechnology. *International Journal of Pharmacy and Pharmaceutical Sciences.* **4**:11–15.

3879. Bawarski W.E. et al. (2008). Emerging nanopharmaceuticals. *Nanomedicine: Nanotechnology, Biology and Medicine.* **4**(4): 273–282.

3880. Bhargavi C. et al. (2013). Nanotherapeutics: an era of drug delivery system in nanoscience. *Indian Journal of Research in Pharmacy and Biotechnology.* **1**(2):210–214.

3881. Bonifácio B.V. et al. (2014). Nanotechnology-based drug delivery systems and herbal medicines: a review. *International Journal of Nanomedicine.* **9**:1–15.

3882. Borm P.J. and Kreyling W. (2004). Toxicological hazards of inhaled nanoparticles—potential implications for drug delivery. *Journal of Nanoscience and Nanotechnology.* **4**(5):521–531.

3883. Cavadas M. et al. (2011). Pathogen-mimetic stealth nanocarriers for drug delivery: a future possibility. *Nanomedicine: Nanotechnology, Biology and Medicine.* **7**(6):730–743.

3884. Cevc G. and Vierl U. (2010). Nanotechnology and the transdermal route: a state of the art review and critical appraisal. *Journal of Controlled Release.* **141**(3):277–299.

3885. De Jong W.H. and Borm P.J. (2008). Drug delivery and nanoparticles: applications and hazards. *International Journal of Nanomedicine.* **3**(2):133.

3886. Devadasu V.R. et al. (2012). Can controversial nanotechnology promise drug delivery? *Chemical Reviews.* **113**(3):1686–1735.

3887. De Villiers M.M. et al. (eds.) (2008). *Nanotechnology in Drug Delivery.* New York: Springer Science & Business Media.

3888. Dou H. et al. (2006). Development of a macrophage-based nanoparticle platform for antiretroviral drug delivery. *Blood.* **108**(8):2827–2835.

3889. Farokhzad O.C. and Langer R. (2009). Impact of nanotechnology on drug delivery. *ACS Nano.* **3**(1):16–20.

3890. Ganta S. et al. (2008). A review of stimuli-responsive nanocarriers for drug and gene delivery. *Journal of Controlled Release.* **126**(3):187–204.

3891. Garcia-Fuentes M. and Alonso M.J. (2012). Chitosan-based drug nanocarriers: where do we stand? *Journal of Controlled Release.* **161**(2):496–504.

3892. Gill S. et al. (2007). Nanoparticles: characteristics, mechanisms of action, and toxicity in pulmonary drug delivery—a review. *Journal of Biomedical Nanotechnology.* **3**(2):107–119.

3893. Gupta A. et al. (2012). Nanotechnology and its applications in drug delivery: a review. *Medical Education.* **3**(1):1–9.

3894. Hamidi M. (2012). Nanotechnology-based drug delivery systems: a general overview of the pharmacokinetic aspects. *Research in Pharmaceutical Sciences.* **7**(5):S952.

3895. Han G. et al. (2007). Functionalized gold nanoparticles for drug delivery. *Future Medicine.* **2**(1):113–123.

3896. He Q. and Shi J. (2011). Mesoporous silica nanoparticle based nano drug delivery systems: synthesis, controlled drug release and delivery, pharmacokinetics and biocompatibility. *Journal of Materials Chemistry.* **21**(16):5845–5855.

3897. Jabr-Milane L. et al. (2008). Multi-functional nanocarriers for targeted delivery of drugs and genes. *Journal of Controlled Release.* **130**(2):121–128.

3898. Hilt J.Z. and Peppas N.A. (2005). Microfabricated drug delivery devices. *International Journal of Pharmaceutics.* **306**(1):15–23.

3899. Jain S. et al. (2011). Nanotechnology in advanced drug delivery. *Journal of Drug Delivery.* **2011**: Article ID 343082.

3900. Jain K. et al. (2015). Nanotechnology in drug delivery: safety and toxicity issues. *Current Pharmaceutical Design.* **21**(29):4252–4261.

3901. Jiang W. et al. (2007). Advances and challenges of nanotechnology-based drug delivery systems. *Expert Opinion on Drug Delivery.* **4**(6). doi:10.1517/1745247.4.6.621

3902. Kim S. et al. (2009). Nanotechnology in drug delivery: past, present, and future. In *Nanotechnology in Drug Delivery.* New York: Springer, pp. 581–596.

3903. Kingsley J.D. et al. (2006). Nanotechnology: a focus on nanoparticles as a drug delivery system. *Journal of Neuroimmune Pharmacology.* **1**(3):340–350.

3904. Koo O.M. et al. (2005). Role of nanotechnology in targeted drug delivery and imaging: a concise review. *Nanomedicine: Nanotechnology, Biology and Medicine.* **1**(3):193–212.

3905. Lakhal S. and Wood M.J. (2011). Exosome nanotechnology: an emerging paradigm shift in drug delivery. *Bioessays.* **33**(10):737–741.

3906. Lamprecht A. (ed.) (2008). *Nanotherapeutics: Drug Delivery Concepts in Nanoscience.* Boca Raton, FL: CRC Press.

3907. Levy-Nissenbaum E. et al. (2008). Nanotechnology and aptamers: applications in drug delivery. *Trends in Biotechnology.* **26**(8):442–449.

3908. Mainardes R.M. and Silva L.P. (2004). Drug delivery systems: past, present, and future. *Current Drug Targets.* **5**(5):449–455.

3909. Makadia H.A. et al. (2013). Self-nano emulsifying drug delivery system (SNEDDS): future aspects. *Asian Journal of Pharmaceutical Research.* **3**(1):21–27.

3910. Marchal S. et al. (2015). Anticancer drug delivery: an update on clinically applied nanotherapeutics. *Drugs.* **75**(14):1601–1611.

3911. Masareddy R.S. et al. (2011). Nano drug delivery systems: a review. *International Journal of Pharmaceutical Sciences and Research.* **2**(2):203–216.

3912. Mirjalili F. et al. (2012). Nanotechnology in drug delivery systems. *International Journal of Drug Delivery.* **4**(3):275.

3913. Mishra B. et al. (2010). Colloidal nanocarriers: a review on formulation technology, types and applications toward targeted drug delivery. *Nanomedicine: Nanotechnology, Biology and Medicine.* **6**(1):9–24.

3914. Moghassemi S. and Hadjizadeh A. (2014). Nano-niosomes as nanoscale drug delivery systems: an illustrated review. *Journal of Controlled Release.* **185**:22–36.

3915. Nayak A.K. and Dhara A.K. (2010). Nanotechnology in drug delivery applications: a review. *Archives of Applied Science Research.* **2**(2):284–293.

3916. Nayar P.G. (2009). Developments in drug delivery with special reference to nanotechnology: a review. *Journal of the Indian Society of Toxicology.* **5**(1):32–36.

3917. Ochekpe N.A. et al. (2009). Nanotechnology and drug delivery. Part 1: Background and applications. *Tropical Journal of Pharmaceutical Research.* **8**(3):265–274.

3918. Ochekpe N.A. et al. (2009). Nanotechnology and drug delivery. Part 2: Nanostructures for drug delivery. *Tropical Journal of Pharmaceutical Research.* **8**(3):275–287.

3919. Park K. (2013). Facing the truth about nanotechnology in drug delivery. *ACS Nano.* **7**(9):7442–7447.

3920. Parveen S. et al. (2012). Nanoparticles: a boon to drug delivery, therapeutics, diagnostics and imaging. *Nanomedicine: Nanotechnology, Biology and Medicine.* **8**(2):147–166.

3921. Patel K.P. et al. (2014). Formulation consideration of nanotechnology based drug delivery systems. *Drug Invention Today.* **6**(1):46–51.

3922. Paul W. and Sharma C.P. (1995). Bioceramics, towards nano-enabled drug delivery: a mini review. *Journal of Materials Science Letters.* **14**:1792.

3923. Pegoraro C. et al. (2012). Transdermal drug delivery: from micro to nano. *Nanoscale.* **4**(6):1881–1894.

3924. Petkar K.C. et al. (2011). Nanostructured materials in drug and gene delivery: a review of the state of the art. *Critical Reviews in Therapeutic Drug Carrier Systems.* **28**(2):101–164.

3925. Petrak K. (2006). Nanotechnology and site-targeted drug delivery. *Journal of Biomaterials Science, Polymer Edition.* **17**(11):1209–1219.

3926. Plapied L. et al. (2011). Fate of polymeric nanocarriers for oral drug delivery. *Current Opinion in Colloid & Interface Science.* **16**(3):228–237.

3927. Qureshi S.R. (2014). Nanotechnology based drug delivery system. *Journal of Pharmaceutical Research & Opinion.* **1**(6):161–165.

3928. Rawat M. et al. (2006). Nanocarriers: promising vehicle for bioactive drugs. *Biological and Pharmaceutical Bulletin.* **29**(9):1790–1798.

3929. Ren Y. et al. (2010). Application of plant viruses as nano drug delivery systems. *Pharmaceutical Research.* **27**(11):2509–2513.

3930. Sahoo S.K. and Labhasetwar V. (2003). Nanotech approaches to drug delivery and imaging. *Drug Discovery Today.* **8**(24):1112–1120.

3931. Sahoo S.K. et al. (2008). Nanotechnology in ocular drug delivery. *Drug Discovery Today.* **13**(3):144–151.

3932. Saxena S.K. et al. (2012). Nanotherapeutics: emerging competent technology in neuroAIDS and CNS drug delivery. *Nanomedicine.* **7**(7):941.

3933. te Kulve H. and Rip A. (2013). Economic and societal dimensions of nanotechnology-enabled drug delivery. *Expert Opinion on Drug Delivery.* **10**(5):611–622.

3934. Torchilin V.P. (2007). Micellar nanocarriers: pharmaceutical perspectives. *Pharmaceutical Research.* **24**(1):1–16.

3935. Torchilin V.P. (2012). Multifunctional nanocarriers. *Advanced Drug Delivery Reviews.* **64**:302–315.

3936. Yatoo M.I. et al. (2014). Nanotechnology based drug delivery at cellular level: a review. *Journal of Animal Science Advances.* **4**(2):705–709.

3937. Yih T.C. and Al-Fandi M. (2006). Engineered nanoparticles as precise drug delivery systems. *Journal of Cellular Biochemistry.* **97**(6):1184–1190.

3938. Yoo J.W. et al. (2011). Bio-inspired, bioengineered and biomimetic drug delivery carriers. *Nature Reviews Drug Discovery.* **10**(7):521–535.

3939. Zhou X. et al. (2014). Nano-enabled drug delivery: a research profile. *Nanomedicine: Nanotechnology, Biology and Medicine.* **10**(5):889–896.

Fullerenes

3940. Anilkumar P. et al. (2011). Fullerenes for applications in biology and medicine. *Current Medicinal Chemistry.* **18**(14): 2045–2059.

3941. Dellinger A. et al. (2013). Application of fullerenes in nanomedicine: an update. *Nanomedicine.* **8**(7):1191–1208.

3942. Djordjević A. et al. (2006). Fullerenes in biomedicine. *Journal of Balkan Union of Oncology.* 11(4):391–404.

3943. Kepley C. (2012). Fullerenes in medicine; will it ever occur? *Journal of Nanomedicine & Nanotechnology.* **3**:e111.

3944. Kepley C. (2013). Application of fullerenes in nanomedicine: an update. *Nanomedicine.* **8**(7):1191–1208.

3945. Schur D.V. et al. (2012). Fullerenes: prospects of using in medicine, biology and ecology. *Visnyk of Dnipropetrovsk University. Biology, Ecology.* **20**(1):139–145.

Gastroenterology

3946. Agudelo Zapata Y. et al. (2008). Nanotechnology in the gastrohepatology. *Revista Colombiana de Gastroenterologia.* **23**(4):361–368.

3947. Brakmane G. et al. (2012). Systematic review: the applications of nanotechnology in gastroenterology. *Alimentary Pharmacology & Therapeutics.* **36**(3):213–221.

3948. Lamprecht A. (2015). Nanomedicines in gastroenterology and hepatology. *Nature Reviews: Gastroenterology & Hepatology.* 12:195–204.

Genetics

3949. Alex S.M. and Sharma C.P. (2013). Nanomedicine for gene therapy. *Drug Delivery and Translational Research.* **3**(5):437–445.

3950. Coccia M. (2012). Converging genetics, genomics and nanotechnologies for groundbreaking pathways in biomedicine and nanomedicine. *International Journal of Healthcare Technology and Management.* **13**(4):184–197.

3951. Guo P. (2005). RNA nanotechnology: engineering, assembly and applications in detection, gene delivery and therapy. *Journal of Nanoscience and Nanotechnology.* **5**(12): 1964.

3952. Martis E.A. et al. (2012). Nanotechnology based devices and applications in medicine: an overview. *Chronicles of Young Scientists.* **3**(1):68.

3953. Moore F.N. (2002). Implications of nanotechnology applications: using genetics as a lesson. *Health Law Review.* **10**(3):9–15.

3954. Nicolini C. (2009). Nanogenomics in medicine. *Wiley Interdisciplinary Reviews: Nanomedicine and Nanobiotechnology.* **2**(1):59–76.

3955. Nicolini C. and Pechkova E. (2010). Nanoproteomics for nanomedicine. *Nanomedicine.* **5**(5):677–682.

3956. Wei C. et al. (2007). Genetic nanomedicine and tissue engineering. *Medical Clinics of North America.* **91**(5):889–898.

3957. Zahid M. et al. (2013). DNA nanotechnology: a future perspective. *Nanoscale Research Letters.* **8**(1):1–13.

Immunology

3958. Dobrovolskaia M.A. and McNeil S.E. (2007). Immunological properties of engineered nanomaterials. *Nature Nanotechnology.* **2**(8):469–478.

3959. Klippstein R. and Pozo D. (2010). Nanotechnology-based manipulation of dendritic cells for enhanced immunotherapy strategies. *Nanomedicine: Nanotechnology, Biology and Medicine.* **6**(4):523–529.

3960. Prasad L.K. et al. (2015). Nanomedicine delivers promising treatments for rheumatoid arthritis. *Nanomedicine.* **10**(13):2063–2074.

3961. Smith D.M. et al. (2013). Applications of nanotechnology for immunology. *Nature Reviews Immunology.* **13**(8):592–605.

Implants and Tissue Engineering

3962. Liu H. and Webster T.J. (2007). Nanomedicine for implants: a review of studies and necessary experimental tools. *Biomaterials.* **28**(2):354–369.

3963. McMahon R.E. et al. (2010). Development of nanomaterials for bone repair and regeneration. *Journal of Biomedical Materials Research Part B: Applied Biomaterials.* **101B**(2):387–397.

3964. Sahoo N.G. et al. (2013). Nanocomposites for bone tissue regeneration. *Nanomedicine.* **8**(4):639–653.

3965. Salgado A.J. et al. (2013). Tissue engineering and regenerative medicine: past, present, and future. *International Review of Neurobiology.* **108**:1–33.

3966. Streicher R.M. et al. (2007). Nanosurfaces and nanostructures for artificial orthopedic implants. *Nanomedicine (London).* **2**(6):861–874.

3967. Tran N. and Webster T.J. (2009). Nanotechnology for bone materials. *Wiley Interdisciplinary Reviews: Nanomedicine and Nanobiotechnology.* **1**(3):336–351.

3968. Webster T.J. (2007). Nanotechnology: better materials for all implants. *Materials Science Forum.* **539**:511–516.

3969. Zhang L. and Webster T.J. (2009). Nanotechnology and nanomaterials: promises for improved tissue regeneration. *Nano Today.* **4**(1):66–80.

Inorganic Nanomedicine

3970. Sekhon B.S. and Kamboj S.R. (2010). Inorganic Nanomedicine. Part 1. *Nanomedicine.* **6**(4): 516–522.
3971. Sekhon B.S. and Kamboj S.R. (2010). Inorganic Nanomedicine. Part 2. *Nanomedicine.* **6**(5): 612–618.

Molecular Medicine

3972. Boenink M. (2009). Tensions and opportunities in convergence: shifting concepts of disease in emerging molecular medicine. *NanoEthics.* **3**(3):243–255.
3973. Boenink M. (2010). Molecular medicine and concepts of disease: the ethical value of a conceptual analysis of emerging biomedical technologies. *Medicine, Health Care and Philosophy.* **13**(1):11–23.
3974. Caskey C.T. (1993). Molecular medicine: a spin-off from the helix. *JAMA.* **269**(15):1986–1992.
3975. Cerami A. and Warren K.S. (1994). Molecular medicine: the future of biomedical science and clinical practice. *Molecular Medicine.* **1**(1):1.
3976. Clark W.R. (1997). *The New Healers: The Promise and Problems of Molecular Medicine in The Twenty-First Century.* New York: Oxford University Press.
3977. Culliton B.J. (1995). Molecular medicine in a changing world. *Nature Medicine.* **1**(1):1
3978. Donner A. (2010). Nanotechnology in molecular medicine. *Trends in Molecular Medicine.* **16**(12):551–552.
3979. Gannon F. (2003). Molecular medicine: trendy title or new reality? *EMBO Reports.* **4**(8):733–733.
3980. Ganten D. (1995). The Journal of Molecular Medicine: tradition, continuity, and renaissance. *Journal of Molecular Medicine.* **73**(1):1–3.

3981. Hoffman E.P. (2007). Skipping toward personalized molecular medicine. *New England Journal of Medicine.* **357**(26): 2719–2722.

3982. Jameson J.L. and Collins F.S. (eds.) (1998). *Principles of Molecular Medicine.* New York: Springer Science & Business Media.

3983. Karp J.E. and Broder S. (1994). New directions in molecular medicine. *Cancer Research.* **54**(3):653–665.

3984. Nathan D.G. (1996). Molecular medicine milestones. *Molecular Medicine.* **2**(2):163.

3985. Ostrowski J. (2006). Molecular medicine of the future-applications and pitfalls. *Acta Poloniae Pharmaceutica.* **63**: 329–332.

3986. Semsarian C. and Seidman C.E. (2001). Molecular medicine in the 21st century. *Internal Medicine Journal.* **31**(1):53–59.

3987. Gannon F. (2003). Molecular medicine: trendy title or new reality? *EMBO Reports.* **4**(8):733.

3988. Short N. (1994). The challenge of molecular medicine. *Nature.* **371**:373.

3989. Sobie E.A. et al. (2003). The challenge of molecular medicine: complexity versus Occam's razor. *Journal of Clinical Investigation.* **111**(6):801.

3990. Steel M. (2005). Molecular medicine: promises, promises? *Journal of the Royal Society of Medicine.* **98**(5):197–199.

3991. Trent R.J. (2005). *Molecular Medicine: An Introductory Text.* Burlington, MA: Elsevier Academic Press.

3992. Weatherall D.J. (2010). Molecular medicine; the road to the better integration of the medical sciences in the twenty-first century. *Notes and Records of the Royal Society.* **64**(Suppl 1):S5–S15.

3993. Weinberger D.R. (2001). Anxiety at the frontier of molecular medicine. *New England Journal of Medicine.* **344**(16):1247–1249.

3994. Woodcock J. (2007). Molecular medicine: how, what, and when? *Clinical Pharmacology & Therapeutics.* **82**(4):376–378.

Nanocapsules

3995. Dineshkumar B. (2013). Nanocapsules: a novel nano-drug delivery system. *International Journal of Research in Drug Delivery.* **3**(1):1–3.

3996. Falqueiro A.M. et al. (2011). Selol-loaded magnetic nanocapsules: a new approach for hyperthermia cancer therapy. *Journal of Applied Physics.* **109**:07B306.

3997. Huynh N.T. et al. (2009). Lipid nanocapsules: a new platform for nanomedicine. *International Journal of Pharmaceutics.* **379**(2):201–209.

3998. Mora-Huertas C.E. et al. (2010). Polymer-based nanocapsules for drug delivery. *International Journal of Pharmaceutics.* **385**:113–142.

3999. Prabhakar V. et al. (2013). Magic bullets: nanocapsules in future medicine. *International Journal of Pharma Sciences.* **3**(4):303–308.

4000. Quintanar-Guerrero D. et al. (1998). Preparation and characterization of nanocapsules from preformed polymers by a new process based on emulsification-diffusion technique. *Pharmaceutical Research.* **15**:1056–1062.

4001. Rübe A. (2006). Development and physico-chemical characterization of nanocapsules. Dissertation: Wittenberg, Martin Luther University.

4002. Ruysschaert T. (2004). Liposome-baesd nanocapsules. *IEEE Transactions on Nanobioscience.* **3**(1):49–55.

4003. Sablon K. (2008). Single-component polymer nanocapsules for drug delivery application. *Nanoscale Research Letters.* **3**:265–267.

4004. Sánchez-Moreno P. et al. (2012). Characterization of different functionalized lipidic nanocapsules as potential drug carriers. *International Journal of Molecular Sciences.* **13**:2405–2424.

4005. Valente I. (2012). Polymeric Nanocapsules for Pharmaceutical Applications. Dissertation: Politecnico di Torino.

4006. Wadhwa J. et al. (2014). Development and optimization of polymeric self-emulsifying nanocapsules for localized drug delivery: design of experiment approach. *The Scientific World Journal.* **2014**: Article ID 516060.

4007. Yurgel V. et al. (2013). Developments in the use of nanocapsules in oncology. *Brazilian Journal of Medical & Biological Research.* **46**(6):486–501.

Nanocomposites

4008. Balogh L. et al. (2002). Dendrimer nanocomposites in medicine. *Chimica Oggi.* **20**(5):35–40.

4009. Boccaccini A.R. et al. (2010). Polymer/bioactive glass nanocomposites for biomedical applications: a review. *Composites Science and Technology.* **70**(13):1764–1776.

4010. Cherian B.M. et al. (2012). Protein based polymer nanocomposites for regenerative medicine. *Natural Polymers.* **2**:255.

4011. Curtis J.M. and Lipp E.D. (2008). Silsesquioxane nanocomposites as tissue implants. *Plastic and Reconstructive Surgery.* **122**(5):1599–1600.

4012. Fukushima K. et al. (2012). PBAT based nanocomposites for medical and industrial applications. *Materials Science and Engineering: C.* **32**(6):1331–1351.

4013. Hule R.A. and Pochan D.J. (2007). Polymer nanocomposites for biomedical applications. *MRS Bulletin.* **32**(04):354–358.

4014. Hussein-Al-Ali S.H. et al. (2014). Novel kojic acid-polymer-based magnetic nanocomposites for medical applications. *International Journal of Nanomedicine.* **9**:351.

4015. Kannan R.Y. et al. (2005). Polyhedral oligomeric silsesquioxane nanocomposites: the next generation material for biomedical applications. *Accounts of Chemical Research.* **38**(11):879–884.

4016. Murugan R. and Ramakrishna S. (2005). Development of nanocomposites for bone grafting. *Composites Science and Technology.* **65**(15):2385–2406.

4017. Sahoo N.G. et al. (2013). Nanocomposites for bone tissue regeneration. *Nanomedicine.* **8**(4):639–653.

4018. Stodolak E. et al. (2009). Nanocomposite fibers for medical applications. *Journal of Molecular Structure.* **924**:208–213.

Nanodiamonds

4019. Passeri D. et al. (2015). Biomedical applications of nanodiamonds: an overview. *Journal of Nanoscience and Nanotechnology.* **15**(2):972–988.

4020. Perevedentseva E. et al. (2013). Biomedical applications of nanodiamonds in imaging and therapy. *Nanomedicine.* **8**(12):2041–2060.

4021. Pramatarova L. et al. (2007). Artificial bones through nanodiamonds. *Journal of Optoelectronics and Advanced Materials.* **9**(1):236–239.

4022. Say J.M. et al. (2011). Luminescent nanodiamonds for biomedical applications. *Biophysical Reviews.* **3**(4):171–184.

4023. Vaijayanthimala V. and Chang H.C. (2009). Functionalized fluorescent nanodiamonds for biomedical applications. *Future Medicine.* **2009**:47–55.

4024. Xing Y. and Dai L. (2009). Nanodiamonds for nanomedicine. *Future Medicine.* **2009**:207–218.

Nanomedicines, Nanodrugs, and Nanopharmaceuticals

4025. Agrawal U. et al. (2013). Multifunctional nanomedicines: potentials and prospects. *Drug Delivery and Translational Research.* **3**(5):479–497.

4026. Ahmad U. and Faiyazuddin M. (2014). Patented nanopharmaceuticals: a hope for patent expired formulations. *Intellectual Property Rights.* **2**:e104.

4027. Ali J. (2011). Nanopharmaceutics. *International Journal of Pharmaceutical Investigation.* **1**(2):61.

4028. Alonso M.J. (2004). Nanomedicines for overcoming biological barriers. *Biomedicine & Pharmacotherapy.* **58**(3):168–172.

4029. Ansari S.H. and Farha Islam M. (2012). Influence of nano-technology on herbal drugs: a review. *Journal of Advanced Pharmaceutical Technology & Research.* **3**(3):142.

4030. Antunes A. et al. (2013). Trends in nanopharmaceutical patents. *International Journal of Molecular Sciences.* **14**(4): 7016–7031.

4031. Barenholz Y. (2012). Doxil R—the first FDA-approved nano-drug: lessons learned. *Journal of Controlled Release.* **160**:117–134.

4032. Bawa R. (2010). Nanopharmaceuticals. *European Journal of Nanomedicine.* **3**(1):34

4033. Bawa R. (2011). Regulating nanomedicine: can the FDA handle it? *Current Drug Delivery.* **8**(3):227–234.

4034. Bawa R. et al. (2008). Nanopharmaceuticals – patenting issues and FDA regulatory challenges. *American Bar Association SciTech Lawyer.* **5**:10–15.

4035. Bawarski W.E. et al. (2008). Emerging nanopharmaceuticals. *Nanomedicine.* **4**(4):273–282.

4036. Berkner S. et al. (2016). Nanopharmaceuticals: tiny chal-lenges for the environmental risk assessment of pharmaceu-ticals. *Environmental Toxicology and Chemistry.* **35**(4):780–787.

4037. Bhattacharya S. et al. (2012). Nanomedicine: pharmacological perspectives. *Nanotechnology Reviews.* **1**(3):235–253.

4038. Brown P.D. and Patel P.R. (2015). Nanomedicine: a pharma perspective. *Wiley Interdisciplinary Reviews: Nanomedicine and Nanobiotechnology.* **7**(2):125–130.

4039. Chan V.S. (2006). Nanomedicine: an unresolved regulatory issue. *Regulatory Toxicology and Pharmacology.* **46**(3):218–224.

4040. Chekman I.S. (2009). [Pharmacological and pharmaceutical foundation of nanodrugs]. *Likars' ka sprava/Ministerstvo okhorony zdorov'ia Ukrainy.* 3–10 (Russian).

4041. Chen J.F. et al. (2004). Feasibility of preparing nanodrugs by high-gravity reactive precipitation. *International Journal of Pharmaceutics.* **269**(1):267–274.

4042. Chen K. and Guan J. (2011). A bibliometric investigation of research performance in emerging nanobiopharmaceuticals. *Journal of Informetrics.* **5**(2):233–247.

4043. Cheng T.F. et al. (2009). Attention on research of pharmacology and toxicology of nanomedicines. *Asian Journal of Pharmacodynamics and Pharmacokinetics.* **9**:27–49.

4044. Debbage P. (2009). Targeted drugs and nanomedicine: present and future. *Current Pharmaceutical Design.* **15**(2): 153–172.

4045. Devalapally H. et al. (2007). Role of nanotechnology in pharmaceutical product development. *Journal of Pharmaceutical Sciences.* **96**(10):2547–2565.

4046. Dobrovolskaia M.A. and McNeil S.E. (2013). Understanding the correlation between in vitro and in vivo immunotoxicity tests for nanomedicines. *Journal of Controlled Release.* **172**(2):456–466.

4047. Duan-yun S. and Chang-xiao L. (2007). Biomedical evaluation of nanomedicines. *Research Letters.* **5**:1010–1019.

4048. Duncan R. (2004). Nanomedicines in action. *The Pharmaceutical Journal.* **273**:485–488.

4049. Duncan R. (2011). Polymer therapeutics as nanomedicines: new perspectives. *Current Opinion in Biotechnology.* **22**(4): 492–501.

4050. Duncan R. and Gaspar R. (2011). Nanomedicine(s) under the microscope. *Molecular Pharmaceutics.* **8**(6):2101–2141.

4051. Eaton M. (2009). Implementing nanomedicine: the importance of the industrial—academic interface for innovation in the pharmaceutical sector. *European Journal of Nanomedicine.* **2**(2):22.

4052. Eaton A.W. (2011). How do we develop nanopharmaceuticals under open innovation? *Nanomedicine: Nanotechnology, Biology, and Medicine.* **7**:371–375.

4053. Eaton A.W. et al. (2015). Delivering nanomedicines to patients: a practical guide. *Nanomedicine.* **11**(4):983–992.

4054. Fenske D.B. and Cullis P.R. (2008). Liposomal nanomedicines. *Expert Opinion on Drug Delivery.* **5**(1):25–44.

4055. Fenske D.B. et al. (2008). Liposomal nanomedicines: an emerging field. *Toxicologic Pathology.* **36**(1):21–29.

4056. Ferrari M. and Downing G. (2005). Medical nanotechnology. *BioDrugs.* **19**(4):203–210.

4057. Finch G. et al. (2014). Nanomedicine drug development: a scientific symposium entitled 'charting a roadmap to commercialization'. *AAPS Journal.* **16**(4):698–704.

4058. Garnett M. (2005). Nanomedicines: delivering drugs using bottom up nanotechnology. *International Journal of Nanoscience.* **4**(5–6):855–861.

4059. Gaspar R. (2007). Regulatory issues surrounding nanomedicines: setting the scene for the next generation of nanopharmaceuticals. *Nanomedicine.* **2**(2):143–147.

4060. Guan J. and Zhao Q. (2013). The impact of university–industry collaboration networks on innovation in nanobiopharmaceuticals. *Technological Forecasting and Social Change.* **80**(7):1271–1286.

4061. Hobson D.W. (2009). Opportunities and challenges in pharmaceutical nanotechnology. *Pharmaceutical Technology.* **33**(7):128–130.

4062. Hoet P. et al. (2009). Do nanomedicines require novel safety assessments to ensure their safety for long-term human use? *Drug Safety.* **32**(8):625–636.

4063. Khan A.U. (2012). Nanodrugs: optimism for emerging trend of multidrug resistance. *International Journal of Nanomedicine.* **7**:4323–4324.

4064. Lagarce F. (2015). Nanomedicines: are we lost in translation? *European Journal of Nanomedicine.* **7**(2):77.

4065. Liang X.J. (2013). *Nanopharmaceutics: The Potential Application of Nanomaterials.* Singapore: World Scientific.

4066. Marcato P.D. and Durán N. (2008). New aspects of nanopharmaceutical delivery systems. *Journal of Nanoscience and Nanotechnology.* **8**(5):2216–2229.

4067. Marchant G.E. and Lindor R.A. (2012). Prudent precaution in clinical trials of nanomedicines. *The Journal of Law, Medicine & Ethics.* **40**(4):831–840.

4068. McKendry R.A. (2012). Nanomechanics of superbugs and superdrugs: new frontiers in Nanomedicine. *Biochemical Society Transactions.* **40**(4):603–608.

4069. Moghimi S.M. et al. (2011). Reshaping the future of nanopharmaceuticals: ad iudicium. *ACS Nano.* **5**(11):8454–8458.

4070. Muthu M.S. and Wilson B. (2012). Challenges posed by the scale-up of nanomedicines. *Nanomedicine.* **7**(3):307–309.

4071. Nagaich U. (2014). Nanomedicine: revolutionary trends in drug delivery and diagnostics. *Journal of Advanced Pharmaceutical Technology & Research.* **5**(1):1.

4072. Nazarov G.V. et al. (2009). Nanosized forms of drugs: a review. *Pharmaceutical Chemistry Journal.* **43**(3):163–170.

4073. O'Malley P. (2010). Nanopharmacology: for the future-think small. *Clinical Nurse Specialist.* **24**(3):123–124.

4074. Onoue S. et al. (2014). Nanodrugs: pharmacokinetics and safety. *International Journal of Nanomedicine.* **9**:1025.

4075. Prasad P.N. (2013). Bhasma: traditional concept of nanomedicine and their modern era prospective. *International Journal of Pharmaceutical and Clinical Research.* **5**(4):150–154.

4076. Raghvendra T. et al. (2010). Clinical applications of nanomedicines: a review. *International Journal of Applied Biology and Pharmaceutical Technology.* **1**(2):660–665.

4077. Saiyed M.A. et al. (2011). Toxicology perspective of nanopharmaceuticals: a critical review. *International Journal of Pharmaceutical Sciences and Nanotechnology.* **4**(1):1287–1295.

4078. Santos-Oliveria R. (2011). Nanoradiopharmacetucials: is that the future of nuclear medicine? *Current Radiopharmaceuticals.* **4**(2):140–143.

4079. Sebastian M. et al. (eds.) (2013). *Nanomedicine and Drug Delivery (Advances in Nanoscience and Nanotechnology, Volume 1).* Point Pleasant, NJ: Apple Academic Press.

4080. Shaalan M. et al. (2015). Recent progress in applications of nanoparticles in fish medicine: a review. *Nanomedicine: Nanotechnology, Biology and Medicine.* **12**(3):701–710.

4081. Shekhawat G.S. and Arya V. (2008). Nanomedicines: emergence of a new era in biomedical sciences. *NanoTrends.* 5:9–20.

4082. Siew A. (2014). Current issues with nanomedicines. *Pharmaceutical Technology.* **38**(3):33.

4083. Sikyta B. (2001). [Nanobiotechnology in pharmacology and medicine]. *Ceska Slov Farm.* **50**(6):263–266 (Czech).

4084. Singh S.K. et al. (2013). Current scope of nanomedicines: an overview. *Current Trends in Biotechnology and Chemical Research.* **3**(1):55–62.

4085. Singh T.G. et al. (2013). Formulation and evaluation of nanopharmaceuticals in drug delivery. *Future Medicine.* 80–95.

4086. Timmermans J. et al. (2011). Ethics and nanopharmacy: value sensitive design of new drugs. *NanoEthics.* **5**:269–283.

4087. Tong R. and Cheng J. (2007). Anticancer polymeric nanomedicines. *Journal of Macromolecular Science: Part C: Polymer Reviews.* **47**(3):345–381.

4088. Torchilin V.P. (2007). Nanocarriers. *Pharmaceutical Research.* **24**(12):2333–2334.

4089. Torchilin V. (2008). *Multifunctional Pharmaceutical Nanocarriers.* New York: Springer Science & Business Media.

4090. Vauthier C. and Couvreur P. (2007). Nanomedicines: a new approach for the treatment of serious diseases. *Journal of Biomedical Nanotechnology.* **3**(3):223–234.

4091. Vincent B.B. and Loeve S. (2014). Metaphors in nanomedicine: the case of targeted drug delivery. *NanoEthics.* **8**:1–17.

4092. Vine W. et al. (2006). Nanodrugs: fact, fiction and fantasy. *Drug Delivery Technology.* **6**(5):35–38.

4093. Wu X. and Mansour H.M. (2011). Nanopharmaceuticals II. The application of nanoparticles and nanocarrier systems in pharmaceutics and nanomedicine. *International Journal of Nanotechnology.* **8**(1–2):115–145.

4094. Yang Z. et al. (2014). Nanomedicine: de novo design of nanodrugs. *Nanoscale.* **6**:663–677.

4095. Youn H. et al. (2008). Nanomedicine: drug delivery systems and nanoparticle targeting. *Nuclear Medicine and Molecular Imaging.* **42**(5):337–346.

Nanoparticles

4096. Alexiou C. et al. (2006). Medical applications of magnetic nanoparticles. *Journal of Nanoscience and Nanotechnology.* **6**(9–10):2762–2768.

4097. Ascencio J.A. et al. (2005). Synthesis and theoretical analysis of samarium nanoparticles: perspectives in nuclear medicine. *The Journal of Physical Chemistry B.* **109**(18):8806–8812.

4098. Doane T.L. and Burda C. (2012). The unique role of nanoparticles in nanomedicine: imaging, drug delivery and therapy. *Chemical Society Reviews.* **41**(7):2885–2911.

4099. Donbrow M. (1991). *Microcapsules and Nanoparticles in Medicine and Pharmacy.* Boca Raton, FL: CRC Press.

4100. Dreaden E.C. et al. (2012). The golden age: gold nanoparticles for biomedicine. *Chemical Society Reviews.* **41**(7):2740–2779.

4101. Duguet E. et al. (2006). Magnetic nanoparticles and their applications in medicine. *Nanomedicine.* **1**(2):157–168.

4102. Dusinska M. et al. (2011). Safety of nanoparticles in medicine. In *Nanomedicine in Health and Disease.* USA: Science Publishers, pp. 203–226.

4103. Lane L.A. et al. (2015). Physical chemistry of nanomedicine: understanding the complex behaviors of nanoparticles in vivo. *Annual Review of Physical Chemistry.* **66**:521–547.

4104. Lascialfari A. and Sangregorio C. (2011). Magnetic Nanoparticles in biomedicine recent advances. *Chimica Oggi-Chemistry Today.* **29**(2).

4105. Medina C. et al. (2007). Nanoparticles: pharmacological and toxicological significance. *British Journal of Pharmacology.* **150**(5):552–558.

4106. Murthy S.K. (2007). Nanoparticles in modern medicine: state of the art and future challenges. *International Journal of Nanomedicine.* **2**:129–141.

4107. Nasimi P. and Haidari M. (2013). Medical uses of nanoparticles: drug delivery and diagnosis. *International Journal of Green Nanotechnology.* **1**:1–5.

4108. Pankhurst Q.A. et al. (2003). Applications of magnetic nanoparticles in biomedicine. *Journal of Physics D: Applied Physics.* **36**:R167–R181.

4109. Parveen S. et al. (2012). Nanoparticles: a boon to drug delivery, therapeutics, diagnostics and imaging. *Nanomedicine: Nanotechnology, Biology and Medicine.* **8**(2):147–166.

4110. Rai M. et al. (2015). Strategic role of selected noble metal nanoparticles in medicine. *Critical Reviews in Microbiology.* 1–24. doi:10.3109/1040841X.2015.1018131

4111. Salata O.V. (2004). Applications of nanoparticles in biology and medicine. *Journal of Nanobiotechnology.* **2**(1):3.

4112. Subbiah R. et al. (2010). Nanoparticles: functionalization and multifunctional applications in biomedical sciences. *Current Medicinal Chemistry.* **17**(36):4559–4577.

4113. Taylor E. and Webster T.J. (2011). Reducing infections through nanotechnology and nanoparticles. *International Journal of Nanomedicine.* **6**:1463–1473.

4114. Wolfram J. et al. (2014). Safety of nanoparticles in medicine. *Current Drug Targets.* **15**(10).

4115. Yildiz I. et al. (2011). Applications of viral nanoparticles in medicine. *Current Opinion in Biotechnology.* **22**(6):901–908.

4116. Yohan D. and Chithrani B.D. (2014). Applications of nanoparticles to nanomedicine. *Journal of Biomedical Nanotechnology.* **10**(9):2371–2392.

4117. Zhang L. et al. (2008). Nanoparticles in medicine: therapeutic applications and developments. *Clinical Pharmacology & Therapeutics.* **83**(5):761–769.

Nanopharmacology

4118. Checkman I.S. (2008). [Nanopharmacology: experimental and clinic aspect]. *Lik Sprava.* 1049 (Latvian).

4119. Jain K. et al. (2014). Potentials and emerging trends in nanopharmacology. *Current Opinion in Pharmacology.* **15**:97–106.

4120. O'Malley P. (2010). Nanopharmacology: for the future—think small. *Clinical Nurse Specialist.* **24**(3):123–124.

4121. Tomuleasa C. et al. (2014). Nanopharmacology in translational hematology and oncology. *International Journal of Nanomedicine.* **9**:3465–3479.

4122. Ullman D. (2006). Let's have a serious discussion of nanopharmacology and homeopathy. *FASEB Journal.* 20(14): 2661.

Nanorobots and Nanorobotics

4123. Abhilash M. (2010). Nanorobots. *International Journal of Pharma and Bio Sciences.* **1**(1).

4124. Al-Hudhud G. (2012). On swarming medical nanorobots. *International Journal of Bio-Science and Bio-Technology.* **4**(1): 75–90.

4125. Anonymous (2007). Manufacturing technology for medical nanorobots. *APNF.* **6**(1):8–13.

4126. Arai F. and Maruyama H. (2013). Nanorobotic manipulation and sensing for biomedical applications. In *Nanorobotics.* New York: Springer, pp. 169–190.

4127. Bhat A.S. (2004). Nanobots: the future of medicine. *International Journal of Engineering and Management Sciences.* **5**(1):44–49.

4128. Cavalcanti A. and Freitas R. (2005). Nanorobotics control design: a collective behavior approach for medicine. *IEEE Transactions on NanoBioscience.* **4**(2):133–140.

4129. Cavalcanti A. et al. (2008). Medical nanorobotics for diabetes control. *Nanomedicine: Nanotechnology, Biology and Medicine.* **4**(2):127–138.

4130. Cavalcanti A. et al. (2008). Nanorobot architecture for medical target identification. *Nanotechnology.* **19**(1): 015103.

4131. Cerofolini G. et al. (2010). A surveillance system for early-stage diagnosis of endogenous diseases by swarms of nanobots. *Advanced Science Letters.* **3**(4):345–352.

4132. Dong L. and Nelson B.J. (2007). Tutorial-robotics in the small. Part II: Nanorobotics. *IEEE Robotics & Automation Magazine.* **14**(3):111–121.

4133. Freitas Jr. R.A. (2005). Current status of nanomedicine and medical nanorobots. *Journal of Computational and Theoretical Nanoscience.* **2**:1–25.

4134. Freitas Jr. R.A. (2007). Medical nanorobotics: breaking the trance of futility in life extension research (a reply to de grey). *Studies in Ethics, Law, and Technology.* **1**(1).

4135. Freitas Jr. R.A. (2009). Computational tasks in medical nanorobotics. *Bio-inspired and Nano-scale Integrated Computing.* 391–428.

4136. Hariharan R. and Manohar J. (2010). Nanorobotics as medicament: (perfect solution for cancer). In *IEEE 2010 International Conference on Emerging Trends in Robotics and Communication Technologies (INTERACT)*, Chennai, pp. 4–7.

4137. Hill C. et al. (2011). Nano-and microrobotics: how far is the reality? *Expert Review of Anticancer Therapy.* **8**(12):1891–1897.

4138. Jacob T. et al. (2011). A nanotechnology-based delivery system: nanobots. Novel vehicles for molecular medicine. *The Journal of Cardiovascular Surgery.* **52**(2):159–167.

4139. Jani P. et al. (2013). DNA nanorobots: the complete healthcare package. *International Journal for Chemical and Pharmaceutical Sciences.* **4**(3):1–6.

4140. Kai K. (2013). Nanotechnology and medical robotics: legal and ethical responsibility. *Waseda Bulletin of Comparative Law.* **30**:1–6.

4141. Khulbe P. (2014). Nanorobots: a review. *International Journal of Pharmaceutical Sciences and Research.* **5**(6):2164.

4142. Kostarelos K. (2010). Nanorobots for medicine: how close are we? *Nanomedicine.* **5**(3):341–342.

4143. Kostarelos K. (2010). Nanorobots for medicine: how close are we? *Nanomedicine (Lond)*. **5**(3):341.

4144. Kroeker K.L. (2009). Medical nanobots. *Communications of the ACM*. **52**(9).

4145. Lenaghan S.C. et al. (2013). Grand challenges in bioengineered nanorobotics for cancer therapy. *IEEE Transactions on Biomedical Engineering*. **60**(3):667–673.

4146. Mali S. (2014). Nanorobots: changing face of healthcare system. *Austin Journal of Biomedical Engineering*. **1**(3):1012.

4147. Manjunath A. and Kishore V. (2014). The promising future in medicine: nanorobots. *Biomedical Science and Engineering*. **2**(2):42–47.

4148. Mavroidis C. and Ferreira A. (2013). Nanorobotics: past, present, and future. In *Nanorobotics*. New York: Springer, pp. 3–27.

4149. Morrison S. (2008). Unmanned voyage: an examination of nanorobotic liability. *Albany Law Journal of Science and Technology*. **18**:229.

4150. Patel G.M. et al. (2006). Nanorobot: a versatile tool in nanomedicine. *Journal of Drug Targeting*. **14**(2):63–67.

4151. Prajapati P.M. et al. (2012). Importance of nanorobots in health care. *International Research Journal of Pharmacy*. **3**(3):122–124.

4152. Rao V.N. (2014). Nanorobots in medicine: a new dimension in bio nanotechnology. *Transactions on Networks and Communications*. **2**(2):46–57.

4153. Salunkeh S.S. (2013). Nanorobots: novel emerging technology in the development of pharmaceuticals for drug delivery applications. *World Journal of Pharmacy and Pharmaceutical Sciences*. **2**(6):4728–4744.

4154. Schürle S. et al. (2013). Generating magnetic fields for controlling nanorobots in medical applications. In *Nanorobotics*. New York: Springer, pp. 275–299.

4155. Sharafi A. et al. (2015). MRI-based communication for untethered intelligent medical microrobots. *Journal of Micro-Bio Robotics*. **10**(1–4):27–35.

4156. Thangavel K. et al. (2014). A survey on nano-robotics in nanomedicine. *Journal of Nanoscience and Nanotechnology.* **2**(1): 525–528.

Nephrology

4157. Fissell W.H. et al. (2007). Dialysis and nanotechnology: now, 10 years, or never? *Blood Purification.* **25**(1):12–17.
4158. Kim S. and Roy S. (2013). Microelectromechanical systems and nephrology: the next frontier in renal replacement technology. *Advances in Chronic Kidney Disease.* **20**(6):516–535.
4159. Lee S.H. et al. (2015). Current progress in nanotechnology applications for diagnosis and treatment of kidney diseases. *Advanced Healthcare Materials.* **4**(13):2037–2045.
4160. Nissenson A.R. et al. (2005). Continuously functioning artificial nephron system: the promise of nanotechnology. *Hemodialysis International.* **9**(3):210–217.
4161. Pisignano D. (2007). Nanotechnology and nephrology. *Giornale italiano di nefrologia : organo ufficiale della Società italiana di nefrologia.* **24**(Suppl 40):S80–S86.
4162. Ronco C. and Nissenson A.R. (2001). Does nanotechnology apply to dialysis? *Blood Purification.* **19**(4):347–352.
4163. Ronco C. et al. (2006). Hemodialysis membranes for high-volume hemodialytic therapies: the application of nanotechnology. *Hemodialysis International.* **10**(s1):S48–S50.
4164. Saini R. et al. (2012). Nanotechnology in nephrology. *Saudi Journal of Kidney Diseases and Transplantation.* **23**(2): 367.

Neurology

4165. Ambesh P. and Angeli D.G. (2015). Nanotechnology in neurology: genesis, current status, and future prospects. *Annals of Indian Academy of Neurology.* **18**(4):382.
4166. Cetin M. et al. (2012). Nanotechnology applications in neuroscience: advances, opportunities and challenges. *Klinik*

Psikofarmakoloji Bulteni-Bulletin of Clinical Psychopharmacology. **22**(2):101–103.

4167. Chhabra R. et al. (2015). Emerging use of nanotechnology in the treatment of neurological disorders. *Current Pharmaceutical Design.* **21**(22):3111–3130.

4168. Ellis-Behnke R. (2007). Nano neurology and the four P's of central nervous system regeneration: preserve, permit, promote, plasticity. *Medical Clinics of North America.* **91**(5):937–962.

4169. Iacob G. and Craciun M. (2011). Update in neurosurgery-nanotechnology in neurosurgery. *Mædica.* **6**(4):345.

4170. Jain K.K. (2009). Current status and future prospects of nanoneurology. *Journal of Nanoneuroscience.* **1**(1):56–64.

4171. Kabanov A.V. and Gendelman H.E. (2007). Nanomedicine in the diagnosis and therapy of neurodegenerative disorders. *Progress in Polymer Science.* **32**(8):1054–1082.

4172. Provenzale J.M. and Mohs A.M. (2010). Nanotechnology in neurology: current status and future possibilities. *U.S. Neurology.* **6**:12–17.

4173. Schaller B. et al. (2008). Molecular medicine successes in neuroscience. *Molecular Medicine.* **14**(7–8):361.

4174. Silva G.A. (2007). What impact will nanotechnology have on neurology? *Nature Clinical Practice Neurology.* **3**(4):180–181.

4175. Srikanth M. and Kessler J.A. (2012). Nanotechnology: novel therapeutics for CNS disorders. *Nature Reviews Neurology.* **8**(6):307–318.

4176. Sriramoju B. et al. (2014). Nanomedicine based nanoparticles for neurological disorders. *Current Medicinal Chemistry.* **21**(36):4154–4168.

4177. Sharma H.S. (2009). Nanoneuroscience is now an established discipline now. *Journal of Nanoneuroscience.* **1**(2):95–96.

4178. Silva G.A. (2007). What impact will nanotechnology have on neurology? *Nature Clinical Practice Neurology.* **3**(4):180–181.

4179. Tosi G. et al. (2012). The bridge between nanotechnology and neuroscience: neuro-nanomedicine. *Journal of Nanoneuroscience.* **2**(1):20–26.

4180. Tosi G. et al. (2015). Nanomedicine and neurodegenerative disorders: so close yet so far. *Expert Opinion on Drug Delivery.* **12**(7):1041–1044.

Nuclear Medicine

4181. Assadi M. et al. (2010). Nanotechnology and nuclear medicine; research and preclinical applications. *Hellenic Journal of Nuclear Medicine.* **14**(2):149–159.

4182. Bhatia A.L. (2008). Nuclear medicine in 21st millennium: an approach via nanotechnology. *Asian Journal of Experimental Science.* **22**(2):12.

4183. Mitra A. et al. (2006). Nanocarriers for nuclear imaging and radiotherapy of cancer. *Current Pharmaceutical Design.* **12**(36): 4729–4749.

4184. Wagenaar D.J. et al. (2006). Rationale for the combination of nuclear medicine with magnetic resonance for preclinical imaging. *Technology in Cancer Research & Treatment.* **5**(4):343–350.

Nursing

4185. Goyette P. (2016). Nanotechnology: an educational program for nurses. Amherst: University of Massachusetts, College of Nursing.

4186. Huston C. (2013). The impact of emerging technology on nursing care: warp speed ahead. *The Online Journal of Issues in Nursing.* Available online at http://nursingworld.org/ MainMenuCategories/ANAMarketplace/ANAPeriodicals/ OJIN/TableofContents/Vol-18-2013/No2-May-2013/Impact -of-Emerging-Technology.html.

4187. McCauley L.A. and McCauley R.D. (2005). Nanotechnology: are occupational health nurses ready? *AAOHN Journal.* **53**(12):517–521.

4188. Stagger N. et al. (2008). Nanotechnology: the coming revolution and its implications for consumers, clinicians, and informatics. *Nursing Outlook.* **56**(5):268–274.

4189. Swadner R.L. (2011). Nanotechnology education in nursing curricula: a critical review of the literature. Doctoral Dissertation: St. Paul, MN, Bethel University.

Nutrition

4190. Nickols-Richardson S.M. and Piehowski K.E. (2008). Nanotechnology in nutritional sciences. *Minerva Biotecnologica.* **20**(3):117.

4191. Ramsden J.J. (2008). The potential contribution of nanotechnology to nutritional well-being. *Journal of Biological Physics and Chemistry.* **8**(2):55–60.

4192. Samah N.A. et al. (2014). The Role of nanotechnology application in antioxidant from herbs and spices for improving health and nutrition: a review. *UNISEL Journal of Science, Engineering and Technology.* **1**(1).

4193. Scrinis G. and Lyons K. (2013). *Nano-Functional Foods: Nanotechnology, Nutritional Engineering and Nutritionally-Reductive Food Marketing.* Boca Raton, FL: CRC Press.

4194. Sonkaria S. et al. (2012). Nanotechnology and its impact on food and nutrition: a review. *Recent Patents on Food, Nutrition & Agriculture.* **4**(1):8–18.

4195. Srinivas P.R. et al. (2010). Nanotechnology research: applications in nutritional sciences. *The Journal of Nutrition.* **140**(1):119–124.

Oncology

4196. Aliosmanoglu A. and Basaran I. (2012). Nanotechnology in cancer treatment. *Journal of Nanomedicine and Biotherapeutic Discovery.* **2**(4):107.

4197. Anajwala C.C. et al. (2010). Current trends of nanotechnology for cancer therapy. *International Journal of Pharmaceutical Sciences and Nanotechnology.* **3**(3):1043–1056.

4198. Balaji S. and Balaji P. (2010). Nanotechnology and cancer: an overview. *International Journal of Pharma & Bio Sciences.* **1**(4):186–201.

4199. Bell I.R. et al. (2014). Integrative nanomedicine: treating cancer with nanoscale natural products. *Global Advances in Health and Medicine.* **3**(1):36–53.

4200. Bhandare N. (2014). Applications of nanotechnology in cancer: a literature review of imaging and treatment. *Journal of Nuclear Medicine & Radiation Therapy.* **5**:195. doi: 10.4172/2155-9619.1000195

4201. Blanco E. et al. (2011). Nanomedicine in cancer therapy: innovative trends and prospects. *Cancer Science.* **102**(7):1247–1252.

4202. Chapman S. et al. (2013). Nanoparticles for cancer imaging: the good, the bad, and the promise. *Nano Today.* **8**(5):454–460.

4203. Chauhan V.P. and Jain R.K. (2013). Strategies for advancing cancer nanomedicine. *Nature Materials.* **12**(11):958–962.

4204. Ediriwickrema A. and Saltzman W.M. (2015). Nanotherapy for cancer: targeting and multifunctionality in the future of cancer therapies. *ACS Biomaterials Science & Engineering.* **1**(2):64–78.

4205. Feng S-S. et al. (2011). Nanomedicine for oral chemotherapy. *Nanomedicine.* **6**(3):407–410.

4206. Ferrari M. (2005). Cancer nanotechnology: opportunities and challenges. *Nature Reviews Cancer.* **5**(3):161–171.

4207. Gao J. et al. (2014). Nanomedicine for treatment of cancer stem cells. *Nanomedicine.* **9**(2):181–184.

4208. Grim J. (2015). Cancer nanomedicine: therapy from within. *Nature Nanotechnology.* **10**(4):299–300.

4209. Grossman J.H. and McNeil S.E. (2012). Nanotechnology in cancer medicine. *Physics Today.* **65**(8):38.

4210. Haque N. et al. (2010). Nanotechnology in cancer therapy: a review. *Journal of Chemical and Pharmaceutical Research.* **2**(5):161–168.

4211. He C. et al. (2015). Nanomedicine for combination therapy of cancer. *EBioMedicine.* **2**(5):366–367.

4212. Heath J.R. et al. (2009). Nanomedicine targets cancer. *Scientific American.* **300**(2):44–51.

4213. Herreros E. et al. (2014). Advances in nanomedicine towards clinical application in oncology and immunology. *Current Pharmaceutical Biotechnology.* **15**(9):864–879.

4214. Institute of Medicine (2010). *Nanotechnology and Oncology: Workshop Summary.* Washington, D.C.: National Academies Press.

4215. Johnson L. et al. (2010). Applications of nanotechnology in cancer. *Discovery Medicine.* **9**(47):374–379.

4216. Kathmann W. (2012). [Cancer nanotherapy]. *Deutsch Mediznische Wochenschrrift.* **137**(Suppl 1):S11–S13.

4217. Kawasaki E.S. and Player A. (2005). Nanotechnology, nanomedicine, and the development of new, effective therapies for cancer. *Nanomedicine.* **1**(2):101–109.

4218. Kolhe S. and Parikh K. (2012). Application of nanotechnology in cancer: a review. *International Journal of Bioinformatics Research & Applications.* **8**(1–2):112–125.

4219. Kurkemeyer M.G. and Wagner W. (2013). Nanomedicine in cancer treatment. *Journal of Nanomedicine & Nanotechnology.* **4**(2):166.

4220. Landesman-Milo D. et al. (2015). Nanomedicine as an emerging platform for metastatic lung cancer therapy. *Cancer and Metastasis Reviews.* **34**(2):291–301.

4221. Lee P.Y. and Wong K.K. (2011). Nanomedicine: new frontier in cancer therapeutics. *Current Drug Delivery.* **8**(3):245–253.

4222. Lian X.J. (2014). Nanotechnology and cancer nanomedicine. *Biotechnology Advances.* **32**(4):665.

4223. Liang R. and Fang F. (2014). The application of nanomaterials in diagnosis and treatment for malignant primary brain tumors. *Nano.* **9**(1). doi:10.1142/S1793292014300011

4224. Liu Y. et al. (2013). Perspectives and potential applications of nanomedicine in breast and prostate cancer. *Medicinal Research Reviews.* **33**(1):3–32.

4225. McCarron P.A. and Faheen A.M. (2010). Nanomedicine-based cancer targeting: a new weapon in an old war. *Nanomedicine.* **5**(1):3–5.

4226. Mody H.R. (2011). Cancer nanotechnology: recent trends and developments. *Internet Journal of Medical Update.* **6**(1):3–7.

4227. Naguib Y.W. and Cui Z. (2014). Nanomedicine: the promise and challenges in cancer chemotherapy. *Advances in Experimental Medicine & Biology.* **811**:207–233.

4228. Nazir S. et al. (2014). Nanomaterials in combatting cancer: therapeutic applications and developments. *Nanomedicine.* **10**(1):19–34.

4229. Nie S. (2010). Understanding and overcoming major barriers in cancer nanomedicine. *Nanomedicine.* **5**(4):532–538.

4230. Pope-Harman A. et al. (2007). Biomedical nanotechnology for cancer. *Medical Clinics of North America.* **91**(5):899–927.

4231. Reddy A.K. (2011). Emerging trends of nanotechnology in cancer therapy. *International Journal of Pharmaceutical & Biological Archive.* **2**(1).

4232. Rosenblum D. and Peer D. (2014). Omics-based Nanomedicine: the future of personalized oncology. *Cancer Letters.* **352**(1): 126–136.

4233. Sebastian M. et al. (eds.) (2013). *Nanomedicine and Cancer Therapies (Advances in Nanoscience and Nanotechnology, Volume 2).* Point Pleasant, NJ: Apple Academic Press.

4234. Song S.Y. (2007). Future direction of nanomedicine in gastrointestinal cancer. *Korean Journal of Gastroenterology.* **49**(5):271–279.

4235. Sukumar U.K. et al. (2013). Emerging applications of nanoparticles for lung cancer diagnosis and therapy. *International Nano Letters.* **3**:45.

4236. Sullivan D.C. and Ferrari M. (2004). Nanotechnology and tumor imaging: seizing an opportunity. *Molecular Imaging.* **3**:364–369.

4237. Tachung C.Y. and Wei C. (2005). Nanomedicine in cancer treatment. *Nanomedicine.* **1**(2):191–192.

4238. Tan A. et al. (2011). Quantum dots and carbon nanotubes in oncology: a review on emerging theranostic applications in nanomedicine. *Nanomedicine (Lond).* **6**:1101–1114.

4239. Venditto V.J. and Szoka F.C. Jr. (2013). Cancer nanomedicines: so many papers and so few drugs! *Advanced Drug Delivery Reviews.* **65**(1):80–88.

4240. Virginie S. (2009). Oncology nanomedicine: study of interactions between nanoparticles activated by external electromagnetic energy sources and cancer cells for enhancement of the therapeutic window. PhD Dissertation: Université Pierre Et Marie Curie.

4241. Virupakshappa B. (2012). Applications of Nanomedicine in oral cancer. *Oral Health & Dental Management.* **11**(2):62–68.

4242. Wang R. et al. (2013). Nanomedicine in action: an overview of cancer nanomedicine on the market and in clinical trials. *Journal of Nanomaterials.* **2013**: Article ID 629681.

4243. Wicki A. et al. (2015). Nanomedicine in cancer therapy: challenges, opportunities and clinical applications. *Journal of Controlled Release.* **200**:138–157.

4244. Yih T.C. and Wei C. (2005). Nanomedicine in cancer treatment. *Nanomedicine.* **1**(2):191–192.

Ophthalmology

4245. Ali J. (2011). Scope of nano ophthalmology. *International Journal of Pharmaceutical Investigation.* **1**(4):199.

4246. Chaurasia S.S. et al. (2015). Nanomedicine approaches for corneal diseases. *Journal of Functional Biomaterials.* **6**(2):277–298.

4247. Cordeiro Sousa D. et al. (2015). Nanomedicine and ophthalmology: looking forward. *Acta Ophthalmologica.* **93**(S255). doi:10.111/j.1755-3768.2015.0669

4248. Gonzalez L. et al. (2013). Nanotechnology in corneal neovascularization therapy: a review. *Journal of Ocular Pharmacology and Therapeutics.* **29**(2):124–134.

4249. Kim N.J. et al. (2014). Nanotechnology and glaucoma: a review of the potential implications of glaucoma nanomedicine. *British Journal of Ophthalmology.* **98**(4):427–431.

4250. Matilda A. et al. (2015). A review on ophthalmology using nanotechnology. *Journal of Nanomedicine & Nanotechnology.* **6**(2):272.

4251. Pasechnikova N.V. et al. (2009). Nanotechnology, nanomedicine, nano ophthalmology (Report 1). *Ophthalmological Journal.* **5**:69–76.

4252. Qazi Y. et al. (2009). Nanoparticles in ophthalmic medicine. *International Journal of Green Nanotechnology: Biomedicine.* **1**(1):B3–B8.

4253. Thomson H. and Lotery A. (2009). The promise of Nanomedicine for ocular disease. *Nanomedicine.* **4**(6):599–604.

4254. Venkatesh S. (2014). Nanotechnology in ophthalmology. *Stanley Medical Journal.* **1**(2):41.

4255. Zarbin M.A. et al. (2010). Nanomedicine in ophthalmology: the new frontier. *American Journal of Ophthalmology.* **150**(2):144–162.

Orthopedics

4256. Balasundaram G. and Webster T.J. (2006). Nanotechnology and biomaterials for orthopedic medical applications. *Nanomedicine.* **1**(2):169–176.

4257. Laurencin C.T. et al. (2008). Nanotechnology and orthopedics: a personal perspective. *Wiley Interdisciplinary Reviews: Nanomedicine and Nanobiotechnology.* **1**(1):6–10.

4258. Mazaheri M. et al. (2015). Nanomedicine applications in orthopedic medicine: state of the art. *International Journal of Nanomedicine.* **10**:6039.

4259. Sato M. and Webster T.J. (2004). Nanobiotechnology: implications for the future of nanotechnology in orthopedic applications. *Expert Review in Medical Devices.* **1**(1):105–114.

4260. Tasker L.H. et al. (2007). Applications of nanotechnology in orthopaedics. *Clinical Orthopaedics and Related Research.* **456**:243–249.

Otolaryngology

4261. Dürr S. (2012). [Nanomedicine in otorhinolaryngology-future prospects]. *Laryngorhinootologie.* **91**(1):6–12 (German).

4262. Poe D.S. and Pyykko I. (2011). Nanotechnology and the treatment of inner ear diseases. *Wiley Interdisciplinary Reviews: Nanomedicine and Nanobiotechnology.* **3**(2):212–221.

Pediatrics

4263. Goldschmidt K. (2011). Nanotechnology in pediatrics: science fiction or reality? *Journal of Pediatric Nursing.* **26**(4):379–382.

4264. Machado M.C. et al. (2010). Nanotechnology: pediatric applications. *Pediatric Research.* **67**:500–504.

4265. Singer D.B. (2015). Pediatric pathology in the year 2050. *Pediatric & Developmental Pathology.* **18**(6):512–518.

4266. Sosnik A. and Carcaboso A.M. (2014). Nanomedicines in the future of pediatric therapy. *Advanced Drug Delivery Reviews.* **73**:140–161.

Personalized Medicine

4267. Jain K.K. (2011). Nanobiotechnology and personalized medicine. *Progress in Molecular Biology and Translational Science.* **104**: 325–354.

4268. Jain K.K. (2011). The role of nanobiotechnology in the development of personalized medicine. *Medical Principles and Practice.* **20**(1):1–3.

4269. Lammers T. et al. (2012). Personalized nanomedicine. *Clinical Cancer Research.* **18**(18):4889–4894.

4270. Marchant G.E. (2009). Small is beautiful: what can nanotechnology do for personalized medicine? *Current Pharmacogenomics and Personalized Medicine.* **7**:231–237.

Photonics

4271. Halas N.J. (2009). The photonic nanomedicine revolution: let the human side of nanotechnology emerge. *Nanomedicine.* **4**(4):369–371.

Regenerative Medicine

4272. Armstead A.L. and Li B. (2011). Nanomedicine as an emerging approach against intracellular pathogens. *International Journal of Nanomedicine.* **6**:3281–3293.
4273. Bahadori M. and Mohammadi F. (2012). Regenerative nanomedicine: ethical, legal and social issues. *Methods in Molecular Biology.* **811**:303–316.
4274. Chaudhury K. et al. (2014). Regenerative nanomedicine: current perspectives and future directions. *International Journal of Nanomedicine.* **9**:4153–4167.
4275. Glenn L.M. and Boyce J.S. (2012). Regenerative Nanomedicine: ethical, legal and social issues. *Methods in Molecular Biology.* **811**:303–316.
4276. Khang D. et al. (2010). Nanotechnology for regenerative medicine. *Biomedical Microdevices.* **12**(4):575–587.
4277. Verma S. et al. (2011). Nanomaterials for regenerative medicine. *Nanomedicine.* **6**(1):157–181.

Reproductive Medicine

4278. Barkalina N. et al. (2014). Nanotechnology in reproductive medicine: emerging applications of nanomaterials. *Nanomedicine.* **10**(5):921–938.
4279. Razi M. et al. (2015). The peep of nanotechnology in reproductive medicine: a mini-review. *International Journal of Medical Laboratory.* **2**(1):1–15.

Respiratory Medicine

4280. Bahadori M. and Mohammadi F. (2012). Nanomedicine for respiratory diseases. *Tanaffos.* **11**(4):18–22.

4281. Garcia Fde M. (2014). Nanomedicine and therapy of lung diseases. *Einstein (Sao Paulo).* **12**(4):531–533.

4282. Omlor A.J. et al. (2015). Nanotechnology in respiratory medicine. *Respiratory Research.* **16**:64. doi:10.1186/s12931-015-0223-5

4283. Pison U. et al. (2006). Nanomedicine for respiratory diseases. *European Journal of Pharmacology.* **533**(1–3):341–350.

4284. Swai H. et al. (2009). Nanomedicine for respiratory diseases. *Wiley Interdisciplinary Reviews: Nanomedicine and Nanobiotechnology.* **1**(3):255–263.

Stem Cells

4285. Ji J. et al. (2010). Advances of nanotechnology in the stem cells research and development. *Nano Biomedicine and Engineering.* **2**(1):67–90.

4286. Liu H. et al. (2015). Stem cells: microenvironment, micro/nanotechnology, and application. *Stem Cells International.* **2015**: Article ID 398510.

4287. Wu K.C. et al. (2013). Nanotechnology in the regulation of stem cell behavior. *Science and Technology of Advanced Materials.* **14**(5):054401.

Supramolecular Nanomedicine

4288. Sekhon B.S. (2015). Supramolecular nanomedicine: an overview. *Current Drug Targets.* **16**(13):1407–1428.

Surgery

4289. Andrews R.J. (2009). Nanotechnology and neurosurgery. *Journal of Nanoscience and Nanotechnology.* **9**(8):5008–5013.

4290. Brenner S.A. and Ling J. F. (2012). Nanotechnology applications in orthopedic surgery. *Journal of Nanotechnology in Engineering and Medicine.* **3**(2):024501.

4291. Ibrahim A.M. et al. (2012). Nanotechnology in plastic surgery. *Plastic and Reconstructive Surgery.* **130**(6):879e–887e.

4292. Jaimes S. et al. (2012). Nanotechnology: advances and expectations in surgery. *Revista Colombiana de Cirugía.* **27**(2):158–166 (Spanish).

4293. Leary S.P. et al. (2005). Toward the emergence of nanoneurosurgery. Part I. Progress in nanoscience, nanotechnology, and the comprehension of events in the mesoscale realm. *Neurosurgery.* **57**(4):606–634.

4294. Leary S.P. et al. (2006). Toward the emergence of nanoneurosurgery. Part II. Nanomedicine: diagnostics and imaging at the nanoscale level. *Neurosurgery.* **58**(5):1009–1025.

4295. Leary S.P. et al. (2006). Toward the emergence of nanoneurosurgery. Part III. Nanomedicine: targeted nanotherapy, nanosurgery, and progress toward the realization of nanoneurosurgery. *Neurosurgery.* **58**(6):1009–1025.

4296. Loizidou M. and Seifalian A.M. (2010). Nanotechnology and its applications in surgery. *British Journal of Surgery.* **97**(4):463–465.

4297. Michler R.E. (2001). Nanotechnology: potential applications in cardiothoracic surgery. *Biomedical Microdevices.* 3(2):119–124.

4298. Parks IV, J. et al. (2012). Nanotechnology applications in plastic and reconstructive surgery: a review. *Plastic Surgical Nursing.* **32**(4):156–164.

4299. Petersen D.K. et al. (2014). Current and future applications of nanotechnology in plastic and reconstructive surgery. *Plastic and Aesthetic Research.* **1**(2):43.

4300. Rebello K.J. (2000). Application of MEMS in surgery. *Proceedings of the IEEE.* **92**(1):43–55.

4301. Schulz M.D. et al. (2010). Nanotechnology in thoracic surgery. *The Annals of Thoracic Surgery.* **89**(6):S2188–S2190.

4302. Singh S. and Singh A. (2013). Current status of nanomedicine and nanosurgery. *Anesthesia Essays Research.* **7**(2):237–242.

4303. Singhal S. et al. (2010). Nanotechnology applications in surgical oncology. *Annual Review of Medicine.* **61**:359.

4304. Valdivia U.J. (2005). Nanotechnology, medicine, and minimally invasive surgery. *Archivos Españoles de Urología.* **58**(9):845.

4305. Weldon C. et al. (2011). Nanotechnology and surgeons. *Wiley Interdisciplinary Reviews: Nanomedicine and Nanobiotechnology.* **3**(3):223–228.

4306. Wong K.K. and Liu X.L. (2012). Nanomedicine: a primer for surgeons. *Pediatric Surgery International.* **28**(10):943–951.

Theranostics

4307. Elinav E. and Peer D. (2013). Harnessing nanomedicine for mucosal theranostics: a silver bullet at last? *ACS Nano.* **7**(4):2883–2890.

4308. Lammers T. et al. (2011). Theranostic nanomedicine. *Accounts of Chemical Research.* **44**(10):1029–1038.

4309. Muthu M.S. et al. (2014). Nanotheranostics: application and further development of nanomedicine strategies for advanced theranostics. *Theranostics.* **4**(6):660–677.

4310. Prabhu P. and Patravale V. (2012). The upcoming field of theranostic nanomedicine: an overview. *Journal of Biomedical Nanotechnology.* **8**(6):859–882.

4311. Wang S.L. and Chuang M.C. (2012). Nanotheranostics: a review of recent publications. *International Journal of Nanomedicine.* **7**:4679–4695.

Therapeutics

4312. Bawa R. (2008). Nanoparticle-based therapeutics in humans: a survey. *Nanotechnology Law & Business.* **5**:135–155.

4313. Duncan R. (2003). The dawning era of polymer therapeutics. *Nature Reviews Drug Discovery.* **2**:347–360.

4314. Emerich D.F. (2005). Nanomedicine: prospective therapeutic and diagnostic applications. *Expert Opinion on Biological Therapy.* **5**(1):1–5.

4315. Farokhzad O.C. and Langer R. (2006). Nanomedicine: developing smarter therapeutic and diagnostic modalities. *Advanced Drug Delivery Reviews.* **58**(14):1456–1459.

4316. Jain R.K. (1998). The next frontier of molecular medicine: delivery of therapeutics. *Nature Medicine*. **4**(6):655–657.

4317. Ye Z. and Mahato R.I. (2008). Role of nanomedicines in cell-based therapeutics. *Nanomedicine*. **3**(1):5–8.

Toxicology

4318. Ahmad I. (2011). Nanotoxicity of natural minerals: an emerging area of nanotoxicology. *Journal of Biomedical Nanotechnology*. **7**(1):32–33.

4319. Ai J. et al. (2011). Nanotoxicology and nanoparticle safety in biomedical design. *International Journal of Nanomedicine*. **6**:1117–1127.

4320. Ai J. et al. (2011). Nanotoxicology and nanoparticle safety in biomedical design. *International Journal of Nanomedicine-icine*. **6**:1117–1127.

4321. Ambrosone A. and Tortiglione C. (2013). Methodological approaches for nanotoxicology using Cnidarian models. *Toxicology Mechanisms and Methods*. **23**(3):207–216.

4322. Ambrosone A. et al. (2014). Nanotoxicology using the sea anemone Nematostella vectensis: from developmental toxicity to genotoxicology. *Nanotoxicology*. **8**(5):508–520.

4323. Arora S. et al. (2012). Nanotoxicology and in vitro studies: the need of the hour. *Toxicology and Applied Pharmacology*. **258**(12):151–165.

4324. Aruguete D.M. et al. (2013). Antimicrobial nanotechnology: its potential for the effective management of microbial drug resistance and implications for research needs in microbial nanotoxicology. *Environmental Science: Processes & Impacts*. **15**(1):93–102.

4325. Azhdarzadeh M. et al. (2015). Nanotoxicology: advances and pitfalls in research methodology. *Nanomedicine*. **10**(18): 2931–2952.

4326. Bacanli M. and Basaran N. (2014). Nanotoxicology-new research area in toxicology. *Turkish Journal of Pharmaceutical Sciences*. **11**(2):231–240.

4327. Baker D. et al. (eds.) (2012). *Essentials of Toxicology for Health Protection: A Handbook for Field Professionals.* Oxford, UK: Oxford University Press.

4328. Bang-wei Z. (2006). Preliminary study of nanotoxicology. *Nanoscience & Nanotechnology.* **6**:20.

4329. Bhattacharjee S. and Brayden D.J. (2015). Development of nanotoxicology: implications for drug delivery and medical devices. *Nanomedicine.* **10**(14):2289–2305.

4330. Boczkowski H. and Hoet P. (2010). What's new in nanotoxicology? Implications for public health from a brief review of the 2008 literature. *Nanotoxicology.* **4**(1):1–14.

4331. Bolt H.M. et al. (2012). Nanotoxicology and oxidative stress control: cutting-edge topics in toxicology. *Archives of Toxicology.* **86**(11):1629–1635.

4332. Bolt H.M. et al. (2013). Recent developments in nanotoxicology. *Archives of Toxicology.* **87**(6):927–928.

4333. Borak J. (2009). Nanotoxicology: characterization, dosing, and health effects. *Journal of Occupational and Environmental Medicine.* **51**(5):620–621.

4334. Borm P.J.A. (2002). Particle toxicology: from coal mining to nanotechnology. *Inhalation Toxicology.* **14**:311–324.

4335. Brayner R. (2008). The toxicological impact of nanoparticles. *Nano Today.* **3**(1–2):48–55.

4336. Bregoli L. et al. (2012). Molecular methods for nanotoxicology. In *Toxic Effects of Nanomaterials.* Bentham Books, pp. 97–120.

4337. Bruinink A. et al. (2015). Effect of particle agglomeration in nanotoxicology. *Archives of Toxicology.* **89**(5):659–675.

4338. Burcham P. (2010). Nanotoxicology: a primer for chemists. *Chemistry in Australia.* **77**(8):18.

4339. Burello E. and Worth A. (2011). Computational nanotoxicology: predicting toxicity of nanoparticles. *Nature & Nanotechnology.* **6**(3):138–139.

4340. Canady R.A. (2010). The uncertainty of nanotoxicology: report of a society for risk analysis workshop. *Risk Analysis.* **30**(11):1663–1670.

4341. Casals E. et al. (2009). Nanotoxicology: exploring new paradigms in toxicology. A poster presented at NanoSpain 2009, March 9–12. Zaragoza, Spain.

4342. Castanova V. (2009). The nanotoxicology program in NIOSH. *Journal of Nanoparticle Research.* **11**(1):5–13.

4343. Catalan-Figueroa J. et al. (2015). Nanomedicine and nanotoxicology: the pros and cons for neurodegeneration and brain cancer. *Nanomedicine.* **11**(2):171–187.

4344. Cattaneo A. G. Systematic review of papers in the field of nanotoxicology. Available online at https://www.researchgate. net/profile/Anna_Giulia_Cattaneo/publication/277138685_ Systematic_review_of_papers_in_the_field_of_nanotoxicology /links/5562fb4d08ae86c06b6605e7.pdf

4345. Chey O.C. et al. (2013). Impact of nanotoxicology towards technologists to end users. *Advanced Materials Letters.* **4**(8):591–597.

4346. Chidambaram M. and Krishnasamy K. (2012). Nanotoxicology: toxicity of engineered nanoparticles and approaches to produce safer nanotherapeutics. *International Journal of Pharmacy and Pharmaceutical Sciences.* **2**(4):117–122.

4347. Clark K.A. et al. (2011). Predictive models for nanotoxicology: current challenges and future opportunities. *Regulatory Toxicology & Pharmacology.* **59**(3):361–363.

4348. Clift M.J. et al. (2011). Nanotoxicology: a perspective and discussion of whether or not in vitro testing is a valid alternative. *Archives of Toxicology.* **85**(7):723–731.

4349. Culha M. and Altunbek M. (2011). Nanotoxicology: how to test the safety of engineered nanomaterials? *Current Opinion in Biotechnology.* **1**(22):S28–S29.

4350. Curtis J. et al. (2006). Nanotechnology and nanotoxicology: a primer for clinicians. *Toxicological Reviews.* **25**(4):245–260.

4351. Czerwinski F. (2009). Nanotoxikologie-es gibt immer zwei Seiten review about nanotoxicology as an emerging field of research, includes interviews with Willie Peijnenburg, Priska Hinz, and Jens Otte. Junge Wissenschaft. **83**.

4352. Dai X. and Cui D. (2012). Advances in the toxicity of nanomaterials. *Nano Biomedicine and Engineering.* **4**(3):150–156.

4353. Dailey L.A. (2007). Inhalation nanotoxicology and how it may impact drug delivery to the lung. *Journal of Pharmacy & Pharmacology.* **59**:A68–A68.

4354. Dhawan A. et al. (2009). Nanomaterials: a challenge for toxicologists. *Nanotoxicology.* **3**(1):1–9.

4355. Dhawan A. et al. (2011). NanoLINEN: nanotoxicology link between India and European nations. *Journal of Biomedical Nanotechnology.* **7**(1):203–204.

4356. Dietz K.J. and Herth S. (2011). Plant nanotoxicology. *Trends in Plant Science.* **16**(11):582–589.

4357. Donaldson K. et al. (2004). Nanotoxicology. *Occupational and Environmental Medicine.* **61**:727–728.

4358. Donaldson K. and Poland C.A. (2013). Nanotoxicity: challenging the myth of nano-specific toxicity. *Current Opinion in Biotechnology.* **24**:724–734.

4359. Drobne D. (2007). Nanotoxicology for safe and sustainable nanotechnology. *Archives of Industrial Hygiene & Toxicology.* **58**(4): 471–478.

4360. Duran N. et al. (eds.) (2014). *Nanotoxicology: Materials, Methodologies, and Assessments.* New York: Springer Science.

4361. Ek S. (2009). A nanobiological approach to nanotoxicology. *Human and Experimental Toxicology.* **28**(6–7):393–400.

4362. Elliott K.C. (2007). Varieties of exploratory experimentation in nanotoxicology. *History & Philosophy of the Life Sciences.* **29**(3):313–336.

4363. Elliott K.C. (2014). Ethical and societal values in nanotoxicology. In *In Pursuit of Nanoethics.* Netherlands: Springer, pp. 147–163.

4364. Fadeel B. et al. (2013). Nanotoxicology. *Toxicology.* **313**(1):1–2.

4365. Fadeel B. et al. (2015). Keeping it real: the importance of material characterization in nanotoxicology. *Biochemical and Biophysical Research Communications.* **468**(3):498–503.

4366. Farhan M. et al. (2014). Nanotoxicology and its implications. *Research Journal of Pharmaceutical, Biological and Chemical Sciences.* **5**(1):470–479.

4367. Feliu N. and Fadeel B. (2010). Nanotoxicology: no small matter. *Nanoscale.* **2**(12):2514–2520.

4368. Fischer H.C. and Chan W.C. (2007). Nanotoxicity: the growing need for in vivo study. *Current Opinion in Biotechnology.* **18**:565–571.

4369. Gagné F. et al. (2007). Aquatic nanotoxicology: a review. *Current Topics in Toxicology.* **4**:51–64.

4370. Gallud A. and Fadeel B. (2015). Keeping it small: towards a molecular definition of nanotoxicology. *European Journal of Nanomedicine.* **7**(3):143–151.

4371. Gangwal S. and Hubal E. (2012). Nanotoxicology: Gangwal et al. respond. *Environmental Health Perspectives.* **120**(1):a13.

4372. García-Remesal M. et al. (2012). Using nanoinformatics methods for automatically identifying relevant nanotoxicology entities from the literature. *BioMed Research International.* **2013**: Article ID 410294.

4373. Gebel T. et al. (2013). The nanotoxicology revolution. *Archives of Toxicology.* **87**(12):2057–2062.

4374. Glushkova A.V. et al. (2007). Nanotechnologies and nanotoxicology: a view on the problem. *Toksikologich Vestn.* **6**:4–8 (Russian).

4375. Godugu C. et al. (2015). Nanotoxicology: contemporary issues and future directions. In *Targeted Drug Delivery: Concepts and Design.* Switzerland: Springer International Publishing, pp. 733–781.

4376. Gomes S.I. (2009). Soil nanotoxicology: effect assessment using biomarkers and gene analysis. PhD Dissertation: Aveiro, Portugal, University de Aveiro.

4377. Gonzalez L. et al. (2008). Genotoxicity of engineered nanomaterials: a critical review. *Nanotoxicology.* **2**(4):252–273.

4378. Gornati R. et al. (2009). In vivo and in vitro models for nanotoxicology testing. In *Nanotoxicity.* Chichester: John Wiley and Sons, pp. 279–302.

4379. Greish K., Thiagarajan G. and Ghandehari H. (2012). In vivo methods of nanotoxicology. In *Nanotoxicity*. Humana Press, pp. 235–253.

4380. Günday N. and Schneider M. (2012). Conference scene— nanomedicine and nanotoxicology: future prospects and the need for translational factors for the combination of both. *Nanomedicine*. **7**(6):811–814.

4381. Gunsolus I.L. and Haynes C.L. (2015). Analytical aspects of nanotoxicology. *Analytical Chemistry*. **88**(1):451–479.

4382. Hanumanthu S.C. (2015). Emerging discipline nanotoxicology: sources, challenges and strategies for addressing risk. *International Journal of Engineering Science and Innovative Technology*. **4**(5):102–110.

4383. Harthorn B. (2009). Nanotoxicology: characterizing the scientific literature, 2000–2007. *Journal of Nanoparticle Research*. **11**(2):251–257.

4384. Haynes C.L. (2009). The growth of nanotoxicology. *Dalton Transactions*. **15**:C29.

4385. Haynes C.L. (2010). The emerging field of nanotoxicology. *Analytical & Bioanalytical Chemistry*. **398**(2):587–588.

4386. Haynes C.L. (2010). *Nanotoxicology*. Berlin: Springer.

4387. Hoet P. and Boczkowski J. (2008). What's new in nanotoxicology? Brief review of the 2007 literature. *Nanotoxicology*. **2**(3):171–182.

4388. Holden P.A. et al. (2012). Ecological nanotoxicology: integrating nanomaterial hazard considerations across the subcellular, population, community, and ecosystems levels. *Accounts of Chemical Research*. **46**(3):813–822.

4389. Holl M.M. (2009). Nanotoxicology: a personal perspective. *Wiley Interdisciplinary Reviews: Nanomedicine and Nanobiotechnology*. **1**(4):353–359.

4390. Hondow N. et al. (2011). STEM mode in the SEM: a practical tool for nanotoxicology. *Nanotoxicology*. **5**(2):215–227.

4391. Houdy P. et al. (eds.) (2012). *Nanoethics and Nanotoxicology*. New York: Springer-Verlag.

4392. Hubs A.F. et al. (2011). Nanotoxicology: a pathologist's perspective. *Toxicologic Pathology.* **39**(2):301–324.

4393. Hull M. et al. (2012). Moving beyond mass: the unmet need to consider dose metrics in environmental nanotoxicology studies. *Environmental Science & Technology.* **46**(20):10881–10882.

4394. Hussain S.M. et al. (2015). At the crossroads of nanotoxicology in vitro: past achievements and current challenges. *Toxicological Sciences.* **147**(1):5–16.

4395. Jain S.K. et al. (2011). Nanotoxicology: an emerging discipline. *Veterinary World.* **4**(1):35–40.

4396. Jones R. (2009). It's not just about nanotoxicology. *Nature Nanotechnology.* **4**(10):615.

4397. Kagan V.E. et al. (2005). Nanomedicine and nanotoxicology: two sides of the same coin. *Nanomedicine.* **1**(4):313–316.

4398. Kahru A. and Dubourguier H-C. (2010). From ecotoxicology to nanoecotoxicology. *Toxicology.* **269**(2–3):105–119.

4399. Kalantari H. (2013). Nanotoxicology. *Jundishapur Journal of Natural Pharmaceutical Products.* **8**(1):1–2.

4400. Kane A.B. and Hurt R.H. (2008). Nanotoxicology: the asbestos analogy revisited. *Nature & Nanotechnology.* **3**(7):378–379.

4401. Karlsson H.L. (2010). The comet assay in nanotoxicology research. *Analytical and Bioanalytical Chemistry.* **398**(2):651–666.

4402. Kendall M. et al. (2011). Particle and nanoparticle interactions with fibrinogen: the importance of aggregation in nanotoxicology. *Nanotoxicology.* **5**(1):55–65.

4403. Kipen H.M. et al. (2005). Smaller is not always better: nanotechnology yields nanotoxicology. *American Journal of Physiology: Lung Cellular & Molecular Physiology.* **33**(5):L696–L697.

4404. Krug H.F. and Wick P. (2011). Nanotoxicology: an interdisciplinary challenge. *Angewandte Chemie International Edition in English.* **50**(6):1260–1278.

4405. Kurath M. and Maasen S. (2006). Toxicology as a nanoscience? *Disciplinary Identities Reconsidered.* **3**. doi:10.1186/1743-8977-3-6

4406. Laurent S. et al. (2012). Crucial ignored parameters on nanotoxicology: the importance of toxicity assay modifications and "cell vision". *PLoS One.* **7**(1):e29997.

4407. Lehr C.M. et al. (2011). Biological barriers: a need for novel tools in nanotoxicology and nanomedicine. *European Journal of Pharmaceutics and Biopharmaceutics.* **77**(3):337.

4408. Li Y. et al. (2014). Nanotoxicity overview: nano-threat to susceptible populations. *International Journal of Molecular Sciences.* **15**(3):3671–3697.

4409. Li Y.F. et al. (2015). Synchrotron radiation techniques for nanotoxicology. *Nanomedicine: Nanotechnology, Biology and Medicine.* **11**(6):1531–1549.

4410. Linkov I. et al. (2008). Nanotoxicology and Nanomedicine: making hard decisions. *Nanomedicine.* **4**(2):167–171.

4411. Lison D. (2013). Nanotoxicology: where are we now? *Acta Clinica Belgica.* **68**(6):466.

4412. Lison D. et al. (2014). Paracelsus in nanotoxicology. *Particle & Fibre Toxicology.* **11**(1):35.

4413. Majumder D.D. et al. (2013). Nanotoxicology: a threat to the environment and to human beings. In *Proceedings of the International Symposium on Engineering under Uncertainty: Safety Assessment and Management (ISEUSAM-2012),* Springer India, pp. 385–400.

4414. Malysheva A. et al. (2015). Bridging the divide between human and environmental nanotoxicology. *Nature Nanotechnology.* **10**(10):835–844.

4415. Marnett L.J. (2009). Nanotoxicology. A new frontier. *Chemical Research in Toxicology.* **22**(9):1491.

4416. Maurer-Jones M.A. and Haynes C.L. (2012). Toward correlation in in vivo and in vitro nanotoxicology studies. *The Journal of Law, Medicine & Ethics.* **40**(4):795–801.

4417. Maynard A.D. (2007). Laying a firm foundation for sustainable nanotechnologies. In *Nanotoxicology: Characterization, Dosing, and Health Effects*. Boca Raton, FL: CRC Press, pp. 1–6.

4418. Maynard A.D. et al. (2011). The new toxicology of sophisticated materials: nanotoxicology and beyond. *Toxicological Sciences*. **120**(Suppl 1):S109–S129.

4419. Meng H. et al. (2009). A predictive toxicological paradigm for the safety assessment of nanomaterials. *ACS Nano*. **3**(7): 1620–1627.

4420. Messing M.E. et al. (2012). Gas-borne particles with tunable and highly controlled characteristics for nanotoxicology studies. *Nanotoxicology*. **7**(6):1052–1063.

4421. Messing M.E. et al. (2014). Nanotoxicology: aerosol surface area determination. A paper presented at the *International Aerosol Conference*. Busan, Korea.

4422. Minocha S. (2012). Investigations into the mechanisms of cell death: the common link between anticancer nanotherapeutics and nanotoxicology. PhD Dissertation: Chapel Hill, University of North Carolina.

4423. Monteiro-Riviere N.A. and Tran C.L. (2007). *Nanotoxicology: Characterization, Dosing and Health Effects*. Boca Raton, FL: CRC Press.

4424. Monterio-Riviere N.A. and Tran C.L. (2012). *Nanotoxicology: Progress toward Nanomedicine*, 2nd edition. Boca Raton, FL: CRC Press.

4425. Müller M. et al. (2008). Nanotoxicology. *Zentralblatt für Arbeitsmedizin, Arbeitsschutz und Ergonomie*. **58**:238–252.

4426. Munoz B. (2013). Perspectives and approaches in nanotoxicology research. *Toxicology Mechanisms & Methods*. **23**(2):151–152.

4427. Mura S. et al. (2015). Latest developments of nanotoxicology in plants. In *Nanotechnology and Plant Sciences*. Switzerland: Springer International Publishing, pp. 125–151.

4428. Myllynen P. (2009). Nanotoxicology: damaging DNA from a distance. *Nature Nanotechnology*. **4**(12):795–796.

4429. Nel A. (2010). Nanotoxicology as a predictive science that can be explored by high content screening and the use of computer-assisted hazard ranking (NSF Cooperative Agreement # NSF-EF9830117). University of California.

4430. Nyland J.F and Silbergeld E.K. (2009). A nanobiological approach to nanotoxicology. *Human & Experimental Toxicology.* **28**(6/7):393–400.

4431. Oberdörster G. (2010). Safety assessment for nanotechnology and nanomedicine: concepts of nanotoxicology. *Journal of Internal Medicine.* **267**(1):89–105.

4432. Oberdörster G. (2012). Nanotoxicology: in vitro–in vivo dosimetry. *Studies.* **7**(3):163–176.

4433. Oberdörster G. et al. (2005). Nanotoxicology: an emerging discipline evolving from studies of ultrafine particles. *Environmental Health Perspectives.* **113**(7):823–839.

4434. Ostrowski A.D. et al. (2009). Nanotoxicology: characterizing the scientific literature, 2000–2007. *Journal of Nanoparticle Research.* **11**(2):251–257.

4435. Papp T. et al. (2008). Human health implications of nanomaterial exposure. *Nanotoxicology.* **2**(1):9–27.

4436. Patel T. et al. (2013). Hierarchical rank aggregation with applications to nanotoxicology. *Journal of Agricultural, Biological, and Environmental Statistics.* **18**(2):159–177.

4437. Peixe T. and Schofield K. (2008). Trends in nanotoxicology exposure at the occupational exposure level. *Toxicology Letters.* **180**:S225.

4438. Philbert M.A. (2012). The new toxicology of sophisticated material: nanotoxicology and beyond. *Toxicology Letters.* **211**:S3.

4439. PourGashtasbi G. (2015). Nanotoxicology and challenges of translation. *Nanomedicine.* **10**(20):3121–3129.

4440. Pourmand A. and Abhdollahi M. (2012). Current opinion on nanotoxicology. *Daru.* **20**(1):95.

4441. Pumera M. (2011). Nanotoxicology: the molecular science point of view. *Chemistry, an Asian Journal.* **6**(2):340–348.

4442. Purzner M. et al. (2008). Nanotoxicology and alternative testing methods–development, chances and needs on OECD and EU level. *ALTEX*. **25**(Suppl 1).

4443. Puzyn T. (2015). Achievements and perspectives of computational nanotoxicology. *Toxicology Letters*. **238**(Suppl):S34–S35.

4444. Reineke J. (2012). *Nanotoxicity: Methods and Protocols*. New York: Humana Press.

4445. Robinson R.L. et al. (2015). An ISA-TAB-Nano based data collection framework to support data-driven modelling of nanotoxicology. *Beilstein Journal of Nanotechnology*. **6**(1):1978-1999.

4446. Rodea-Palomares I. et al. (2014). From basel: biophysical interactions at the bio-nano interface: relevance for aquatic nanotoxicology. *Globe*. **15**(6):1–2.

4447. Sahu S.C. and Casciano D.A. (eds.) (2009). *Nanotoxicity: From In Vivo and In Vitro Models to Health Risks*. West Sussex, UK: John Wiley & Sons.

4448. Sahu S.C. and Casciano D.A. (eds.) (2014). *Hand Book of Nanotoxicology, Nanomedicine and Stem Cell Use in Toxicology*. Wiley Online Library at http://onlinelibrary.wiley.com/doi/10.1002/9781118856017

4449. Santamaria A. (2012). Historical overview of nanotechnology and nanotoxicology. In *Nanotoxicity*. Humana Press, pp. 1–12.

4450. Sauer U.G. et al. (2011). A knowledge-based search engine to navigate the information thicket of nanotoxicology. *Regulatory Toxicology & Pharmacology*. **59**(1):47–52.

4451. Schirmer K. and Auffan M. (2015). Nanotoxicology in the environment. *Environmental Science: Nano*. **2**(6):561–563.

4452. Schrader-Frechette K. (2007). Nanotoxicology and ethical conditions for informed consent. *NanoEthics*. **1**(1):47–56.

4453. Schnackenberg L.K. et al. (2012). Metabolomics techniques in nanotoxicology studies. In *Nanotoxicity*. Humana Press, pp. 141–156.

4454. Seaton A. (2005). Nanotoxicology: hazard and risk. *Nano-Biotechnology*. **1**(3):316.

4455. Seaton A. and Donaldson K. (2005). Nanoscience, nanotoxicology and the need to think small. *Lancet.* **365**(9463):923–924.

4456. Service R.F. (2004). Nanotoxicology. Nanotechnology grows up. *Science.* **304**(5678):1732–1734.

4457. Silbergeld E. et al. (2010). Nanoscale methods for nanotoxicology. *Toxicology Letters.* **196**:S283.

4458. Silbergeld E. et al. (2011). Nanotoxicology: "The End of the Beginning"—signs on the roadmap to a strategy for assuring the safe application and use of nanomaterials. *ALTEX.* **28**:236–241.

4459. Singh N. (2009). Conference scene—nanotoxicology: health and environmental impacts. *Nanomedicine (Lond).* **4**(4):385–390.

4460. Song Y. and Ning B. (2014). [Focus on study of nanotoxicology in China]. *Zhongua Yu Fang Yi Xue Za Zhi [Chinese Journal of Preventive Medicine].* **48**(7):552–554 (Chinese).

4461. Stone V. (2008). Current trends in nanotoxicology—view from the editor. *Toxicology Letters.* **180**:S20–S21.

4462. Stone V. and Donaldson K. (2006). Nanotoxicology: signs of stress. *Nature Nanotechnology.* **1**(1):23–24.

4463. Suh W.H. et al. (2009). Nanotechnology, nanotoxicology, and neuroscience. *Progress in Neurobiology.* **87**(3):133–170.

4464. Sundaram S.K. et al. (2013). Fourier-transform infrared spectroscopy for rapid screening and live-cell monitoring: application to nanotoxicology. *Nanomedicine.* **8**(1):145–156.

4465. Tantra R. et al. (2015). Nano(Q)SAR: challenges, pitfalls and perspectives. *Nanotoxicology.* **9**(5):636–642.

4466. Tiwle R. (2012). Nanotoxicology: an emerging tool used for the toxicity of nanomaterials. *Journal of Biomedical and Pharmaceutical Research.* **1**(3). Available at http://jbpr.in/index.php/jbpr/article/view/40

4467. Tyne W. et al. (2013). A new medium for Caenorhabditis elegans toxicology and nanotoxicology studies designed to better reflect natural soil solution conditions. *Environmental Toxicology and Chemistry.* **32**(8):1711–1717.

4468. Valerian E.K. et al. (2005). Nanomedicine and nanotoxicology: two sides of the same coin. *Nanomedicine.* **1**(4):313–316.

4469. Wang B. et al. (2010). New methods for nanotoxicology: synchrotron radiation-based techniques. *Analytical Bioanalytical Chemistry.* **398**(2):667–676.

4470. Warheit D.B. (2010). Debunking some misconceptions about nanotoxicology. *Nano Letters.* **10**(12):4777–4782.

4471. Webster T.J. (2007). NanoTox: hysteria or scientific studies? *Nature Nanotechnology.* **2**(12):732–734.

4472. Wehmas L. and Tanguay R.L. (2013). Nanotoxicology in green nanoscience. In *Innovations in Green Chemistry and Green Engineering.* New York: Springer, pp. 157–178.

4473. White J.W. et al. (2009). Protein interfacial structure and nanotoxicology. *Nuclear Instruments and Methods in Physics Research Section A: Accelerators, Spectrometers, Detectors and Associated Equipment.* **600**(1):263–265.

4474. Wick P. et al. (2013). Knocking at the door of the unborn child: issues in developmental and placental nanotoxicology, and directions for future research. *Reproductive Toxicology.* (41):15.

4475. Winkler D.A. et al. (2013). Applying quantitative structure–activity relationship approaches to nanotoxicology: current status and future potential. *Toxicology.* **313**(1):15–23.

4476. Wittmaack K. (2007). Dose and response metrics in nanotoxicology: Wittmaack responds to Oberdoerster et al. and Stoeger et al. *Environmental Health Perspectives.* **115**(6): A291.

4477. Yildirimer L. et al. (2011). Toxicology and clinical potential of nanoparticles. *Nano Today.* **6**(6):585–607.

4478. Zhao Y. and Ng K.W. (2014). Nanotoxicology in the skin: how deep is the issue? *Nano Life.* **4**(1). doi:10.1142/S1793984414400042

4479. Zhifang Z. (2011). Radioanalytical methods in nanotoxicology studies. *Progress in Chemistry.* **7**:030.

4480. Zhou S.F. (2015). American Journal of Nanotoxicology and Nanoscience aims to publish novel discoveries in nanotoxicology and nanoscience. *American Journal of Nano-toxicology and Nanoscience.* **1**(1). doi: 10.18802/ajnn.28301

Transplantation

4481. Fissell W.H. et al. (2007). Dialysis and nanotechnology: now, 10 years, or never? *Blood Purification.* **25**(1):12–17.
4482. Malchesky P.S. (2004). Nanotechnology and artificial organs: today and for the future. *Artificial Organs.* **28**(11):969–970.
4483. Merina R.M. (2010). Use of nanorobots in heart transplantation. In *IEEE 2010 International Conference on Emerging Trends in Robotics and Communication Technologies (INTERACT)*, Chennai, pp. 265–268.
4484. Patel K.J. et al. (2015). Utilization of machine perfusion and nanotechnology for liver transplantation. *Current Transplantation Reports.* **2**(4):303–311.
4485. Shen L.J. et al. (2007). Nanomedicines in renal transplant rejection–focus on sirolimus. *International Journal of Nanomedicine.* **2**(1):25.

Urology

4486. Álvarez J.D. et al. (2008). Nanotecnología, advances y expectativas en urología. *Urología Colombiana.* **18**:41–48.
4487. Gommersall L. et al. (2007). Nanotechnology and its relevance to the urologist. *European Urology.* **52**(2):368–375.
4488. Jin S. and Labhasetwar V. (2009). Nanotechnology in urology. *Urologic Clinics of North America.* **36**(2):179–188.
4489. Maddox M. et al. (2014). Nanotechnology applications in urology: a review. *BJU International.* **114**(5):653–660.
4490. Shergill I.S. et al. (2006). Nanotechnology: potential applications in urology. *BJU International.* **97**(2):219–220.
4491. Tsiouris A. et al. (2006). Nanotechnology: potential applications in urology. *BJU International.* **98**(1):231–232.

Vaccines, Vaccinations, and Immunizations

4492. Boyapalle S. et al. (2012). Nanotechnology applications to HIV vaccines and microbicides. *Journal of Global Infectious Diseases.* **4**(1):62.

4493. Heller P. et al. (2010). From polymers to nanomedicines: new materials for future vaccines. In Giese M. (ed.), *Molecular Vaccines*, Volume 2. Switzerland: Springer International Publishing. Chapter 41, pp. 644–663.

4494. Fangueiro J.F. et al. (2012). Nanomedicines for immunization and vaccines. In Souto E.B. (ed.), *Patenting Nanomedicine.* Berlin Heidelberg: Springer-Verlag. Chapter 15, pp. 436–445.

4495. Kim M.G. et al. (2014). Nanotechnology and vaccine development. *Asian journal of Pharmaceutical Sciences.* **9**(5):227–235.

4496. Lawson L.B. et al. (2007). Use of nanocarriers for transdermal vaccine delivery. *Clinical Pharmacology & Therapeutics.* **82**(6):641–643.

4497. Nasir A. (2009). Nanotechnology in vaccine development: a step forward. *Journal of Investigative Dermatology.* **129**(5): 1055–1059.

4498. Peek L.J. et al. (2008). Nanotechnology in vaccine delivery. *Advanced Drug Delivery Reviews.* **60**(8):915–928.

4499. van Riet E. et al. (2014). Combatting infectious diseases; nanotechnology as a platform for rational vaccine design. *Advanced Drug Delivery Reviews.* **74**:28–34.

4500. Wilkinson A. (2014). Nanotechnology for the delivery of vaccines. Doctoral Dissertation: Birmingham, UK: Aston University.

4501. Zhao L. et al. (2014). Nanoparticle vaccines. *Vaccine.* **32**(3): 327–337.

Vascular Medicine

4502. Brenner S.A. and Pautler M. (2010). Nanotechnology applications in vascular disease. *Journal of Nanotechnology in Engineering and Medicine.* **1**(4):044501.

4503. Gupta A.S. (2011). Nanomedicine approaches in vascular disease: a review. *Nanomedicine: Nanotechnology, Biology and Medicine.* **7**(6):763–779.

4504. Vorp D.A. et al. (2001). Vascular applications of micro- and nanotechnology. *Journal of Vascular and Interventional Radiology.* **12**(1):P236–P240.

4505. Wickline S.A. et al. (2006). Applications of nanotechnology to atherosclerosis, thrombosis, and vascular biology. *Arteriosclerosis, Thrombosis, and Vascular Biology.* **26**(3):435–441.

Veterinary Medicine

4506. Chakravarthi V.P. and Balji S.N. (2010). Applications of nanotechnology in veterinary medicine. *Veterinary World.* **3**(10):477–480.

4507. Dilbaghi N. et al. (2013). Nanoscale devices for veterinary technology: trends and future prospective. *Advanced Materials Letters.* **4**(3):175–184.

4508. Mohanty M.N. et al. (2014). An overview of nanomedicine in veterinary science. *Veterinary Research International.* **2**(4):90–95.

4509. Num S.M. and Useh N.M. (2013). Nanotechnology applications in veterinary diagnostics and therapeutics. *Sokoto Journal of Veterinary Sciences.* **11**(2):10–14.

4510. Scott N.R. (2007). Nanoscience in veterinary medicine. *Veterinary Research Communications.* **31**(Suppl 1):139–144.

4511. Underwood C. and van Eps A.W. (2012). Nanomedicine and veterinary science: the reality and the practicality. *Veterinary Journal.* **193**(1):12–23.

Nanophotonics

4512. Aravind A. (2014). Nanophotonics and its uses. *IJRECE.* **2**(4):126–128.

4513. Brown C.T.A. et al. (2010). Nanobiophotonics: photons that shine their light on the life at the nanoscale. *Journal of Biophotonics.* **3**(10–11):639–640.

4514. Ghoshal S.K. et al. (2011). Nanophotonics for the 21st century. In *Optoelectronics: Devices and Applications*. InTech Open Access Publisher. Available at http://cdn.intechopen.com/pdfs/20499/ InTech-Nanophotonics_for_21st_century.pdf

4515. Gomes A.S. (2009). Commentary: nanobiophotonics: example of scientific convergence. *Journal of Nanophotonics*. **3**(1):030304–030304.

4516. Hai M., Lin C.B.L.Q.T. and Pei W.A.N.G. (2004). Latest developments in nanophotonics. *Physics* (物理). **33**:636–640.

4517. Halas N.J. (2004). Nanophotonics: an overview. *Materials Today*. **7**(12):49.

4518. Huser T. (2008). Nano-biophotonics: new tools for chemical nano-analytics. *Current Opinion in Chemical Biology*. **12**(5):497–504.

4519. Jenkins A. (2008). The road to nanophotonics. *Nature Photonics*. **2**(5):258–260.

4520. Kamp M. et al. (2006). Recent advances in nanophotonics: from physics to devices. *Current Applied Physics*. **6**:e166–e171.

4521. Kirchain R. and Kimerling L. (2007). A roadmap for nanophotonics. *Nature Photonics*. **1**(6):303–305.

4522. Ohtsu M. et al. (2008). *Principles of Nanophotonics*. Boca Raton, FL: CRC Press.

4523. Popescu G. (2010). *Nanobiophotonics*. New York: McGraw Hill Professional.

4524. Prasad P.N. (2003). Future opportunities in nanophotonics. *Nanocrystals and Organic and Hybrid Nanomaterials*. **5222**: 87–93.

4525. Prasad P.N. (2004). *Nanophotonics*. New York: John Wiley & Sons.

4526. Shen Y. and Prasad P.N. (2002). Nanophotonics: a new multidisciplinary frontier. *Applied Physics B*. **74**(7–8):641–645.

4527. Taney S. et al. (2008). Editorial: biophotonics. *Advances in Optical Technologies*. **2008**: Article ID 134215.

4528. National Research Council (2008). Nanophotonics: accessibility and applicability. Washington, D.C.: Division on Engineering & Physical Sciences.

4529. Yuanwu Q.I.U. (2007). Nanophotonics material review. *Laser & Optoelectronics Progress.* **3**:005.

Nanophysics

4530. Fichthorn K. and Scheffler M. (2004). Nanophysics: a step up to self-assembly. *Nature.* **429**(6992):617–618.

4531. Sattler K.D. (ed.) (2010). *Handbook of Nanophysics: Nanomedicine and Nanorobotics.* Boca Raton, FL: CRC Press.

4532. Tala S. (2011). Enculturation into technoscience: analysis of the views of novices and experts on modeling and learning in nanophysics. *Science & Education.* **20**(7–8):733–760.

Nanorobots and Nanorobotics

4533. Bogue R. (2010). Microrobots and nanorobots: a review of recent developments. *Industrial Robot: An International Journal.* **37**(4):341–346.

4534. Copoţ M. et al. (2001). Achievements and perspectives in the field of nanorobotics. *The Romanian Review Precision Mechanics, Optics & Mechatronics.* **19**(36):61–66.

4535. Dong L. and Nelson B.J. (2007). Tutorial-robotics in the small. Part II. Nanorobotics. *IEEE Robotics & Automation Magazine.* **14**(3):111–121.

4536. Frederick W.C. (2000). Genes, nanobots, and the human future. *Professional Ethics, A Multidisciplinary Journal.* **8**(3/4): 101–122.

4537. Hollis R. (1996). Whither microbots? In *IEEE Proceedings of the Seventh International Symposium on Micro Machine and Human Science*, pp. 9–12.

4538. Klocke V. (2002). Engineering in the nanocosmos: nanorobotics moves kilograms of mass. *Journal of Nanoscience and Nanotechnology.* **2**(3–4):435–440.

4539. Kurzweil R. and Miles K. (2015). Nanobots in our brains will make us godlike. *New Perspectives Quarterly.* **32**(4):24–29.

4540. Mallouk T.E. and Sen A. (2009). Powering nanorobots. *Scientific American.* **300**(5):72–77.

4541. Mavrodis C. and Ferreira A. (eds.) (2013). *Nanorobotics: Current Approaches and Techniques.* New York: Springer.

4542. Mazmuder S. and Bhargava S. (2014). Nanorobots: current state and future perspectives. *International Journal of Development Research.* **4**(6):1234–1239.

4543. Moore A. (2004). Waiter, there's a nanobot in my martini! *EMBO Reports.* **5**(5):448–450.

4544. Nelson B.J. (2009). Towards nanorobots. In *Solid-State Sensors, Actuators and Microsystems Conference,* Transducers 2009, International IEEE, Piscataway, pp. 2155–2159.

4545. Nerlich B. (2008). Powered by imagination: nanobots at the Science Photo Library. *Science as Culture.* **17**(3):269–292.

4546. Neto A.M.J.C. et al. (2010). A review on nanorobotics. *Journal of Computational and Theoretical Nanoscience.* **7**(10):1870–1877.

4547. Petrina A.M. (2012). Nanorobotics: simulation and experiments. *Automatic Documentation and Mathematical Linguistics.* **46**(4): 159–169.

4548. Requicha A.A.G. (2003). Nanorobots, NEMS and nanoassembly. *Proceedings IEEE.* **91**(11):1922–1933.

4549. Sánchez S. and Pumera M. (2009). Nanorobots: the ultimate wireless self-propelled sensing and actuating devices. *Chemistry–An Asian Journal.* **4**(9):1402–1410.

4550. Sengupta S. et al. (2012). Fantastic voyage: designing self-powered nanorobots. *Angewandte Chemie International Edition.* 51(34):8434–8445.

4551. Sharma N.N. and Mittal R.K. (2008). Nanorobot movement: challenges and biologically inspired solutions. *International Journal on Smart Sensing and Intelligent Systems.* **1**(1):87–109.

4552. Sujatha V., Suresh M. and Mahalaxmi S. (2010). Nanorobotics: a futuristic approach. *SRM University Journal of Dental Sciences.* **1**(1):86–90.

4553. Unciti-Broceta A. (2015). Bioorthogonal catalysis: rise of the nanobots. *Nature Chemistry.* **7**(7):538–539.

4554. Wier N.A. et al. (2005). A review of research in the field of nanorobotics (SAND2005-6808). Albuquerque: Sandia National Laboratories.

4555. Winder R. (2003). Nanobots: will they rule? *Chemistry and Industry.* (10):18–19.

Nanoscale

4556. Adams D.M. et al. (2003). Charge transfer on the nanoscale: current status. *The Journal of Physical Chemistry B.* **107**(28):6668–6697.

4557. Amendola V. and Meneghetti M. (2009). Self-healing at the nanoscale. *Nanoscale.* **1**(1):74–88.

4558. Baird D. et al. (eds.) (2004). *Discovering the Nanoscale.* Amsterdam: IOS Press.

4559. Billinge S.J. and Levin I. (2007). The problem with determining atomic structure at the nanoscale. *Science.* **316**(5824): 561–565.

4560. Brites C.D. et al. (2012). Thermometry at the nanoscale. *Nanoscale.* **4**(16):4799–4829.

4561. Datta S. (2000). Nanoscale device modeling: the Green's function method. *Superlattices and Microstructures.* **28**(4):253–278.

4562. Di Ventra M. et al. (2004). *Introduction to Nanoscale Science and Technology.* New York: Springer Science & Business Media.

4563. French R.H. et al. (2010). Long range interactions in nanoscale science. *Reviews of Modern Physics.* **82**(2):1887.

4564. Gavankar S. et al. (2012). Life cycle assessment at nanoscale: review and recommendations. *International Journal of Life Cycle Assessment.* **17**(3):295–303.

4565. Gimzewski J.K. and Joachim C. (1999). Nanoscale science of single molecules using local probes. *Science.* **283**(5408): 1683–1688.

4566. Goodsell D.S. (2006). Seeing the nanoscale. *Nano Today.* **1**(3):44–49.

4567. Gruverman A. and Kholkin A. (2006). Nanoscale ferroelectrics: processing, characterization and future trends. *Reports on Progress in Physics.* **69**(8):2443.

4568. Heath J.R. (1999). Nanoscale materials. *Accounts of Chemical Research.* **32**(5):388–388.

4569. Kelsall R. et al. (eds.) (2005). *Nanoscale Science and Technology.* New York: John Wiley & Sons.

4570. Klabunde K.J. and Richards R.M. (eds.) (2009). *Nanoscale Materials in Chemistry.* New York: John Wiley & Sons.

4571. Lieber C.M. (2003). Nanoscale science and technology: building a big future from small things. *MRS Bulletin.* **28**(07):486–491.

4572. Mann S. (2008). Life as a nanoscale phenomenon. *Angewandte Chemie International Edition.* **47**(29):5306–5320.

4573. Mo Y. et al. (2009). Friction laws at the nanoscale. *Nature.* **457**(7233):1116–1119.

4574. Mody C. and Lynch M. (2010). Test objects and other epistemic things: a history of a nanoscale object. *The British Journal for the History of Science.* **43**(03):423–458.

4575. Roco M.C. (2003). Converging science and technology at the nanoscale: opportunities for education and training. *Nature Biotechnology.* **21**(10):1247–1249.

4576. Roco M.C. (2004). Nanoscale science and engineering: unifying and transforming tools. *AIChE Journal.* **50**:890–897.

4577. Rousseau E. et al. (2009). Radiative heat transfer at the nanoscale. *Nature Photonics.* **3**(9):514–517.

4578. Samuelson L. (2003). Self-forming nanoscale devices. *Materials Today.* **6**(10):22–31.

4579. Tretter T. (2006). Conceptualizing nanoscale. *Science Teacher.* **73**(9):50–53.

Recreation and Sports

4580. Gong Z.G. (2013). Nanotechnology application in sports. *Advanced Materials Research.* **662**:186–189.

4581. Li C.Y. and Chen W. (2004). Opportunity and challenge of competitive sports in nanometer era. *Journal of Hubei Sports Science.* **23**(2):179–181.

4582. Li G. and Cheng Y. (2013). The application prospects of nanotechnology in future competitive sports development. *Advanced Materials Research.* **662**:190–193.

4583. Maney K. (2004). Nanotech could put a new spin on sports. *USA Today.* **17**:1–4.

4584. Qi Z. (2014). The ethical reflection on nanotechnology applications in the field of sports. *Applied Mechanics and Materials.* **556**:40–42.

4585. Ren Y.Z. et al. (2006). Application of nanotechnology to sport costumes. *Sports Science Research.* **27**(3):36.

4586. Shen F. and Wang H.J. (2012). The application of nano-materials and technologies in sports physical sciences. *Advanced Materials Research.* **485**:478–481.

4587. Wang P. and Wang J.Y. (2014). Development and application of nanotechnology in sports. *Advanced Materials Research.* **918**:54–58.

4588. Yu W.D. et al. (2006). Applying nanotechnology to athletic sports. *Sports Science Research.* **27**(3):33.

4589. Zhao H.E. and Shen F. (2012). The applied research of nanophase materials in sports engineering. *Advanced Materials Research.* **496**:126–129.

Transportation

4590. Coelho M. C., Torrão G. and Emami N. (2012). Nanotechnology in automotive industry: research strategy and trends for the future—small objects, big impacts. *Journal of Nanoscience and Nanotechnology.* **12**(8):6621–6630.

4591. Khan M.S. (2011). Nanotechnology in transportation: evolution of a revolutionary technology. TR News. Issue No. 277.

4592. Presting H. and König U. (2003). Future nanotechnology developments for automotive applications. *Materials Science and Engineering: C.* **23**(6):737–741.

4593. Wallner E.D. et al. (2010). Nanotechnology applications in future automobiles. SEA Technical Paper No. 2010-01-1149.

Tribology

4594. Belak J.F. (1993). Nanotribology. *MRS Bulletin.* **18**(05):15–19.

4595. Bennewitz R. et al. (2010). Nanotribology: fundamental studies of friction and plasticity. *Advanced Engineering Materials.* **12**(5):362–367.

4596. Bhushan B. (1995). Micro/nanotribology and its applications. *Journal of the Korean Society of Tribologists and Lubrication Engineers.* **11**(5):128–135.

4597. Bhushan B. (1998). *Handbook of Micro/Nanotribology.* Boca Raton, FL: CRC Press.

4598. Bhushan B. (2005). Nanotribology and nanomechanics. *Wear.* **259**(7):1507–1531.

4599. Bhushan B. (2008). Nanotribology, nanomechanics and nanomaterials characterization. *Philosophical Transactions of the Royal Society of London A: Mathematical, Physical and Engineering Sciences.* **366**(1869):1351–1381.

4600. Bhushan B. (ed.) (2011). *Nanotribology and Nanomechanics I: Measurement Techniques and Nanomechanics,* vol. 1. New York: Springer Science & Business Media.

4601. Bhushan B. et al. (1995). Nanotribology: friction, wear and lubrication at the atomic scale. *Nature.* **374**(6523):607–616.

4602. Braun O.M. and Naumovets A.G. (2006). Nanotribology: microscopic mechanisms of friction. *Surface Science Reports.* **60**(6):79–158.

4603. Canter N. (2004). Nanotribology: the science of thinking small. *Tribology & Lubrication Technology.* **60**(6):42.

4604. Carpick R.W. (2008). On the scientific and technological importance of nanotribology. In *STLE/ASME 2008 International*

Joint Tribology Conference, American Society of Mechanical Engineers, pp. 95–97.

4605. Dedkov G.V. (2000). Nanotribology: experimental facts and theoretical models. *Physics-Uspekhi.* **43**(6):541–572.

4606. Dickinson T.J. (2005). Nanotribology: rubbing on a small scale. *Journal of Chemical Education.* **82**(5):734–742.

4607. Elango B. et al. (2013). Global nanotribology research output (1996–2010): a scientometric analysis. *PLoS One.* **8**(12):e81094.

4608. Gnecco E. et al. (2002). Nanotribology. *CHIMIA International Journal for Chemistry.* **56**(10):562–565.

4609. Gulseren O. et al. (2010). New trends in nanotribology. *Tribology Letters.* **39**(3):227–227.

4610. Guoyu S. and Kaihe C. (2003). The study condition of nanotribology and processing of nanotribology samples. *Machinery Manufacturing Engineer.* **12**:001.

4611. Hsu S.M. and Ying Z.C. (eds.) (2012). *Nanotribology: Critical Assessment and Research Needs.* New York: Springer Science & Business Media.

4612. Jianning D. (1997). Progress in the research of nanotribology. *Journal of Jiangsu University of Science and Technology.* **6**: 012.

4613. Landman U. et al. (1993). Nanotribology and the stability of nanostructures. *Japanese Journal of Applied Physics.* **32**(3S):1444.

4614. Liu E. et al. (1998). Comparative study between macrotribology and nanotribology. *Journal of Applied Physics.* **84**(9): 4859–4865.

4615. Luo J. et al. (1998). Progresses and problems in nanotribology. *Chinese Science Bulletin.* **43**(5):369–378.

4616. Marti O. (1993). Nanotribology: friction on a nanometer scale. *Physica Scripta.* **1993**(T49B):599.

4617. Rymuza Z. (2010). Advanced techniques for nanotribological studies. *Scientific Problems of Machines Operation and Maintenance.* **1**(161):33–43.

4618. Shizhu W. (2007). Progress of research on nanotribology. *Chinese Journal of Mechanical Engineering.* **10**(43):1–8.

4619. Shore A.K. (2009). *Nanotribology and Nanomechanics*, 2nd edition. London: Taylor & Francis.

4620. Urbakh M. and Meyer E. (2010). Nanotribology: the renaissance of friction. *Nature Materials.* **9**(1):8–10.

4621. Van Rensselar J. (2009). Nanotribology from lab ... to real world. *Tribology & Lubrication Technology.* **65**(1):34.

Appendix 1

Countries with Minimal Nanotechnology Footprints

- Afghanistan
- Albania
- Algeria
- Angola
- Antigua/Barbuda
- Azerbaijan
- Bahamas, The
- Bahrain
- Barbados
- Belize
- Benin
- Bhutan
- Bosnia and Herzegovina
- Botswana
- Brunei
- Burundi
- Cambodia
- Cape Verde
- Chad
- Comoros
- Congo, Democratic Republic of the
- Congo, Republic of the
- Cook Islands

- Côte d'Ivoire (Ivory Coast)
- Cyprus
- Djibouti
- Dominica, Commonwealth of
- Dominican Republic
- East Timor
- Eretria
- Ethiopia
- Fiji
- Gabon
- Gambia
- Georgia
- Guinea
- Guinea-Bissau
- Guyana
- Haiti
- Honduras
- Iceland
- Jordon
- Korea, North
- Kuwait
- Kyrgyzstan
- Laos
- Lebanon
- Lesotho
- Liberia
- Liechtenstein
- Luxembourg
- Maldives
- Mali
- Madagascar
- Marshall Islands
- Mauritania
- Mauritius
- Micronesia
- Moldova
- Montenegro
- Mozambique

- Myanmar
- Namibia
- Nauru
- Nicaragua
- Niger
- Oman
- Palau
- Palestine
- Panama
- Paraguay
- Rwanda
- Saint Kitts and Nevis
- Saint Lucia
- Saint Vincent and the Grenadines
- Samoa
- San Marino
- São Tomé and Principe
- Seychelles
- Sierra Leone
- Solomon Islands
- Somalia
- South Sudan
- Sudan
- Suriname
- Syria
- Tajikistan
- Tanzania
- Togo
- Tonga
- Trinidad and Tobago
- Turkmenistan
- Tuvalu
- Vanuatu

Appendix 2

Subject Index

Appendix 3

Geographic Index

Note: The numbers in the index designate the numerically ordered references in the book.